Data Clustering in C++
An Object-Oriented Approach

Chapman & Hall/CRC
Data Mining and Knowledge Discovery Series

SERIES EDITOR
Vipin Kumar
University of Minnesota
Department of Computer Science and Engineering
Minneapolis, Minnesota, U.S.A

AIMS AND SCOPE

This series aims to capture new developments and applications in data mining and knowledge discovery, while summarizing the computational tools and techniques useful in data analysis. This series encourages the integration of mathematical, statistical, and computational methods and techniques through the publication of a broad range of textbooks, reference works, and handbooks. The inclusion of concrete examples and applications is highly encouraged. The scope of the series includes, but is not limited to, titles in the areas of data mining and knowledge discovery methods and applications, modeling, algorithms, theory and foundations, data and knowledge visualization, data mining systems and tools, and privacy and security issues.

PUBLISHED TITLES

UNDERSTANDING COMPLEX DATASETS: DATA MINING WITH MATRIX DECOMPOSITIONS
David Skillicorn

COMPUTATIONAL METHODS OF FEATURE SELECTION
Huan Liu and Hiroshi Motoda

CONSTRAINED CLUSTERING: ADVANCES IN ALGORITHMS, THEORY, AND APPLICATIONS
Sugato Basu, Ian Davidson, and Kiri L. Wagstaff

KNOWLEDGE DISCOVERY FOR COUNTERTERRORISM AND LAW ENFORCEMENT
David Skillicorn

MULTIMEDIA DATA MINING: A SYSTEMATIC INTRODUCTION TO CONCEPTS AND THEORY
Zhongfei Zhang and Ruofei Zhang

NEXT GENERATION OF DATA MINING
Hillol Kargupta, Jiawei Han, Philip S. Yu, Rajeev Motwani, and Vipin Kumar

DATA MINING FOR DESIGN AND MARKETING
Yukio Ohsawa and Katsutoshi Yada

THE TOP TEN ALGORITHMS IN DATA MINING
Xindong Wu and Vipin Kumar

GEOGRAPHIC DATA MINING AND KNOWLEDGE DISCOVERY, SECOND EDITION
Harvey J. Miller and Jiawei Han

TEXT MINING: CLASSIFICATION, CLUSTERING, AND APPLICATIONS
Ashok N. Srivastava and Mehran Sahami

BIOLOGICAL DATA MINING
Jake Y. Chen and Stefano Lonardi

INFORMATION DISCOVERY ON ELECTRONIC HEALTH RECORDS
Vagelis Hristidis

TEMPORAL DATA MINING
Theophano Mitsa

RELATIONAL DATA CLUSTERING: MODELS, ALGORITHMS, AND APPLICATIONS
Bo Long, Zhongfei Zhang, and Philip S. Yu

KNOWLEDGE DISCOVERY FROM DATA STREAMS
João Gama

STATISTICAL DATA MINING USING SAS APPLICATIONS, SECOND EDITION
George Fernandez

INTRODUCTION TO PRIVACY-PRESERVING DATA PUBLISHING: CONCEPTS AND TECHNIQUES
Benjamin C. M. Fung, Ke Wang, Ada Wai-Chee Fu, and Philip S. Yu

HANDBOOK OF EDUCATIONAL DATA MINING
Cristóbal Romero, Sebastian Ventura, Mykola Pechenizkiy, and Ryan S.J.d. Baker

DATA MINING WITH R: LEARNING WITH CASE STUDIES
Luís Torgo

MINING SOFTWARE SPECIFICATIONS: METHODOLOGIES AND APPLICATIONS
David Lo, Siau-Cheng Khoo, Jiawei Han, and Chao Liu

DATA CLUSTERING IN C++: AN OBJECT-ORIENTED APPROACH
Guojun Gan

Chapman & Hall/CRC
Data Mining and Knowledge Discovery Series

Data Clustering in C++
An Object-Oriented Approach

Guojun Gan

CRC Press
Taylor & Francis Group
Boca Raton London New York

CRC Press is an imprint of the
Taylor & Francis Group, an **informa** business

A CHAPMAN & HALL BOOK

CRC Press
Taylor & Francis Group
6000 Broken Sound Parkway NW, Suite 300
Boca Raton, FL 33487-2742

First issued in paperback 2019

© 2011 by Taylor & Francis Group, LLC
CRC Press is an imprint of Taylor & Francis Group, an Informa business

No claim to original U.S. Government works

ISBN-13: 978-1-4398-6223-0 (hbk)
ISBN-13: 978-0-367-38295-7 (pbk)

This book contains information obtained from authentic and highly regarded sources. Reasonable efforts have been made to publish reliable data and information, but the author and publisher cannot assume responsibility for the validity of all materials or the consequences of their use. The authors and publishers have attempted to trace the copyright holders of all material reproduced in this publication and apologize to copyright holders if permission to publish in this form has not been obtained. If any copyright material has not been acknowledged please write and let us know so we may rectify in any future reprint.

Except as permitted under U.S. Copyright Law, no part of this book may be reprinted, reproduced, transmitted, or utilized in any form by any electronic, mechanical, or other means, now known or hereafter invented, including photocopying, microfilming, and recording, or in any information storage or retrieval system, without written permission from the publishers.

For permission to photocopy or use material electronically from this work, please access www.copyright.com (http://www.copyright.com/) or contact the Copyright Clearance Center, Inc. (CCC), 222 Rosewood Drive, Danvers, MA 01923, 978-750-8400. CCC is a not-for-profit organization that provides licenses and registration for a variety of users. For organizations that have been granted a photocopy license by the CCC, a separate system of payment has been arranged.

Trademark Notice: Product or corporate names may be trademarks or registered trademarks, and are used only for identification and explanation without intent to infringe.

Visit the Taylor & Francis Web site at
http://www.taylorandfrancis.com

and the CRC Press Web site at
http://www.crcpress.com

Dedication

To my grandmother and my parents

Dedication

Contents

List of Figures	xv
List of Tables	xix
Preface	xxi

I Data Clustering and C++ Preliminaries   1

1 Introduction to Data Clustering   3
- 1.1 Data Clustering 3
  - 1.1.1 Clustering versus Classification 4
  - 1.1.2 Definition of Clusters 5
- 1.2 Data Types 7
- 1.3 Dissimilarity and Similarity Measures 8
  - 1.3.1 Measures for Continuous Data 9
  - 1.3.2 Measures for Discrete Data 10
  - 1.3.3 Measures for Mixed-Type Data 10
- 1.4 Hierarchical Clustering Algorithms 11
  - 1.4.1 Agglomerative Hierarchical Algorithms 12
  - 1.4.2 Divisive Hierarchical Algorithms 14
  - 1.4.3 Other Hierarchical Algorithms 14
  - 1.4.4 Dendrograms 15
- 1.5 Partitional Clustering Algorithms 15
  - 1.5.1 Center-Based Clustering Algorithms 17
  - 1.5.2 Search-Based Clustering Algorithms 18
  - 1.5.3 Graph-Based Clustering Algorithms 19
  - 1.5.4 Grid-Based Clustering Algorithms 20
  - 1.5.5 Density-Based Clustering Algorithms 20
  - 1.5.6 Model-Based Clustering Algorithms 21
  - 1.5.7 Subspace Clustering Algorithms 22
  - 1.5.8 Neural Network-Based Clustering Algorithms 22
  - 1.5.9 Fuzzy Clustering Algorithms 23
- 1.6 Cluster Validity 23
- 1.7 Clustering Applications 24
- 1.8 Literature of Clustering Algorithms 25
  - 1.8.1 Books on Data Clustering 25

	1.8.2 Surveys on Data Clustering	26
1.9	Summary	28

2 The Unified Modeling Language 29
2.1	Package Diagrams	29
2.2	Class Diagrams	32
2.3	Use Case Diagrams	36
2.4	Activity Diagrams	38
2.5	Notes	39
2.6	Summary	40

3 Object-Oriented Programming and C++ 41
3.1	Object-Oriented Programming	41
3.2	The C++ Programming Language	42
3.3	Encapsulation	45
3.4	Inheritance	48
3.5	Polymorphism	50
	3.5.1 Dynamic Polymorphism	51
	3.5.2 Static Polymorphism	52
3.6	Exception Handling	54
3.7	Summary	56

4 Design Patterns 57
4.1	Singleton	58
4.2	Composite	61
4.3	Prototype	64
4.4	Strategy	67
4.5	Template Method	69
4.6	Visitor	72
4.7	Summary	75

5 C++ Libraries and Tools 77
5.1	The Standard Template Library	77
	5.1.1 Containers	77
	5.1.2 Iterators	82
	5.1.3 Algorithms	84
5.2	Boost C++ Libraries	86
	5.2.1 Smart Pointers	87
	5.2.2 Variant	89
	5.2.3 Variant versus Any	90
	5.2.4 Tokenizer	92
	5.2.5 Unit Test Framework	93
5.3	GNU Build System	95
	5.3.1 Autoconf	96
	5.3.2 Automake	97
	5.3.3 Libtool	97

		5.3.4 Using GNU Autotools	98
	5.4	Cygwin	98
	5.5	Summary	99

II A C++ Data Clustering Framework 101

6 The Clustering Library 103
6.1 Directory Structure and Filenames 103
6.2 Specification Files 105
    6.2.1 configure.ac 105
    6.2.2 Makefile.am 106
6.3 Macros and typedef Declarations 109
6.4 Error Handling 111
6.5 Unit Testing 112
6.6 Compilation and Installation 113
6.7 Summary 114

7 Datasets 115
7.1 Attributes 115
    7.1.1 The Attribute Value Class 115
    7.1.2 The Base Attribute Information Class 117
    7.1.3 The Continuous Attribute Information Class 119
    7.1.4 The Discrete Attribute Information Class 120
7.2 Records 122
    7.2.1 The Record Class 122
    7.2.2 The Schema Class 124
7.3 Datasets 125
7.4 A Dataset Example 127
7.5 Summary 130

8 Clusters 131
8.1 Clusters 131
8.2 Partitional Clustering 133
8.3 Hierarchical Clustering 135
8.4 Summary 138

9 Dissimilarity Measures 139
9.1 The Distance Base Class 139
9.2 Minkowski Distance 140
9.3 Euclidean Distance 141
9.4 Simple Matching Distance 142
9.5 Mixed Distance 143
9.6 Mahalanobis Distance 144
9.7 Summary 147

10 Clustering Algorithms   149
10.1 Arguments 149
10.2 Results 150
10.3 Algorithms 151
10.4 A Dummy Clustering Algorithm 154
10.5 Summary 158

11 Utility Classes   161
11.1 The Container Class 161
11.2 The Double-Key Map Class 164
11.3 The Dataset Adapters 167
    11.3.1 A CSV Dataset Reader 167
    11.3.2 A Dataset Generator 170
    11.3.3 A Dataset Normalizer 173
11.4 The Node Visitors 175
    11.4.1 The Join Value Visitor 175
    11.4.2 The Partition Creation Visitor 176
11.5 The Dendrogram Class 177
11.6 The Dendrogram Visitor 179
11.7 Summary 180

III   Data Clustering Algorithms   183

12 Agglomerative Hierarchical Algorithms   185
12.1 Description of the Algorithm 185
12.2 Implementation 187
    12.2.1 The Single Linkage Algorithm 192
    12.2.2 The Complete Linkage Algorithm 192
    12.2.3 The Group Average Algorithm 193
    12.2.4 The Weighted Group Average Algorithm 194
    12.2.5 The Centroid Algorithm 194
    12.2.6 The Median Algorithm 195
    12.2.7 Ward's Algorithm 196
12.3 Examples 197
    12.3.1 The Single Linkage Algorithm 198
    12.3.2 The Complete Linkage Algorithm 200
    12.3.3 The Group Average Algorithm 202
    12.3.4 The Weighted Group Average Algorithm 204
    12.3.5 The Centroid Algorithm 207
    12.3.6 The Median Algorithm 210
    12.3.7 Ward's Algorithm 212
12.4 Summary 214

13 DIANA — 217
13.1 Description of the Algorithm 217
13.2 Implementation 218
13.3 Examples 223
13.4 Summary 227

14 The k-means Algorithm — 229
14.1 Description of the Algorithm 229
14.2 Implementation 230
14.3 Examples 235
14.4 Summary 240

15 The c-means Algorithm — 241
15.1 Description of the Algorithm 241
15.2 Implementaion 242
15.3 Examples 246
15.4 Summary 253

16 The k-prototypes Algorithm — 255
16.1 Description of the Algorithm 255
16.2 Implementation 256
16.3 Examples 258
16.4 Summary 263

17 The Genetic k-modes Algorithm — 265
17.1 Description of the Algorithm 265
17.2 Implementation 267
17.3 Examples 274
17.4 Summary 277

18 The FSC Algorithm — 279
18.1 Description of the Algorithm 279
18.2 Implementation 281
18.3 Examples 284
18.4 Summary 290

19 The Gaussian Mixture Algorithm — 291
19.1 Description of the Algorithm 291
19.2 Implementation 293
19.3 Examples 300
19.4 Summary 306

20	**A Parallel k-means Algorithm**	**307**
	20.1 Message Passing Interface	307
	20.2 Description of the Algorithm	310
	20.3 Implementation	311
	20.4 Examples	316
	20.5 Summary	320
A	**Exercises and Projects**	**323**
B	**Listings**	**325**
	B.1 Files in Folder `ClusLib`	325
	B.1.1 Configuration File `configure.ac`	325
	B.1.2 m4 Macro File `acinclude.m4`	326
	B.1.3 Makefile	327
	B.2 Files in Folder `cl`	328
	B.2.1 Makefile	328
	B.2.2 Macros and `typedef` Declarations	328
	B.2.3 Class `Error`	329
	B.3 Files in Folder `cl/algorithms`	331
	B.3.1 Makefile	331
	B.3.2 Class `Algorithm`	332
	B.3.3 Class `Average`	334
	B.3.4 Class `Centroid`	334
	B.3.5 Class `Cmean`	335
	B.3.6 Class `Complete`	339
	B.3.7 Class `Diana`	339
	B.3.8 Class `FSC`	343
	B.3.9 Class `GKmode`	347
	B.3.10 Class `GMC`	353
	B.3.11 Class `Kmean`	358
	B.3.12 Class `Kprototype`	361
	B.3.13 Class `LW`	362
	B.3.14 Class `Median`	364
	B.3.15 Class `Single`	365
	B.3.16 Class `Ward`	366
	B.3.17 Class `Weighted`	367
	B.4 Files in Folder `cl/clusters`	368
	B.4.1 Makefile	368
	B.4.2 Class `CenterCluster`	368
	B.4.3 Class `Cluster`	369
	B.4.4 Class `HClustering`	370
	B.4.5 Class `PClustering`	372
	B.4.6 Class `SubspaceCluster`	375
	B.5 Files in Folder `cl/datasets`	376
	B.5.1 Makefile	376

	B.5.2	Class `AttrValue`	376
	B.5.3	Class `AttrInfo`	377
	B.5.4	Class `CAttrInfo`	379
	B.5.5	Class `DAttrInfo`	381
	B.5.6	Class `Record`	384
	B.5.7	Class `Schema`	386
	B.5.8	Class `Dataset`	388
B.6	Files in Folder `cl/distances`		392
	B.6.1	Makefile	392
	B.6.2	Class `Distance`	392
	B.6.3	Class `EuclideanDistance`	393
	B.6.4	Class `MahalanobisDistance`	394
	B.6.5	Class `MinkowskiDistance`	395
	B.6.6	Class `MixedDistance`	396
	B.6.7	Class `SimpleMatchingDistance`	397
B.7	Files in Folder `cl/patterns`		398
	B.7.1	Makefile	398
	B.7.2	Class `DendrogramVisitor`	399
	B.7.3	Class `InternalNode`	401
	B.7.4	Class `LeafNode`	403
	B.7.5	Class `Node`	404
	B.7.6	Class `NodeVisitor`	405
	B.7.7	Class `JoinValueVisitor`	405
	B.7.8	Class `PCVisitor`	407
B.8	Files in Folder `cl/utilities`		408
	B.8.1	Makefile	408
	B.8.2	Class `Container`	409
	B.8.3	Class `DataAdapter`	411
	B.8.4	Class `DatasetGenerator`	411
	B.8.5	Class `DatasetNormalizer`	413
	B.8.6	Class `DatasetReader`	415
	B.8.7	Class `Dendrogram`	418
	B.8.8	Class `nnMap`	421
	B.8.9	Matrix Functions	423
	B.8.10	Null Types	425
B.9	Files in Folder `examples`		426
	B.9.1	Makefile	426
	B.9.2	Agglomerative Hierarchical Algorithms	426
	B.9.3	A Divisive Hierarchical Algorithm	429
	B.9.4	The k-means Algorithm	430
	B.9.5	The c-means Algorithm	433
	B.9.6	The k-prototypes Algorithm	435
	B.9.7	The Genetic k-modes Algorithm	437
	B.9.8	The FSC Algorithm	439
	B.9.9	The Gaussian Mixture Clustering Algorithm	441

			B.9.10 A Parallel k-means Algorithm	444

B.10 Files in Folder `test-suite` 450
    B.10.1 Makefile 450
    B.10.2 The Master Test Suite 451
    B.10.3 Test of `AttrInfo` 451
    B.10.4 Test of `Dataset` 453
    B.10.5 Test of `Distance` 454
    B.10.6 Test of `nnMap` 456
    B.10.7 Test of Matrices 458
    B.10.8 Test of `Schema` 459

C Software   461

C.1 An Introduction to Makefiles 461
    C.1.1 Rules 461
    C.1.2 Variables 462
C.2 Installing Boost 463
    C.2.1 Boost for Windows 463
    C.2.2 Boost for Cygwin or Linux 464
C.3 Installing Cygwin 465
C.4 Installing GMP 465
C.5 Installing MPICH2 and Boost MPI 466

Bibliography   469

Author Index   487

Subject Index   493

List of Figures

1.1	A dataset with three compact clusters.	6
1.2	A dataset with three chained clusters.	7
1.3	Agglomerative clustering.	12
1.4	Divisive clustering.	13
1.5	The dendrogram of the Iris dataset.	16
2.1	UML diagrams.	30
2.2	UML packages.	31
2.3	A UML package with nested packages placed inside.	31
2.4	A UML package with nested packages placed outside.	31
2.5	The visibility of elements within a package.	32
2.6	The UML dependency notation.	32
2.7	Notation of a class.	33
2.8	Notation of an abstract class.	33
2.9	A template class and one of its realizations.	34
2.10	Categories of relationships.	35
2.11	The UML actor notation and use case notation.	36
2.12	A UML use case diagram.	37
2.13	Notation of relationships among use cases.	37
2.14	An activity diagram.	39
2.15	An activity diagram with a flow final node.	39
2.16	A diagram with notes.	40
3.1	Hierarchy of C++ standard library exception classes.	54
4.1	The singleton pattern	58
4.2	The composite pattern.	62
4.3	The prototype pattern.	65
4.4	The strategy pattern.	67
4.5	The template method pattern.	70
4.6	The visitor pattern.	74
5.1	Iterator hierarchy.	83
5.2	Flow diagram of `Autoconf`.	96
5.3	Flow diagram of `Automake`.	97
5.4	Flow diagram of `configure`.	98

6.1	The directory structure of the clustering library.	104
7.1	Class diagram of attributes.	116
7.2	Class diagram of records.	123
7.3	Class diagram of `Dataset`.	125
8.1	Hierarchy of cluster classes.	132
8.2	A hierarchical tree with levels.	136
10.1	Class diagram of algorithm classes.	153
11.1	A generated dataset with 9 points.	174
11.2	An EPS figure.	177
11.3	A dendrogram that shows 100 nodes.	181
11.4	A dendrogram that shows 50 nodes.	182
12.1	Class diagram of agglomerative hierarchical algorithms.	188
12.2	The dendrogram produced by applying the single linkage algorithm to the Iris dataset.	199
12.3	The dendrogram produced by applying the single linkage algorithm to the synthetic dataset.	200
12.4	The dendrogram produced by applying the complete linkage algorithm to the Iris dataset.	201
12.5	The dendrogram produced by applying the complete linkage algorithm to the synthetic dataset.	203
12.6	The dendrogram produced by applying the group average algorithm to the Iris dataset.	204
12.7	The dendrogram produced by applying the group average algorithm to the synthetic dataset.	205
12.8	The dendrogram produced by applying the weighted group average algorithm to the Iris dataset.	206
12.9	The dendrogram produced by applying the weighted group average algorithm to the synthetic dataset.	207
12.10	The dendrogram produced by applying the centroid algorithm to the Iris dataset.	208
12.11	The dendrogram produced by applying the centroid algorithm to the synthetic dataset.	209
12.12	The dendrogram produced by applying the median algorithm to the Iris dataset.	211
12.13	The dendrogram produced by applying the median algorithm to the synthetic dataset.	212
12.14	The dendrogram produced by applying the ward algorithm to the Iris dataset.	213
12.15	The dendrogram produced by applying Ward's algorithm to the synthetic dataset.	214

13.1	The dendrogram produced by applying the DIANA algorithm to the synthetic dataset.	225
13.2	The dendrogram produced by applying the DIANA algorithm to the Iris dataset.	226

List of Tables

1.1	The six essential tasks of data mining.	4
1.2	Attribute types.	8
2.1	Relationships between classes and their notation.	34
2.2	Some common multiplicities.	35
3.1	Access rules of base-class members in the derived class.	50
4.1	Categories of design patterns.	57
4.2	The singleton pattern.	58
4.3	The composite pattern.	61
4.4	The prototype pattern.	64
4.5	The strategy pattern.	67
4.6	The template method pattern.	70
4.7	The visitor pattern.	73
5.1	STL containers	78
5.2	Non-modifying sequence algorithms.	84
5.3	Modifying sequence algorithms.	84
5.4	Sorting algorithms.	84
5.5	Binary search algorithms.	85
5.6	Merging algorithms.	85
5.7	Heap algorithms.	85
5.8	Min/max algorithms.	85
5.9	Numerical algorithms defined in the header file `numeric`.	85
5.10	Boost smart pointer class templates.	87
5.11	Boost unit test log levels.	95
7.1	An example of class `DAttrInfo`.	121
7.2	An example dataset.	127
10.1	Cluster membership of a partition of a dataset with 5 records.	151
12.1	Parameters for the Lance-Williams formula, where $\Sigma = \|C\| + \|C_{i_1}\| + \|C_{i_2}\|$.	186

12.2 Centers of combined clusters and distances between two clusters for geometric hierarchical algorithms, where $\mu(\cdot)$ denotes a center of a cluster and $D_{euc}(\cdot,\cdot)$ is the Euclidean distance. 187

C.1 Some automatic variables in `make`. 462

Preface

Data clustering is a highly interdisciplinary field whose goal is to divide a set of objects into homogeneous groups such that objects in the same group are similar and objects in different groups are quite distinct. Thousands of papers and a number of books on data clustering have been published over the past 50 years. However, almost all papers and books focus on the theory of data clustering. There are few books that teach people how to implement data clustering algorithms.

This book was written for anyone who wants to implement data clustering algorithms and for those who want to implement new data clustering algorithms in a better way. Using object-oriented design and programming techniques, I have exploited the commonalities of all data clustering algorithms to create a flexible set of reusable classes that simplifies the implementation of any data clustering algorithm. Readers can follow me through the development of the base data clustering classes and several popular data clustering algorithms.

This book focuses on how to implement data clustering algorithms in an object-oriented way. Other topics of clustering such as data pre-processing, data visualization, cluster visualization, and cluster interpretation are touched but not in detail. In this book, I used a direct and simple way to implement data clustering algorithms so that readers can understand the methodology easily. I also present the material in this book in a straightforward way. When I introduce a class, I present and explain the class method by method rather than present and go through the whole implementation of the class.

Complete listings of classes, examples, unit test cases, and GNU configuration files are included in the appendices of this book as well as in the CD-ROM of the book. I have tested the code under Unix-like platforms (e.g., Ubuntu and Cygwin) and Microsoft Windows XP. The only requirements to compile the code are a modern C++ compiler and the Boost C++ libraries.

This book is divided into three parts: Data Clustering and C++ Preliminaries, A C++ Data Clustering Framework, and Data Clustering Algorithms. The first part reviews some basic concepts of data clustering, the unified modeling language, object-oriented programming in C++, and design patterns. The second part develops the data clustering base classes. The third part implements several popular data clustering algorithms. The content of each chapter is described briefly below.

Chapter 1. Introduction to Data Clustering. In this chapter, we review some basic concepts of data clustering. The clustering process, data types, similarity and dissimilarity measures, hierarchical and partitional clustering algorithms, cluster validity, and applications of data clustering are briefly introduced. In addition, a list of survey papers and books related to data clustering are presented.

Chapter 2. The Unified Modeling Language. The Unified Modeling Language (UML) is a general-purpose modeling language that includes a set of standardized graphic notation to create visual models of software systems. In this chapter, we introduce several UML diagrams such as class diagrams, use-case diagrams, and activity diagrams. Illustrations of these UML diagrams are presented.

Chapter 3. Object-Oriented Programming and C++. Object-oriented programming is a programming paradigm that is based on the concept of objects, which are reusable components. Object-oriented programming has three pillars: encapsulation, inheritance, and polymorphism. In this chapter, these three pillars are introduced and illustrated with simple programs in C++. The exception handling ability of C++ is also discussed in this chapter.

Chapter 4. Design Patterns. Design patterns are reusable designs just as objects are reusable components. In fact, a design pattern is a general reusable solution to a problem that occurs over and over again in software design. In this chapter, several design patterns are described and illustrated by simple C++ programs.

Chapter 5. C++ Libraries and Tools. As an object-oriented programming language, C++ has many well-designed and useful libraries. In this chapter, the standard template library (STL) and several Boost C++ libraries are introduced and illustrated by C++ programs. The GNU build system (i.e., GNU Autotools) and the Cygwin system, which simulates a Unix-like platform under Microsoft Windows, are also introduced.

Chapter 6. The Clustering Library. This chapter introduces the file system of the clustering library, which is a collection of reusable classes used to develop clustering algorithms. The structure of the library and file name convention are introduced. In addition, the GNU configuration files, the error handling class, unit testing, and compilation of the clustering library are described.

Chapter 7. Datasets. This chapter introduces the design and implementation of datasets. In this book, we assume that a dataset consists of a set of records and a record is a vector of values. The attribute value class, the attribute information class, the schema class, the record class, and the dataset class are introduced in this chapter. These classes are illustrated by an example in C++.

Chapter 8. Clusters. A cluster is a collection of records. In this chapter, the cluster class and its child classes such as the center cluster class and the subspace cluster class are introduced. In addition, partitional clustering class and hierarchical clustering class are also introduced.

Chapter 9. Dissimilarity Measures. Dissimilarity or distance measures are an important part of most clustering algorithms. In this chapter, the design of the distance base class is introduced. Several popular distance measures such as the Euclidean distance, the simple matching distance, and the mixed distance are introduced. In this chapter, we also introduce the implementation of the Mahalanobis distance.

Chapter 10. Clustering Algorithms. This chapter introduces the design and implementation of the clustering algorithm base class. All data clustering algorithms have three components: arguments or parameters, clustering method, and clustering results. In this chapter, we introduce the argument class, the result class, and the base algorithm class. A dummy clustering algorithm is used to illustrate the usage of the base clustering algorithm class.

Chapter 11. Utility Classes. This chapter, as its title implies, introduces several useful utility classes used frequently in the clustering library. Two template classes, the container class and the double-key map class, are introduced in this chapter. A CSV (comma-separated values) dataset reader class and a multivariate Gaussian mixture dataset generator class are also introduced in this chapter. In addition, two hierarchical tree visitor classes, the join value visitor class and the partition creation visitor class, are introduced in this chapter. This chapter also includes two classes that provide functionalities to draw dendrograms in EPS (Encapsulated PostScript) figures from hierarchical clustering trees.

Chapter 12. Agglomerative Hierarchical Algorithms. This chapter introduces the implementations of several agglomerative hierarchical clustering algorithms that are based on the Lance-Williams framework. In this chapter, single linkage, complete linkage, group average, weighted group average, centroid, median, and Ward's method are implemented and illustrated by a synthetic dataset and the Iris dataset.

Chapter 13. DIANA. This chapter introduces a divisive hierarchical clustering algorithm and its implementation. The algorithm is illustrated by a synthetic dataset and the Iris dataset.

Chapter 14. The k-means Algorithm. This chapter introduces the standard k-means algorithm and its implementation. A synthetic dataset and the Iris dataset are used to illustrate the algorithm.

Chapter 15. The c-means Algorithm. This chapter introduces the fuzzy c-means algorithm and its implementation. The algorithm is also illustrated by a synthetic dataset and the Iris dataset.

Chapter 16. The k-prototype Algorithm. This chapter introduces the k-prototype algorithm and its implementation. This algorithm was designed to cluster mixed-type data. A numeric dataset (the Iris dataset), a categorical dataset (the Soybean dataset), and a mixed-type dataset (the heart dataset) are used to illustrate the algorithm.

Chapter 17. The Genetic k-modes Algorithm. This chapter introduces the genetic k-modes algorithm and its implementation. A brief introduction to the genetic algorithm is also presented. The Soybean dataset is used to illustrate the algorithm.

Chapter 18. The FSC Algorithm. This chapter introduces the fuzzy subspace clustering (FSC) algorithm and its implementation. The algorithm is illustrated by a synthetic dataset and the Iris dataset.

Chapter 19. The Gaussian Mixture Model Clustering Algorithm. This chapter introduces a clustering algorithm based on the Gaussian mixture model.

Chapter 20. A Parallel k-means Algorithm. This chapter introduces a simple parallel version of the k-means algorithm based on the message passing interface and the Boost MPI library.

Chapters 2–5 introduce programming related materials. Readers who are already familiar with object-oriented programming in C++ can skip those chapters. Chapters 6–11 introduce the base clustering classes and some utility classes. Chapter 12 includes several agglomerative hierarchical clustering algorithms. Each one of the last eight chapters is devoted to one particular clustering algorithm. The eight chapters introduce and implement a diverse set of clustering algorithms such as divisive clustering, center-based clustering, fuzzy clustering, mixed-type data clustering, search-based clustering, subspace clustering, mode-based clustering, and parallel data clustering.

A key to learning a clustering algorithm is to implement and experiment the clustering algorithm. I encourage readers to compile and experiment the examples included in this book. After getting familiar with the classes and their usage, readers can implement new clustering algorithms using these classes or even improve the designs and implementations presented in this book. To this end, I included some exercises and projects in the appendix of this book.

This book grew out of my wish to help undergraduate and graduate students who study data clustering to learn how to implement clustering algorithms and how to do it in a better way. When I was a PhD student, there were no books or papers to teach me how to implement clustering algorithms. It took me a long time to implement my first clustering algorithm. The clustering programs I wrote at that time were just C programs written in C++. It has taken me years to learn how to use the powerful C++ language in the right way. With the help of this book, readers should be able to learn how to implement clustering algorithms and how to do it in a better way in a short period of time.

I would like to take this opportunity to thank my boss, Dr. Hong Xie, who taught me how to write in an effective and rigorous way. I would also like to thank my ex-boss, Dr. Matthew Willis, who taught me how to program in C++ in a better way. I thank my PhD supervisor, Dr. Jianhong Wu, who brought me into the field of data clustering. Finally, I would like to thank my wife, Xiaoying, and my children, Albert and Ella, for their support.

Guojun Gan
Toronto, Ontario
December 31, 2010

Part I

Data Clustering and C++ Preliminaries

Chapter 1

Introduction to Data Clustering

In this chapter, we give a review of data clustering. First, we describe what data clustering is, the difference between clustering and classification, and the notion of clusters. Second, we introduce types of data and some similarity and dissimilarity measures. Third, we introduce several popular hierarchical and partitional clustering algorithms. Then, we discuss cluster validity and applications of data clustering in various areas. Finally, we present some books and review papers related to data clustering.

1.1 Data Clustering

Data clustering is a process of assigning a set of records into subsets, called *clusters*, such that records in the same cluster are similar and records in different clusters are quite distinct (Jain et al., 1999). Data clustering is also known as *cluster analysis, segmentation analysis, taxonomy analysis*, or *unsupervised classification*.

The term *record* is also referred to as *data point, pattern, observation, object, individual, item*, and *tuple* (Gan et al., 2007). A record in a multidimensional space is characterized by a set of *attributes, variables*, or *features*.

A typical clustering process involves the following five steps (Jain et al., 1999):

(a) pattern representation;

(b) dissimilarity measure definition;

(c) clustering;

(d) data abstraction;

(e) assessment of output.

In the pattern representation step, the number and type of the attributes are determined. *Feature selection*, the process of identifying the most effective subset of the original attributes to use in clustering, and *feature extraction*, the process of transforming the original attributes to new attributes, are also done in this step if needed.

In the dissimilarity measure definition step, a distance measure appropriate to the data domain is defined. Various distance measures have been developed and used in data clustering (Gan et al., 2007). The most common one among them, for example, is the Euclidean distance.

In the clustering step, a clustering algorithm is used to group a set of records into a number of meaningful clusters. The clustering can be *hard clustering*, where each record belongs to one and only one cluster, or *fuzzy clustering*, where a record can belong to two or more clusters with probabilities. The clustering algorithm can be *hierarchical*, where a nested series of partitions is produced, or *partitional*, where a single partition is identified.

In the data abstraction step, one or more prototypes (i.e., representative records) of a cluster is extracted so that the clustering results are easy to comprehend. For example, a cluster can be represented by a centroid.

In the final step, the output of a clustering algorithm is assessed. There are three types of assessments: *external*, *internal*, and *relative* (Jain and Dubes, 1988). In an external assessment, the recovered structure of the data is compared to the a priori structure. In an internal assessment, one tries to determine whether the structure is intrinsically appropriate to the data. In a relative assessment, a test is performed to compare two structures and measure their relative merits.

1.1.1 Clustering versus Classification

Data clustering is one of the six essential tasks of data mining, which aims to discover useful information by exploring and analyzing large amounts of data (Berry and Linoff, 2000). Table 1.1 shows the six tasks of data mining, which are grouped into two categories: direct data mining tasks and indirect data mining tasks. The difference between direct data mining and indirect data mining lies in whether a variable is singled out as a target.

Direct Data Mining	Indirect Data Mining
Classification	Clustering
Estimation	Association Rules
Prediction	Description and Visualization

TABLE 1.1: The six essential tasks of data mining.

Classification is a direct data mining task. In classification, a set of labeled or preclassified records is provided and the task is to classify a newly encountered but unlabeled record. Precisely, a classification algorithm tries to model a set of labeled data points $(\mathbf{x}_i, y_i)$ $(1 \leq i \leq n)$ in terms of some mathematical function $y = f(\mathbf{x}, \mathbf{w})$ (Xu and Wunsch, II, 2009), where $\mathbf{x}_i$ is a data point, y_i is the label or class of $\mathbf{x}_i$, and $\mathbf{w}$ is a vector of adjustable parameters. An inductive learning algorithm or inducer is used to determine the values of these parameters by minimizing an empirical risk function on the set of labeled data points (Kohavi, 1995; Cherkassky and Mulier, 1998; Xu and Wunsch, II, 2009). Suppose $\mathbf{w}^*$ is the vector of parameters determined by the inducer. Then we obtain an induced classifier $y = f(\mathbf{x}, \mathbf{w}^*)$, which can be used to classify new data points. The set of labeled data points $(\mathbf{x}_i, y_i)$ $(1 \leq i \leq n)$ is also called the training data for the inducer.

Unlike classification, data clustering is an indirect data mining task. In data clustering, the task is to group a set of unlabeled records into meaningful subsets or clusters, where each cluster is associated with a label. As mentioned at the beginning of this section, a clustering algorithm takes a set of unlabeled data points as input and tries to group these unlabeled data points into a finite number of groups or clusters such that data points in the same cluster are similar and data points in different clusters are quite distinct (Jain et al., 1999).

1.1.2 Definition of Clusters

Over the last 50 years, thousands of clustering algorithms have been developed (Jain, 2010). However, there is still no formal uniform definition of the term *cluster*. In fact, formally defining cluster is difficult and may be misplaced (Everitt et al., 2001).

Although no formal definition of cluster exists, there are several operational definitions of cluster. For example, Bock (1989) suggested that a cluster is a group of data points satisfying various plausible criteria such as

(a) Share the same or closely related properties;

(b) Show small mutual distances;

(c) Have "contacts" or "relations" with at least one other data point in the group;

(d) Can be clearly distinguishable from the rest of the data points in the dataset.

Carmichael et al. (1968) suggested that a set of data points forms a cluster if the distribution of the set of data points satisfies the following conditions:

(a) Continuous and relatively dense regions exist in the data space; and

(b) Continuous and relatively empty regions exist in the data space.

Lorr (1983) suggested that there are two kinds of clusters for numerical data: compact clusters and chained clusters. A compact cluster is formed by a group of data points that have high mutual similarity. For example, Figure

1.1 shows a two-dimensional dataset with three compact clusters[1]. Usually, such a compact cluster has a center (Michaud, 1997).

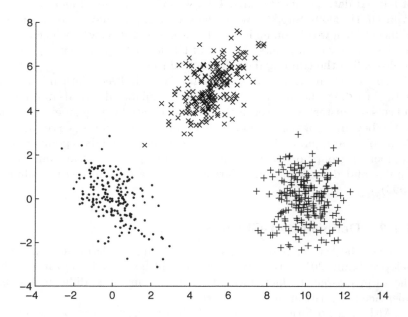

FIGURE 1.1: A dataset with three compact clusters.

A chained cluster is formed by a group of data points in which any two data points in the cluster are reachable through a path. For example, Figures 1.2 shows a dataset with three chained clusters. Unlike a compact cluster, which can be represented by a single center, a chained cluster is usually represented by multiple centers.

Everitt (1993) also summarized several operational definitions of cluster. For example, one definition of cluster is that a cluster is a set of data points that are alike and data points from different clusters are not alike. Another definition of cluster is that a cluster is a set of data points such that the distance between any two points in the cluster is less than the distance between any point in the cluster and any point not in it.

---

[1]This dataset was generated by the dataset generator program in the clustering library presented in this book.

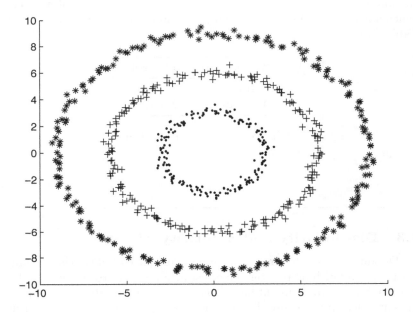

FIGURE 1.2: A dataset with three chained clusters.

1.2 Data Types

Most clustering algorithms are associated with data types. It is important to understand different types of data in order to perform cluster analysis. By data type we mean hereby the type of a single attribute.

In terms of how the values are obtained, an attribute can be typed as *discrete* or *continuous*. The values of a discrete attribute are usually obtained by some sort of counting; while the values of a continuous attribute are obtained by some sort of measuring. For example, the number of cars is discrete and the weight of a person is continuous. There is a gap between two different discrete values and there is always a value between two different continuous values.

In terms of measurement scales, an attribute can be typed as *ratio*, *interval*, *ordinal*, or *nominal*. Nominal data are discrete data without a natural ordering. For example, name of a person is nominal. Ordinal data are discrete data that have a natural ordering. For example, the order of persons in a line is ordinal. Interval data are continuous data that have a specific order and equal intervals. For example, temperature is interval data. Ratio data are continuous data that are interval data and have a natural zero. For example, the annual salary of a person is ratio data. The ratio and interval types are continuous types, while the ordinal and nominal types are discrete types (see Table 1.2).

Continuous	Discrete
Ratio	Ordinal
Interval	Nominal

TABLE 1.2: Attribute types.

1.3 Dissimilarity and Similarity Measures

Dissimilarity or distance is an important part of clustering as almost all clustering algorithms rely on some distance measure to define the clustering criteria. Since records might have different types of attributes, the appropriate distance measures are also different. For example, the most popular Euclidean distance is used to measure dissimilarities between continuous records; i.e., records consist of continuous attributes.

A distance function D on a dataset X is a binary function that satisfies the following conditions (Anderberg, 1973; Zhang and Srihari, 2003; Xu and Wunsch, II, 2009):

(a) $D(\mathbf{x},\mathbf{y}) \geq 0$ (Nonnegativity);

(b) $D(\mathbf{x},\mathbf{y}) = D(\mathbf{y},\mathbf{x})$ (Symmetry or Commutativity);

(c) $D(\mathbf{x},\mathbf{y}) = 0$ if and only if $\mathbf{x} = \mathbf{y}$ (Reflexivity);

(d) $D(\mathbf{x},\mathbf{y}) \leq D(\mathbf{x},\mathbf{z}) + D(\mathbf{z}+\mathbf{y})$ (Triangle inequality),

where $\mathbf{x}$, $\mathbf{y}$, and $\mathbf{z}$ are arbitrary data points in X. A distance function is also called a metric, which satisfies the above four conditions.

If a function satisfies the first three conditions and does not satisfy the triangle inequality, then the function is called a semimetric. In addition, if a metric D satisfies the following condition

$$D(\mathbf{x},\mathbf{y}) \leq \max\{D(\mathbf{x},\mathbf{z}), D(\mathbf{z}+\mathbf{y})\},$$

then the metric is called an ultrametric (Johnson, 1967).

Unlike distance measures, similarity measures are defined in the opposite way. The more the two data points are similar to each other, the larger the similarity is and the smaller the distance is.

1.3.1 Measures for Continuous Data

The most common distance measure for continuous data is the Euclidean distance. Given two data points **x** and **y** in a d-dimensional space, the Euclidean distance between the two data points is defined as

$$D_{euc}(\mathbf{x}, \mathbf{y}) = \sqrt{\sum_{j=1}^{d}(x_j - y_j)^2}, \tag{1.1}$$

where x_j and y_j are the jth components of **x** and **y**, respectively.

The Euclidean distance measure is a metric (Xu and Wunsch, II, 2009). Clustering algorithms that use the Euclidean distance tend to produce hyperspherical clusters. Clusters produced by clustering algorithms that use the Euclidean distance are invariant to translations and rotations in the data space (Duda et al., 2001). One disadvantage of the Euclidean distance is that attributes with large values and variances dominate other attributes with small values and variances. However, this problem can be alleviated by normalizing the data so that each attribute contributes equally to the distance.

The squared Euclidean distance between two data points is defined as

$$D_{squ}(\mathbf{x}, \mathbf{y}) = \sum_{j=1}^{d}(x_j - y_j)^2. \tag{1.2}$$

The Manhattan distance or city block distance between two data points is defined as

$$D_{man}(\mathbf{x}, \mathbf{y}) = \sum_{j=1}^{d}|x_j - y_j|. \tag{1.3}$$

The maximum distance between two data points is defined as

$$D_{max}(\mathbf{x}, \mathbf{y}) = \max_{1 \leq j \leq d}|x_j - y_j|. \tag{1.4}$$

The Euclidean distance and the Manhattan distance are special cases of the Minkowski distance, which is defined as

$$D_{min}(\mathbf{x}, \mathbf{y}) = \left(\sum_{j=1}^{d}|x_j - y_j|^p\right)^{\frac{1}{p}}, \tag{1.5}$$

where $p \geq 1$. In fact, the maximum distance is also a special case of the Minkowski distance when we let $p \to \infty$.

The Mahalanobis distance is defined as

$$D_{mah}(\mathbf{x}, \mathbf{y}) = \sqrt{(\mathbf{x} - \mathbf{y})^T \Sigma^{-1}(\mathbf{x} - \mathbf{y})}, \tag{1.6}$$

where Σ^{-1} is the inverse of a covariance matrix Σ, **x** and **y** are column vectors, and $(\mathbf{x}-\mathbf{y})^T$ denotes the transpose of $(\mathbf{x}-\mathbf{y})$. The Mahalanobis distance can be used to alleviate the distance distortion caused by linear combinations of attributes (Jain and Dubes, 1988; Mao and Jain, 1996).

Some other distance measures for continuous data have also been proposed. For example, the average distance (Legendre and Legendre, 1983), the generalized Mahalanobis distance (Morrison, 1967), the weighted Manhattan distance (Wishart, 2002), the chord distance (Orlóci, 1967), and the Pearson correlation (Eisen et al., 1998), to name just a few. Many other distance measures for numeric data can be found in (Gan et al., 2007).

1.3.2 Measures for Discrete Data

The most common distance measure for discrete data is the simple matching distance. The simple matching distance between two categorical data points $\mathbf{x}$ and $\mathbf{y}$ is defined as (Kaufman and Rousseeuw, 1990; Huang, 1997a,b, 1998)

$$D_{sim}(\mathbf{x}, \mathbf{y}) = \sum_{j=1}^{d} \delta(x_j, y_j), \qquad (1.7)$$

where d is the dimension of the data points and $\delta(\cdot, \cdot)$ is defined as

$$\delta(x_j, y_j) = \begin{cases} 0 & \text{if } x_j = y_j, \\ 1 & \text{if } x_j \neq y_j. \end{cases}$$

Some other matching coefficients for categorical data have also been proposed. For a comprehensive list of matching coefficients, readers are referred to (Gan et al., 2007, Chapter 6). Gan et al. (2007) also contains a comprehensive list of similarity measures for binary data, which is a special case of categorical data.

1.3.3 Measures for Mixed-Type Data

A dataset might contain both continuous and discrete data. In this case, we need to use a measure for mixed-type data. Gower (1971) proposed a general similarity coefficient for mixed-type data, which is defined as

$$S_{gow}(\mathbf{x}, \mathbf{y}) = \frac{1}{\sum\limits_{j=1}^{d} w(x_j, y_j)} \sum_{j=1}^{d} w(x_j, y_j) s(x_j, y_j), \qquad (1.8)$$

where $s(x_j, y_j)$ is a similarity component for the jth components of $\mathbf{x}$ and $\mathbf{y}$, and $w(x_j, y_j)$ is either one or zero depending on whether a comparison for the jth component of the two data points is valid or not.

For different types of attributes, $s(x_j, y_j)$ and $w(x_j, y_j)$ are defined differently. If the jth attribute is continuous, then

$$s(x_j, y_j) = 1 - \frac{|x_j - y_j|}{R_j},$$

$$w(x_j, y_j) = \begin{cases} 0 & \text{if } x_j \text{ or } y_j \text{ is missing}, \\ 1 & \text{otherwise}, \end{cases}$$

where R_j is the range of the jth attribute.

If the jth attribute is binary, then

$$s(x_j, y_j) = \begin{cases} 1 & \text{if both } x_j \text{ and } y_j \text{ are "present"}, \\ 0 & \text{otherwise}, \end{cases}$$

$$w(x_j, y_j) = \begin{cases} 0 & \text{if both } x_j \text{ and } y_j \text{ are "absent"}, \\ 1 & \text{otherwise}. \end{cases}$$

If the jth attribute is nominal or categorical, then

$$s(x_j, y_j) = \begin{cases} 1 & \text{if } x_j = y_j, \\ 0 & \text{otherwise}, \end{cases}$$

$$w(x_j, y_j) = \begin{cases} 0 & \text{if } x_j \text{ or } y_j \text{ is missing}, \\ 1 & \text{otherwise}. \end{cases}$$

A general distance measure was defined similarly in (Gower, 1971). Ichino (1988) and Ichino and Yaguchi (1994) proposed a generalized Minkowski distance, which was also presented in (Gan et al., 2007).

1.4 Hierarchical Clustering Algorithms

A hierarchical clustering algorithm is a clustering algorithm that divides a dataset into a sequence of nested partitions. Hierarchical clustering algorithms can be further classified into two categories: agglomerative hierarchical clustering algorithms and divisive hierarchical clustering algorithms.

An agglomerative hierarchical algorithm starts with every single record as a cluster and then repeats merging the closest pair of clusters according to some similarity criteria until only one cluster is left. For example, Figure 1.3 shows an agglomerative clustering of a dataset with 5 records.

In contrast to agglomerative clustering algorithms, a divisive clustering algorithm starts with all records in a single cluster and then repeats splitting large clusters into smaller ones until every cluster contains only a single record. Figure 1.4 shows an example of divisive clustering of a dataset with 5 records.

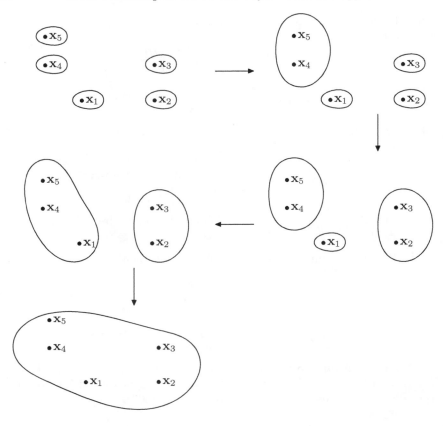

FIGURE 1.3: Agglomerative clustering.

1.4.1 Agglomerative Hierarchical Algorithms

Based on different ways to calculate the distance between two clusters, agglomerative hierarchical clustering algorithms can be classified into the following several categories (Murtagh, 1983):

(a) Single linkage algorithms;

(b) Complete linkage algorithms;

(c) Group average algorithms;

(d) Weighted group average algorithms;

(e) Ward's algorithms;

(f) Centroid algorithms;

(g) Median algorithms;

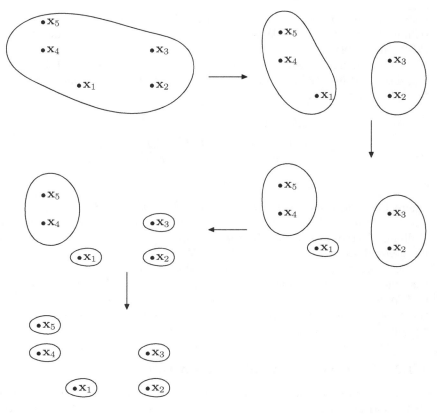

FIGURE 1.4: Divisive clustering.

(h) Other agglomerative algorithms that do not fit into the above categories.

For algorithms in the first seven categories, we can use the Lance-Williams recurrence formula (Lance and Williams, 1967a,b) to calculate the distance between an existing cluster and a cluster formed by merging two existing clusters. The Lance-Williams formula is defined as

$$\begin{aligned}&D(C_k, C_i \cup C_j)\\ =\ &\alpha_i D(C_k, C_i) + \alpha_j D(C_k, C_j)\\ &+\beta D(C_i, C_j) + \gamma |D(C_k, C_i) - D(C_k, C_j)|,\end{aligned}$$

where C_k, C_i, and C_j are three clusters, $C_i \cup C_j$ denotes the cluster formed by merging clusters C_i and C_j, $D(\cdot, \cdot)$ is a distance between clusters, and α_i, α_j, β, and γ are adjustable parameters. Section 12.1 presents various values of these parameters.

When the Lance-Williams formula is used to calculate distances, the single linkage and the complete linkage algorithms induce a metric on the dataset known as the ultrametric (Johnson, 1967). However, other agglomerative algorithms that use the Lance-Williams formula might not produce an ultrametric (Milligan, 1979).

A more general recurrence formula has been proposed by Jambu (1978) and discussed in Gordon (1996) and Gan et al. (2007). The general recurrence formula is defined as

$$\begin{aligned}&D(C_k, C_i \cup C_j)\\ &= \alpha_i D(C_k, C_i) + \alpha_j D(C_k, C_j)\\ &+ \beta D(C_i, C_j) + \gamma |D(C_k, C_i) - D(C_k, C_j)|\\ &+ \delta_i h(C_i) + \delta_j h(C_j) + \epsilon h(C_k),\end{aligned}$$

where $h(C)$ denotes the height of cluster C in the dendrogram, and δ_i, δ_j, and ϵ are adjustable parameters. Other symbols are the same as in the Lance-Williams formula. If we let the three parameters δ_i, δ_j, and ϵ be zeros, then the general formula becomes the Lance-Williams formula.

Some other agglomerative hierarchical clustering algorithms are based on the general recurrence formula. For example, the flexible algorithms (Lance and Williams, 1967a), the sum of squares algorithms (Jambu, 1978), and the mean dissimilarity algorithms (Holman, 1992; Gordon, 1996) are such agglomerative hierarchical clustering algorithms.

1.4.2 Divisive Hierarchical Algorithms

Divisive hierarchical algorithms can be classified into two categories: monothetic and polythetic (Willett, 1988; Everitt, 1993). A monothetic algorithm divides a dataset based on a single specified attribute. A polythetic algorithm divides a dataset based on the values of all attributes.

Given a dataset containing n records, there are $2^n - 1$ nontrivial different ways to split the dataset into two pieces (Edwards and Cavalli-Sforza, 1965). As a result, it is not feasible to enumerate all possible ways of dividing a large dataset. Another difficulty of divisive hierarchical clustering is to choose which cluster to split in order to ensure monotonicity.

Divisive hierarchical algorithms that do not consider all possible divisions and that are monotonic do exist. For example, the algorithm DIANA (DIvisive ANAlysis) is such a divisive hierarchical clustering algorithm (Kaufman and Rousseeuw, 1990).

1.4.3 Other Hierarchical Algorithms

In the previous two subsections, we presented several classic hierarchical clustering algorithms. These classic hierarchical clustering algorithms have drawbacks. One drawback of these algorithms is that they are sensitive to noise and outliers. Another drawback of these algorithms is that they cannot handle large datasets since their computational complexity is at least $O(n^2)$ (Xu and

Wunsch, II, 2009), where n is the size of the dataset. Several hierarchical clustering algorithms have been developed in an attempt to improve these drawbacks. For example, BIRCH (Zhang et al., 1996), CURE (Guha et al., 1998), ROCK (Guha et al., 2000), and Chameleon (Karypis et al., 1999) are such hierarchical clustering algorithms.

Other hierarchical clustering algorithms have also been developed. For example, Leung et al. (2000) proposed an agglomerative hierarchical clustering algorithm based on the scale-space theory in human visual system research. Li and Biswas (2002) proposed a similarity-based agglomerative clustering (SBAC) to cluster mixed-type data. Basak and Krishnapuram (2005) proposed a divisive hierarchical clustering algorithm based on unsupervised decision trees.

1.4.4 Dendrograms

Results of a hierarchical clustering algorithm are usually visualized by dendrograms. A dendrogram is a tree in which each internal node is associated with a height. The heights in a dendrogram satisfy the following ultrametric conditions (Johnson, 1967):

$$h_{ij} \leq \max\{h_{ik}, h_{jk}\} \quad \forall i, j, k \in \{1, 2, \cdots, n\},$$

where n is the number of records in a dataset and h_{ij} is the height of the internal node corresponding to the smallest cluster to which both record i and record j belong.

Figure 1.5 shows a dendrogram of the famous Iris dataset (Fisher, 1936). This dendrogram was created by the single linkage algorithm with the Euclidean distance. From the dendrogram we see that the single linkage algorithm produces two natural clusters for the Iris dataset.

More information about dendrograms can be found in Gordon (1996), Gordon (1987), Sibson (1973), Jardine et al. (1967), Banfield (1976), van Rijsbergen (1970), Rohlf (1974), and Gower and Ross (1969). Gordon (1987) discussed the ultrametric conditions for dendrograms. Sibson (1973) presented a mathematical representation of a dendrogram. Rohlf (1974) and Gower and Ross (1969) discussed algorithms for plotting dendrograms.

1.5 Partitional Clustering Algorithms

A partitional clustering algorithm is a clustering algorithm that divides a dataset into a single partition. Partitional clustering algorithms can be further classified into two categories: hard clustering algorithms and fuzzy clustering algorithms. In hard clustering, each record belongs to one and only one clus-

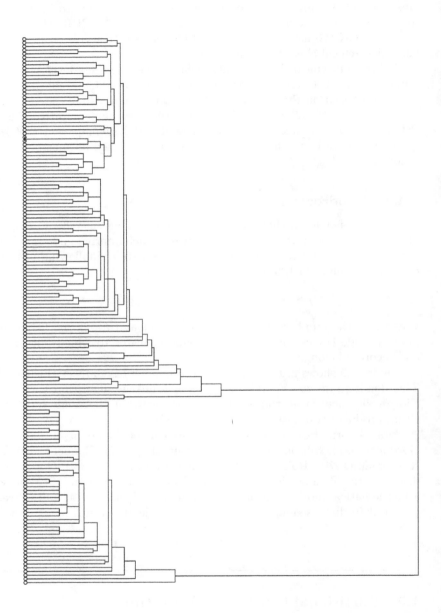

FIGURE 1.5: The dendrogram of the Iris dataset.

ter. In fuzzy clustering, a record can belong to two or more clusters with probabilities.

Suppose a dataset with n records is clustered into k clusters by a partitional clustering algorithm. The clustering result of the partitional clustering algorithm can be represented by a $k \times n$ matrix U defined as

$$U = \begin{pmatrix} u_{11} & u_{12} & \cdots & u_{1n} \\ u_{21} & u_{22} & \cdots & u_{2n} \\ \vdots & \vdots & \ddots & \vdots \\ u_{k1} & u_{k2} & \cdots & u_{kn} \end{pmatrix}. \tag{1.9}$$

The matrix U produced by a hard clustering algorithm has the following properties:

$$u_{ji} = 0 \text{ or } 1, \quad 1 \leq j \leq k, 1 \leq i \leq n, \tag{1.10a}$$

$$\sum_{j=1}^{k} u_{ji} = 1, \quad 1 \leq i \leq n, \tag{1.10b}$$

$$\sum_{i=1}^{n} u_{ji} > 0, \quad 1 \leq j \leq k. \tag{1.10c}$$

The matrix U produced by a fuzzy clustering algorithm has the following properties:

$$0 \leq u_{ji} \leq 1, \quad 1 \leq j \leq k, 1 \leq i \leq n, \tag{1.11a}$$

$$\sum_{j=1}^{k} u_{ji} = 1, \quad 1 \leq i \leq n, \tag{1.11b}$$

$$\sum_{i=1}^{n} u_{ji} > 0, \quad 1 \leq j \leq k. \tag{1.11c}$$

In this section, we present a survey of several partitional clustering algorithms. For a more comprehensive list of partitional clustering algorithms, readers are referred to (Xu and Wunsch, II, 2009) and (Gan et al., 2007).

1.5.1 Center-Based Clustering Algorithms

Center-based clustering algorithms (Zhang, 2003; Teboulle, 2007) are clustering algorithms that use a center to represent a cluster. Center-based clustering algorithms have two important properties (Zhang, 2003):

(a) They have a clearly defined objective function;

(b) They have a low runtime cost.

The standard k-means algorithm is a center-based clustering algorithm and is also one of the most popular and simple clustering algorithms. Although the k-means algorithm was first published in 1955 (Jain, 2010), more than 50 years ago, it is still widely used today.

Let $X = \{\mathbf{x}_1, \mathbf{x}_2, \cdots, \mathbf{x}_n\}$ be a dataset with n records. The k-means algorithm tries to divide the dataset into k disjoint clusters $C_1, C_2, \cdots, C_k$ by minimizing the following objective function

$$E = \sum_{i=1}^{k} \sum_{\mathbf{x} \in C_i} D(\mathbf{x}, \mu(C_i)),$$

where $D(\cdot, \cdot)$ is a distance function and $\mu(C_i)$ is the center of the cluster C_i and is usually defined as

$$\mu(C_i) = \frac{1}{|C_i|} \sum_{\mathbf{x} \in C_i} \mathbf{x}.$$

The standard k-means algorithm minimizes the objective function using an iterative process (Selim and Ismail, 1984; Bobrowski and Bezdek, 1991; Phillips, 2002).

The standard k-means algorithm has several variations (Gan et al., 2007). For example, the continuous k-means algorithm (Faber, 1994), the compare-means algorithm (Phillips, 2002), the sort-means algorithm (Phillips, 2002), the k-means algorithm based on kd-tree (Pelleg and Moore, 1999), and the timmed k-means algorithm (Cuesta-Albertos et al., 1997) are variations of the standard k-means algorithm.

Other center-based clustering algorithms include the k-modes algorithm (Huang, 1998; Chaturvedi et al., 2001), the k-probabilities algorithm (Wishart, 2002), the k-prototypes algorithm (Huang, 1998), the x-means algorithm (Pelleg and Moore, 1999), the k-harmonic means algorithm (Zhang et al., 2001), the mean-shift algorithm (Fukunaga and Hostetler, 1975; Cheng, 1995; Comaniciu and Meer, 1999, 2002), and the maximum-entropy clustering (MEC) algorithm (Rose et al., 1990).

1.5.2 Search-Based Clustering Algorithms

Many data clustering algorithms are formulated as some optimization problems (Dunn, 1974a; Ng and Wong, 2002), which are complicated and have many local optimal solutions. Most of the clustering algorithms will stop when they find a locally optimal partition of the dataset. That is, most of the clustering algorithms may not be able to find the globally optimal partition of the dataset. For example, the fuzzy k-means algorithm (Selim and Ismail, 1984) is convergent but may stop at a local minimum of the optimization problem.

Search-based clustering algorithms are developed to deal with the problem mentioned above. A search-based clustering algorithm aims at finding a globally optimal partition of a dataset by exploring the solution space of the underlying optimization problem. For example, clustering algorithms based on genetic algorithms (Holland, 1975) and tabu search (Glover et al., 1993) are search-based clustering algorithms.

Al-Sultan and Fedjki (1997) proposed a clustering algorithm based on a tabu search technique. Ng and Wong (2002) improved the fuzzy k-means algorithm using a tabu search algorithm. Other search-based clustering algorithms include the J-means algorithm (Mladenović and Hansen, 1997), the genetic k-means algorithm (Krishna and Narasimha, 1999), the global k-means algorithm (Likas and Verbeek, 2003), the genetic k-modes algorithm (Gan et al., 2005), and the SARS algorithm (Hua et al., 1994).

1.5.3 Graph-Based Clustering Algorithms

Clustering algorithms based on graphs have also been proposed. A graph is a collection of vertices and edges. In graph-based clustering, a vertex represents a data point or record and an edge between a pair of vertices represents the similarity between the two records represented by the pair of vertices (Xu and Wunsch, II, 2009). A cluster usually corresponds to a highly connected subgraph (Hartuv and Shamir, 2000).

Several graph-based clustering algorithms have been proposed and developed. The chameleon (Karypis et al., 1999) algorithm is a graph-based clustering algorithm that uses a sparse graph to represent a dataset. The CACTUS algorithm (Ganti et al., 1999) is another graph-based clustering algorithm that uses a graph, called the similarity graph, to represent the inter-attribute and intra-attribute summaries. The ROCK algorithm (Guha et al., 2000) is an agglomerative hierarchical clustering algorithm that uses a graph connectivity to calculate the similarities between data points.

Gibson et al. (2000) proposed a clustering algorithm based on hypergraphs and dynamical systems. Foggia et al. (2007) proposed a graph-based clustering algorithm that is able to find clusters of any size and shape and does not require specifying the number of clusters. Foggia et al. (2009) compared the performance of several graph-based clustering algorithms.

Most of the graph-based clustering algorithms mentioned above use graphs as data structures and do not use graph analysis. Spectral clustering algorithms, which are also graph-based clustering algorithms, first construct a similarity graph and then use graph Laplacian matrices and standard linear algebra methods to divide a dataset into a number of clusters. Luxburg (2007) presented a tutorial on spectral clustering. Filippone et al. (2008) also presented a survey of spectral clustering. Interested readers are referred to these two papers and the book by Ding and Zha (2010).

1.5.4 Grid-Based Clustering Algorithms

Grid-based clustering algorithms are very efficient for clustering very large datasets since these algorithms perform clustering on the grid cells rather than the individual data points. A typical grid-based clustering algorithm consists of the following basic steps (Grabusts and Borisov, 2002):

(a) Construct a grid structure by dividing the data space into a finite number of cells;

(b) Calculate the density for each cell;

(c) Sort the cells based on their densities;

(d) Identify cluster centers;

(e) Traverse neighbor cells.

The STING (STatistical INformation Grid-based) algorithm is a grid-based clustering algorithm proposed by Wang et al. (1997) for clustering spatial datasets. STING was designed for clustering low-dimensional data and cannot be scalable for clustering high-dimensional data. Keim and Hinneburg (1999) proposed a grid-based clustering algorithm, OptiGrid, for clustering high-dimensional data. Schikuta and Erhart (1997) proposed a BANG-clustering algorithm, which is also a grid-based clustering algorithm. Nagesh et al. (2001) proposed a clustering algorithm based on adaptive grids. Other grid-based clustering algorithms include GRIDCLUS (Schikuta, 1996), GDILC (Zhao and Song, 2001), and WaveCluster (Sheikholeslami et al., 2000).

More recently, Qiu et al. (2007) also proposed a grid-based clustering algorithm that is capable of dealing with high dimensional datasets. Park and Lee (2007) proposed a grid-based subspace clustering algorithm to cluster data streams. Lin et al. (2008) proposed a grid-based clustering algorithm that is less influenced by the size of the grid cells than many other grid-based clustering algorithms.

1.5.5 Density-Based Clustering Algorithms

Density-based clustering algorithms are a kind of clustering algorithm that is capable of finding arbitrarily shaped clusters. In density-based clustering, a cluster is defined as a dense region surrounded by low-density regions. Usually, density-based clustering algorithms do not require specifying the number of clusters since these algorithms can automatically detect clusters and the number of clusters (El-Sonbaty et al., 2004). One drawback of most density-based clustering algorithms is that it is hard to determine certain parameters, such as the density threshold.

Popular density-based clustering algorithms include DBSCAN (Ester et al., 1996) and its variations and extensions such as GDBSCAN (Sander et al., 1998), PDBSCAN (Xu et al., 1999), DBCluC (Zaiane and Lee, 2002).

BRIDGE (Dash et al., 2001) is a hybrid clustering algorithm that is based on the k-means algorithm and DBSCAN. Other density-based clustering algorithms include DENCLUE (Keim and Hinneburg, 1999) and CUBN (Wang and Wang, 2003).

1.5.6 Model-Based Clustering Algorithms

Mode-based clustering algorithms are clustering algorithms based on probability models. The term *model* usually refers to the type of constraints and geometric properties of the covariance matrices (Martinez and Martinez, 2005). In model-based clustering, the data are viewed as samples coming from a mixture of probability distributions, each of which represents a cluster.

In model-based clustering, there are two approaches to formulate a model for the composite of clusters: the classification likelihood approach and the mixture likelihood approach (Celeux and Govaert, 1995; Fraley and Raftery, 1998). In the classification likelihood approach, the objective function

$$\mathcal{L}_C(\Theta_1, \Theta_2, \cdots, \Theta_k; \gamma_1, \gamma_2, \cdots, \gamma_n | X) = \prod_{i=1}^{n} f_{\gamma_i}(\mathbf{x}_i | \Theta_{\gamma_i})$$

is maximized, where $\gamma_i = j$ if record $\mathbf{x}_i$ belongs to the jth component or cluster, Θ_j ($j = 1, 2, \cdots, k$) are parameters, and $X = \{\mathbf{x}_1, \mathbf{x}_2, \cdots, \mathbf{x}_n\}$ is a dataset.

In the mixture likelihood approach, the objective function

$$\mathcal{L}_M(\Theta_1, \Theta_2, \cdots, \Theta_k; \tau_1, \tau_2, \cdots, \tau_k | X) = \prod_{i=1}^{n} \sum_{j=1}^{k} \tau_j f_j(\mathbf{x}_i | \Theta_j)$$

is maximized, where $\tau_j \geq 0$ is the probability that a record belongs to the jth component and

$$\sum_{j=1}^{k} \tau_j = 1.$$

Some classical and powerful model-based clustering algorithms are based on Gaussian mixture models (Banfield and Raftery, 1993). Celeux and Govaert (1995) presented sixteen model-based clustering algorithms based on different constraints on the Gaussian mixture model. These algorithms use the EM algorithm (Dempster et al., 1977; Meng and van Dyk, 1997) to estimate the parameters.

Bock (1996) presented a survey of data clustering based on a probabilistic and inferential framework. Edwards and Cavalli-Sforza (1965), Day (1969), Wolfe (1970), Scott and Symons (1971), and Binder (1978) presented some early work on model-based clustering. Other model-based clustering algorithms are discussed in (Gan et al., 2007, Chapter 14).

1.5.7 Subspace Clustering Algorithms

Almost all conventional clustering algorithms do not work well for high dimensional datasets due to the following two problems associated with high dimensional data. First, the distance between any two data points in a high dimensional space becomes almost the same (Beyer et al., 1999). Second, clusters of high dimensional data are embedded in the subspaces of the data space and different clusters may exist in different subspaces (Agrawal et al., 1998). Subspace clustering algorithms are clustering algorithms that are capable of finding clusters embedded in subspaces of the original data space.

Most subspace clustering algorithms can be classified into two major categories (Parsons et al., 2004): top-down algorithms and bottom-up algorithms. In top-down subspace clustering, a conventional clustering is performed and then the subspace of each cluster is evaluated. In bottom-up subspace clustering, dense regions in low dimensional spaces are identified and then these dense regions are combined to form clusters.

Examples of top-down subspace clustering algorithms include PART (Cao and Wu, 2002), PROCLUS (Aggarwal et al., 1999), ORCLUS (Aggarwal and Yu, 2000), and FINDIT (Woo and Lee, 2002), and δ-cluster (Yang et al., 2002). Examples of bottom-up subspace clustering algorithms include CLIQUE (Agrawal et al., 1998), ENCLUS (Cheng et al., 1999), MAFIA (Goil et al., 1999), CLTree (Liu et al., 2000), DOC (Procopiuc et al., 2002), and CBF (Chang and Jin, 2002).

There are also some subspace clustering algorithms that do not fit into the aforementioned categories. For example, the FSC (Fuzzy Subspace Clustering) algorithm (Gan et al., 2006a; Gan and Wu, 2008) is subspace clustering, which is very similar to the k-means algorithm. The FSC algorithm uses a weight to represent the importance of a dimension or attribute to a cluster and incorporates the weights into the optimization problem.

Other subspace clustering algorithms include SUBCAD (Gan and Wu, 2004), the MSSC (Mean Shift for Subspace Clustering) algorithm (Gan, 2007; Gan et al., 2007), and the grid-based subspace clustering algorithm (Park and Lee, 2007). Recent work on subspace clustering is presented in Patrikainen and Meila (2006), Jing et al. (2007), Müller et al. (2009), Kriegel et al. (2009), and Deng et al. (2010).

1.5.8 Neural Network-Based Clustering Algorithms

Neural network-based clustering algorithms are related to the concept of competitive learning (Fukushima, 1975; Grossberg, 1976a,b; Rumelhart and Zipser, 1986). There are two types of competitive learning paradigms: hard competitive learning and soft competitive learning. Hard competitive learning is also known as winner-take-all or crisp competitive learning (Baraldi and Blonda, 1999a,b). In hard competitive learning, only a particular winning neuron that matches best with the given input pattern is allowed to learn. In soft competitive learning, all neurons have the opportunity to learn based on the input pattern. Hence soft competitive learning is also known as winner-take-most competitive learning (Baraldi and Blonda, 1999a,b).

Examples of neural network-based clustering algorithms include PART (Cao and Wu, 2002, 2004), which is also a subspace clustering algorithm. The PARTCAT algorithm (Gan et al., 2006b) is based on PART but was developed for clustering categorical data. Several other neural network-based clustering algorithms are presented and discussed in (Xu and Wunsch, II, 2009).

1.5.9 Fuzzy Clustering Algorithms

Most of the clustering algorithms presented in the previous several subsections are hard clustering algorithms, which require that each record belongs to one and only one cluster. Since Zadeh (1965) introduced the concept of fuzzy sets, fuzzy set theory has been applied to the area of data clustering (Bellman et al., 1966; Ruspini, 1969). In fuzzy clustering, a record is allowed to belong to two or more clusters with probabilities.

Examples of fuzzy clustering algorithms include the fuzzy k-means algorithm (Bezdek, 1974, 1981a), the fuzzy k-modes algorithm (Huang and Ng, 1999), and the c-means algorithm (Dunn, 1974a,b; Selim and Ismail, 1986; Hathaway et al., 1987; Hathaway and Bezdek, 1984; Fukuyama and Sugeno, 1989). For more information about fuzzy clustering, readers are referred to the book by Höppner et al. (1999) and the survey paper by Döring et al. (2006). For recent work on fuzzy clustering, readers are referred to Zhang et al. (2008) and Campello et al. (2009).

1.6 Cluster Validity

Cluster validity is a collection of quantitative and qualitative measures that are used to evaluate and assess the results of clustering algorithms (Jain and Dubes, 1988; Jain, 2010). Cluster validity indices can be defined based on three fundamental criteria: internal criteria, relative criteria, and external criteria (Jain and Dubes, 1988; Theodoridis and Koutroubas, 1999; Halkidi et al., 2002a,b). Both internal and external criteria are related to statistical testing.

In the external criteria approach, the results of a clustering algorithm are evaluated based on a prespecified structure imposed on the underlying dataset. Usually, external criteria require using the Monte Carlo simulation to do the evaluation (Halkidi et al., 2002b). Hence cluster validity based on external criteria is computationally expensive.

In the internal criteria approach, the results of a clustering algorithm are evaluated based only on quantities and features inherited from the underlying dataset. Cluster validity based on internal criteria can be used to assess results of hierarchical clustering algorithms as well as partitional clustering algorithms.

In the relative criteria approach, the results of a clustering algorithm are evaluated with other clustering results, which are produced by different clustering algorithms or the same algorithm but with different parameters. For example, a relative criterion is used to compare the results produced by the k-means algorithm with different parameter k (the number of clusters) in order to find the best clustering of the dataset.

1.7 Clustering Applications

Data clustering has been applied to many fields. According to Scoltock (1982), data clustering has been applied to the following five major groups:

(a) **Biology and zoology**. Clustering algorithms have been used to group animals and plants and develop taxonomies. In fact, data clustering algorithms were first developed in this field in which clustering is known as taxonomy analysis.

(b) **Medicine and psychiatry**. Clustering algorithms have been used to group diseases, including mental and physical diseases.

(c) **Sociology, criminology, anthropology, and archaeology**. In fields in this group, data clustering algorithms have been used to group organizations, criminals, crimes, and cultures.

(d) **Geology, geography, and remote sensing**. In the fields in this group, clustering algorithms have been used to group rock samples, sediments, cities, and land-use patterns.

(e) **Information retrieval, pattern recognition, market research, and economics**. Clustering algorithms have been used to analyze images, documents, industries, consumers, products, and markets.

The above list shows the applications of data clustering about 30 years ago. Nowadays, data clustering has a very broad application. For example, data clustering has been applied to areas such as computational intelligence, machine learning, electrical engineering, genetics, and insurance (Xu and Wunsch, II, 2009; Everitt et al., 2001).

1.8 Literature of Clustering Algorithms

Since the k-means algorithm was published about 50 years ago (Jain, 2010), thousands of papers and a number of books on data clustering have been published. In this section, some books and survey papers related to data clustering are listed. For a list of journals and conference proceedings in which papers on data clustering are published, readers are referred to Gan et al. (2007).

1.8.1 Books on Data Clustering

The following is a list of books related to data clustering that were published in the past 40 years.

(a) *Principles of Numerical Taxonomy* (Sokal and Sneath, 1963). This book reviewed most of the applications of numerical taxonomy in the field of biology at that time. A new edition of this book is titled *Numerical Taxonomy: The Principles and Practice of Numerical Classification* (Sokal and Sneath, 1973). Although the two books were written for researchers in the field of biology, these books reviewed much of the literature of cluster analysis and presented many clustering techniques.

(b) *Cluster Analysis: Survey and Evaluation of Techniques* by Bijnen (1973) selected a number of clustering techniques related to sociological and psychological research.

(c) *Cluster Analysis: A Survey* by Duran and Odell (1974) supplied an exposition of various works in the literature of cluster analysis at that time. Many references that played a role in developing the theory of cluster analysis are included in the book.

(d) *Cluster Analysis for Applications* (Anderberg, 1973). This book collected many clustering techniques and provided many FORTRAN procedures for readers to perform cluster analysis on real data.

(e) *Clustering Algorithms* (Hartigan, 1975). This book presented data clustering from a statistician's point of view. This book includes a wide range of procedures in FORTRAN, methods, and examples.

(f) *Cluster Analysis for Social Scientists* (Lorr, 1983). This book is an elementary book on data clustering and was written for researchers and graduate students in the social and behavioral sciences.

(g) *Algorithms for Clustering Data* (Jain and Dubes, 1988). This book was written for the scientific community and emphasizes informal algorithms for clustering data and interpreting results.

(h) *Introduction to Statistical Pattern Recognition* (Fukunaga, 1990). This book introduces fundamental mathematical tools for classification. This book also discusses data clustering based on statistics.

(i) *Finding Groups in Data—Introduction to Cluster Analysis* (Kaufman and Rousseeuw, 1990). This book was written for general users of data clustering algorithms, especially for people who do not have a strong background in mathematics and statistics.

(j) *Cluster Analysis* (Everitt, 1993). This book introduces data clustering for works in a variety of areas. This book includes many examples of clustering and describes several software packages for clustering.

(k) *Clustering for Data Mining: A Data Recovery Approach* (Mirkin, 2005). This book presents data recovery models based on the k-means algorithm and hierarchical algorithms as well as reviews some clustering algorithms.

(l) *Data Clustering: Theory, Algorithms, and Applications* (Gan et al., 2007). This book presents a comprehensive survey of data clustering and includes some simple C++ and MATLAB clustering programs. This book was written as a reference book.

(m) *Advances in Fuzzy Clustering and Its Applications* (Valente de Oliveira and Pedrycz, 2007). This book presents a comprehensive and in-depth list of many state-of-the-art fuzzy clustering algorithms. Fundamentals of fuzzy clustering, visualization techniques, and real-time clustering are also presented.

(n) *Algorithms for Fuzzy Clustering: Methods in c-Means Clustering with Applications* (Miyamoto et al., 2008). This book presents a family of fuzzy c-means algorithms with an emphasis on entropy-based methods and uncovers the theoretical and methodological differences between the standard method and the entropy-based method.

(o) *Clustering* (Xu and Wunsch, II, 2009). This book is also a reference book on data clustering and includes many classic and recently developed clustering algorithms.

(p) *Spectral Clustering, Ordering and Ranking* (Ding and Zha, 2010). This book presents some recent advances in spectral clustering. Applications to web and text mining and genomics are also included.

1.8.2 Surveys on Data Clustering

The following is a list of survey papers related to data clustering.

(a) *A Review of Classification* (Cormack, 1971)

(b) *Cluster Analysis: Survey and Evaluation of Techniques* (Bijnen, 1973)

(c) *Cluster Analysis - A Survey* (Duran and Odell, 1974)

(d) *A Survey of the Literature of Cluster Analysis* (Scoltock, 1982)

(e) *A Survey of Recent Advances in Hierarchical Clustering Algorithms* (Murtagh, 1983)

(f) *Counting Dendrograms: A Survey* (Murtagh, 1984)

(g) *A Review of Hierarchical Classification* (Gordon, 1987)

(h) *A Survey of Fuzzy Clustering* (Yang, 1993)

(i) *A Survey of Fuzzy Clustering Algorithms for Pattern Recognition (I)* (Baraldi and Blonda, 1999a)

(j) *A Survey of Fuzzy Clustering Algorithms for Pattern Recognition (II)* (Baraldi and Blonda, 1999b)

(k) *Data Clustering: A Review* (Jain et al., 1999)

(l) *Statistical Pattern Recognition: A Review* (Jain et al., 2000)

(m) *Survey of Clustering Data Mining Techniques* (Berkhin, 2002)

(n) *Cluster Analysis for Gene Expression Data: A Survey* (Jiang et al., 2004)

(o) *Subspace Clustering for High Dimensional Data: A Review* (Parsons et al., 2004)

(p) *Mining Data Streams: A Review* (Gaber et al., 2005)

(q) *Survey of Clustering Algorithms* (Xu and Wunsch II, 2005)

(r) *Data Analysis with Fuzzy Clustering Methods* (Döring et al., 2006)

(s) *A Survey of Kernel and Spectral Methods for Clustering* (Filippone et al., 2008)

(t) *Clustering High-Dimensional Data: A Survey on Subspace Clustering, Pattern Based Clustering, and Correlation Clustering* (Kriegel et al., 2009)

(u) *Data Clustering: 50 Years beyond k-means* (Jain, 2010)

1.9 Summary

In this chapter, were introduced some basic concepts of data clustering, some similarity and dissimilarity measure, and some clustering algorithms. The chapter did not go into detail about most of the topics. Instead, it provided references to these topics to which interested readers can refer for more information. In particular, it listed a number of books and survey papers related to data clustering in this chapter. These books and papers are valuable resources to get information on data clustering.

Chapter 2

The Unified Modeling Language

The Unified Modeling Language (UML) is a standardized general-purpose modeling language used to specify, document, and visualize software systems. The UML includes a set of well-defined and expressive notations used to create visual models of software systems. This standard set of notation makes it possible for an architecture to be formulated and communicated unambiguously to others. Since the Object Management Group (OMG), an international not-for-profit consortium that creates ards for the computer industry, adopted the UML as a standard in 1997, the UML has been revised many times. In this chapter, we give a brief introduction to UML 2.0. For further details on UML 2.0, readers are referred to Booch et al. (2007). In what follows, UML means UML 2.0.

There are two types of UML diagrams: structure diagrams and behavior diagrams (see Figure 2.1). The UML structure diagrams are used to show the static structure of elements in a software system. The UML structure diagrams include the following six types of diagrams: package diagram, class diagram, component diagram, deployment diagram, object diagram, and composite structure diagram. The UML behavior diagrams are used to describe the dynamic behavior of a software system. The UML behavior diagrams include the following seven types of diagrams: use case diagram, activity diagram, state machine diagram, sequence diagram, communication diagram, and interaction overview diagram, the last four of which are also called interaction diagrams.

In this chapter, we introduce several types of UML diagrams.

2.1 Package Diagrams

The UML package diagram is used to organize the artifacts of the object-oriented development process to clearly present the analysis of the problem and the associated design. A package diagram has three essential elements: packages, their visibility, and their dependencies.

The notation for a package is a rectangle with a tab on the top left. If a package is empty, i.e., the package contains no UML elements, the name of

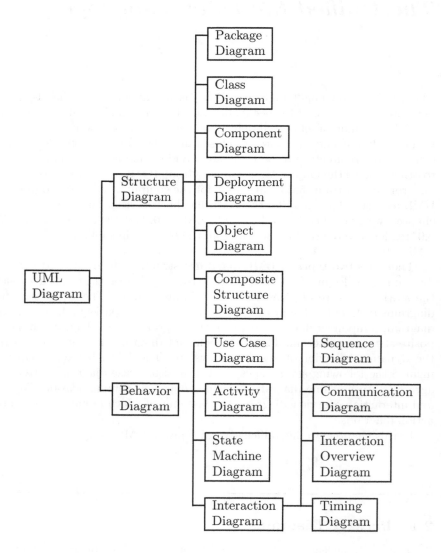

FIGURE 2.1: UML diagrams.

the package is placed in the interior of the rectangle. Figure 2.2(a) shows an empty UML package. If a package contains UML elements, the name of the package is placed within the tab, the small rectangle above the main rectangle. Figure 2.2(b) shows a UML package containing two classes.

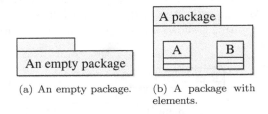

(a) An empty package.   (b) A package with elements.

FIGURE 2.2: UML packages.

Packages can contain other UML elements such as classes and packages. For example, Figure 2.3 shows that Package A contains Package B and Package C. In Figure 2.3, the constituent packages are placed within the containing package. We use this nested notation when there are fewer elements to be shown or because we have a focused concern. Figure 2.4 shows Package A with its constituent packages placed outside it and linked to it with nest links. The tail of the nest link is located at the contained element, and the head of the nest link (a circle with a plus sign inside) is located at the containing element.

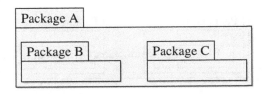

FIGURE 2.3: A UML package with nested packages placed inside.

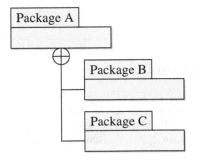

FIGURE 2.4: A UML package with nested packages placed outside.

The visibility of an element in a package is defined by the package and can be either public or private. An element with public visibility is visible to elements within its containing package, including nested packages, and to external elements. An element with private visibility is visible only to elements within its containing package, including nested packages.

The public visibility notation is "+" and the private visibility notation is "-". On a UML diagram, the visibility notation is placed in front of the element name. Figure 2.5 shows a package containing a public element and a private element.

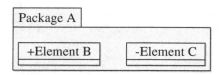

FIGURE 2.5: The visibility of elements within a package.

Dependencies between UML elements are denoted by a dashed arrow with an open arrowhead, where the tail of the arrow is located at the element having the dependency and the head is located at the element supporting the dependency. Figure 2.6 shows that Element D is dependent on Element E.

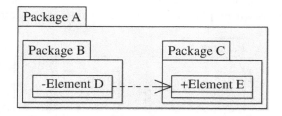

FIGURE 2.6: The UML dependency notation.

2.2 Class Diagrams

A class diagram is a type of static structure diagram and the main building block in object-oriented modeling (Booch et al., 2007). In the UML, a class diagram is used to describe the structure of a system by showing the system's classes and their relationships. Classes and their relationships are the two essential elements of a class diagram.

In a class diagram, a simple class is represented by a box, which consists of three compartments: the upper compartment containing the name of the class, the middle compartment containing the attributes of the class, and the bottom compartment containing the operations of the class. The format of an attribute is

```
visibility AttributeName: Type
```

and the format of an operation is

```
visibility OperationName(parameterName:parameterType):
                     returnType
```

The visibility of an attribute or operation is denoted by a lock. An open, semi-open, or closed lock means that the attribute or operation has public, protected, or private visibility. Figure 2.7 shows an example class in a class diagram. In some books, plus sign (+), pound sign (#), or minus sign (-) is used to indicate public, protected, or private visibility.

Shape
🔒 _name:string
▬ setName(name:string)
▬ getName():string
🔒 checkName():void

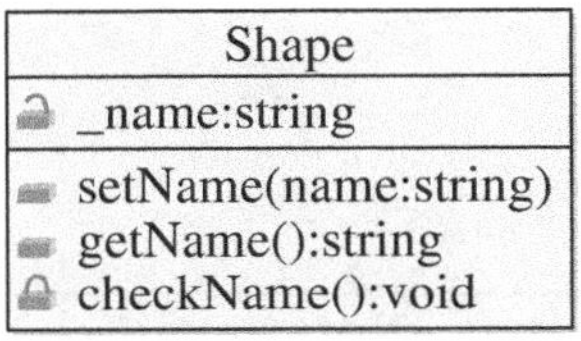

FIGURE 2.7: Notation of a class.

If a class is abstract, then its name in the class icon is typeset in an oblique font. For example, Figure 2.8 shows a **Shape** class, which is abstract.

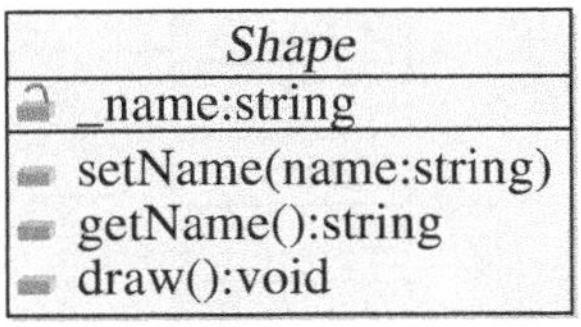

Shape
🔒 _name:string
▬ setName(name:string)
▬ getName():string
▬ draw():void

FIGURE 2.8: Notation of an abstract class.

If a class is a parameterized class or template, then the class is represented by a simple class but with a dashed-line box in the upper right-hand corner, which contains the formal parameters of the template. Figure 2.9 shows a template class and one of its realizations.

In class diagrams, different relationships between classes are represented by different types of arrows. Table 2.1 gives a list of relationships and their notation. Essentially, there are six relationships: association, generalization, aggregation, composition, realization, and dependency. These relationships can be grouped into three categories: instance-level relationships, class-level relationships, and general relationships (see Figure 2.10).

An association represents a semantic connection between two classes and is often labeled with a noun phrase denoting the nature of the relationship.

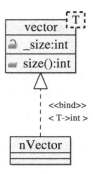

FIGURE 2.9: A template class and one of its realizations.

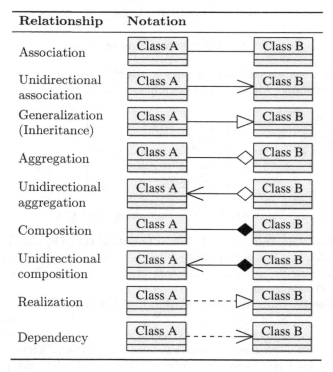

TABLE 2.1: Relationships between classes and their notation.

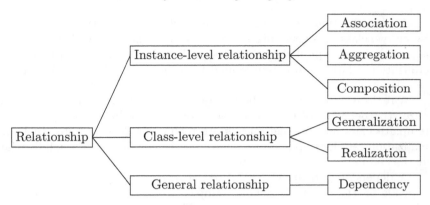

FIGURE 2.10: Categories of relationships.

Bidirectional and unidirectional are the most common types of association. An association can be further adorned with its multiplicity, which is applied to the target end of the association and denotes the number of instances of the target class that participate in the association. Table 2.2 gives some common multiplicities.

Multiplicity	Meaning
0	Zero instances
0 .. 1	Zero or one instance
0 .. *	Zero or more instances
*	Zero or more instances
2 .. *	Two or more instances
2 .. 6	2, 3, 4, 5, or 6 instances

TABLE 2.2: Some common multiplicities.

An aggregation is a type of association that represents whole/part relationship between two classes. Since an aggregation is a type of association, it can have the same adornments that an association can. In UML, an aggregation is represented by a line with a hollow diamond located at the end denoting the aggregate or the whole.

A composition represents a containment relationship between two classes. In UML, a composition is represented by a line with a filled diamond located at the end denoting the containing class. Unlike an aggregation, where the lifetimes of the objects of the containing class and the lifetimes of the objects of the contained class are independent, a composition implies that the object of the containing class owns the objects of the contained class, that is, if the object of the containing class is destroyed, the objects of the contained class are also destroyed.

A generalization represents a generalization/specialization relationship between two classes. A generalization occurs when one class, called the subclass, is a specialized form of the other class, called the superclass. In UML, a generalization relationship is represented by a line with a hollow triangle at the end denoting the superclass.

A realization relationship between two classes indicates that one class realizes the behavior of the other class. For example, relationships between a template class and its realizations are realization relationships. In UML, a realization relationship is represented by a dashed line with a hollow triangle at the end denoting the template.

A dependency relationship between two classes indicates that one class depends on the other class. For example, a dependency relationship occurs when one class is a parameter or local variable of a method of another class.

2.3 Use Case Diagrams

Use case diagrams are used to show the functionality provided by a software system in terms of actors, which are entities interfacing with the system. Use case diagrams depict what interacts with the system and show what actors want the system to do.

On a UML diagram, the actor notation is a stylized stick figure with a name below the figure and the use case notation is an oval with a description inside the oval as shown in Figure 2.11.

FIGURE 2.11: The UML actor notation and use case notation.

A use case diagram shows which actors use which use cases by connecting them via lines as shown in Figure 2.12.

The details of the functionality provided by a use case are described in a use case specification, which includes the name of the use case, its purpose, the optimistic flow (i.e., the flow when everything goes as one intended), and some pragmatic flows (i.e., flows where things do not occur as one intended).

There are three relationships among use cases that are used often in practice: ≪include≫, ≪extend≫, and generalization (see Figure 2.13). The ≪include≫ and ≪extend≫ relationships are used primarily for organizing use case models. The generalization relationship is used to represent use cases with common behaviors.

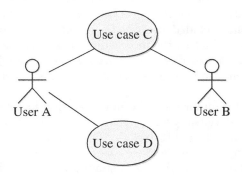

FIGURE 2.12: A UML use case diagram.

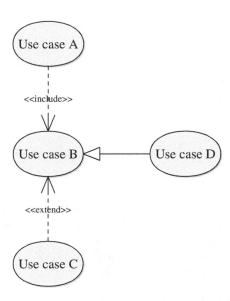

FIGURE 2.13: Notation of relationships among use cases.

The UML notation for the ≪include≫ relationship is a dashed arrow with an open head and the label "≪include≫" in the middle of the arrow as shown in Figure 2.13. The tail of the arrow is located at the including use case and the head is located at the included use case. The ≪include≫ relationship indicates that the included use case is required for the completion of the including use case. That is, the included use case must be executed when the including use case is executed. In Figure 2.13, for example, Use case B must be executed when Use case A is executed because Use case A includes Use case B.

The UML notation for the ≪extend≫ relationship is the same as that for the ≪include≫ relationship except that the label in the middle of the arrow is "≪extend≫". The ≪extend≫ relationship indicates that the extending use case is optional to the extended use case.

Maksimchuk and Naiburg (2005) summarized the following four key differences of the ≪include≫ relationship and the ≪extend≫ relationship:

(a) The extending use case is optional, but the included use case is not;

(b) The base use case (i.e., the included or extended use case) is complete without the extending use case, but is not without the included use case;

(c) The execution of the extending use case is conditional, but that of the included use case is not;

(d) The extending use case changes the behavior of the base use case, but the included case does not.

The UML notation for the generalization relationship is a solid line ending in a hollow triangle drawn from the generalized use case to the more general use case as shown in Figure 2.13.

2.4 Activity Diagrams

In the UML, activity diagrams are used to represent workflows of stepwise activities and actions in a system. There are two primary types of elements in an activity diagram: action nodes and control nodes. Control nodes can be classified into three categories: initial and final, decision and merge, and fork and join. Final control nodes can be further classified into two categories: activity final and flow final.

In activity diagrams, activities and actions are represented by rounded rectangles. Decisions are represented by diamonds. Start and end of concurrent activities are represented by bars. The start of the workflow is represented by a black circle and the end of the workflow is represented by an encircled black circle. An example activity diagram is shown in Figure 2.14.

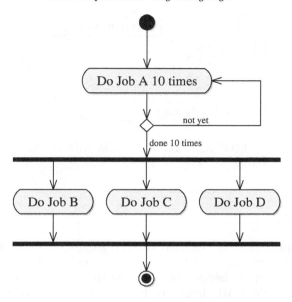

FIGURE 2.14: An activity diagram.

Unlike the notation for activity final, the notation for flow final is an encircled cross. Figure 2.15 shows an activity diagram with a flow final node.

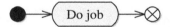

FIGURE 2.15: An activity diagram with a flow final node.

2.5 Notes

A note is non-UML diagram element that is used to add comments or textual information to any element of any diagram. In UML, a note is represented by a rectangle with a folded upper-right corner. Figure 2.16 shows a class diagram with two notes. If a note affects a particular element in a diagram, we connect it and the element by a dashed line. If a note applies to the diagram as a whole, we just leave it unconnected.

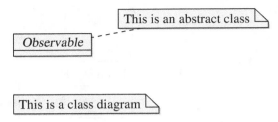

FIGURE 2.16: A diagram with notes.

2.6 Summary

UML diagrams can be classified into two groups: structure diagrams, which emphasize what things must be in a system, and behavior diagrams, which emphasize what must happen in a system. In this chapter, we gave a brief introduction to the UML diagrams. In particular, we introduced package diagrams, class diagrams, use case diagrams, and activity diagrams. More information about UML can be found in Maksimchuk and Naiburg (2005) and Sintes (2001).

Chapter 3

Object-Oriented Programming and C++

In this chapter, we give a brief introduction to object-oriented programming and the C++ programming language. We illustrate some concepts of object-oriented programming using simple C++ programs.

3.1 Object-Oriented Programming

Object-oriented programming (OOP) is a programming paradigm that is based on the concept of *objects*, which are data structures consisting of data and methods (Sintes, 2001).

Prior to the invention of OO programming, programmers used machine language programming, procedural programming, and modular programming to write their programs. In machine language programming, as the name suggests, programmers write their programs in the machine's binary languages and enter the programs directly into the computer's main memory via banks of switches. Obviously, machine language programming is error prone and the resulting code is difficult to read and maintain.

In procedural programming, programmers split their programs into a sequence of small procedures, which are easier to understand and debug. This represents a significant improvement over machine language programming. Procedural programming still has some drawbacks. For example, procedural programming limits code reuse and its data-centric nature requires each procedure to know how to properly manipulate the data.

In modular programming, programmers break their programs down into a number of modules, which combine data and the procedures used to process the data. Clearly, modular programming improves on some of the deficiencies found in procedural programming. However, modules are not extendable, that is, one has to break open the code and make changes directly when making changes to a module.

OO programming improves modular programming by adding inheritance and polymorphism to the module. In OO programming, a program is a collection of interacting objects, each of which models some aspect of the problem that one tries to solve. For example, an OO program implementing an online shopping cart contains item, shopping cart, and cashier objects.

In OO programming, a class defines the attributes and behaviors common to a type of object and an object is an instance of a class. Attributes of a class are visible characteristics of the class. For example, name and address of a person are attributes of some classes. A behavior of a class is an action taken by an object of the class when passed a message or in response to a state change. Some common behaviors of a class are constructors, accessors, and mutators. Constructors of a class are methods used to initialize objects of the class during their instantiation or creation. Accessors of a class are methods that allow one to access an object's internal data. Mutators are methods that allow one to change an object's internal state.

An OO program has the following five characteristics (Sintes, 2001):

(a) Natural. OO programming enables one to model a problem at a functional level rather than at the implementation level.

(b) Reliable. OO programs that are well-designed and carefully written are reliable. In addition, the modular nature of objects allows one to make changes to one part of the program without affecting other parts.

(c) Reusable. Objects can be reused in many different programs and can be extended.

(d) Maintainable. Well-designed OO programs are maintainable. For example, one needs to correct only one place in order to fix a bug.

(e) Extendable. The features of OO programming, such as inheritance and polymorphism, allow one to extend the existing code easily.

3.2 The C++ Programming Language

The C++ programming language is an object-oriented programming language developed by Bjarne Stroustrup starting in 1979 at Bell Labs (Stroustrup, 1994). C++ was built on the base of the C language, which evolved from two previous programming languages: BCPL (Basic Combined Programming Language) and B (Deitel and Deitel, 2009). BCPL was designed by Martin Richards as a language for writing operating systems and compilers. The B language was developed by Ken Thompson at Bell Labs on the base of BCPL and was used to create early versions of the UNIX operating system.

The process of creating and executing a C++ program consists of six phases: creating, preprocessing, compiling, linking, loading, and executing (Deitel and Deitel, 2009).

In Phase 1, a C++ program is created with a text editor such as vi (Lamb and Robbins, 1998) or Emacs (Cameron et al., 1996) and saved to a disk. If the C++ program is small, we can just put the program into a single source file. If the C++ program is big, we might split the program into a set of header files and source files and put these files into different directories. C++ header filenames often have extension ".hpp" and C++ source filenames often have extension ".cpp".

In Phase 2, the C++ program is preprocessed by the preprocessor program invoked by the compiler as the first part of translation. The preprocessor program handles directives, which are lines included in the code of the program that are not program statements but directives for the preprocessor. For example, source file inclusion (#include), macro definitions (#define and #undef), and conditional inclusion (#ifdef, #ifndef, #if, #endif, #else, and #elif) are common C++ directives.

In Phase 3, the C++ program is compiled. In this phase, the C++ program is translated into object code by the compiler. The object code is a sequence of computer instructions in a machine code format, which can be understood directly by a specific type of CPU.

In Phase 4, the object code is linked with the code for the missing functions as C++ programs typically contain references to functions and data defined elsewhere, such as in the C++ standard libraries. If Phase 3 and Phase 4 are completed successfully, an executable image is produced and saved to a disk.

In Phase 5, the executable image of the C++ program is loaded into the memory. This phase is done by the loader program. The loader program also loads additional components from shared libraries that support the C++ program.

In Phase 6, the program is executed by the computer.

We now consider a simple C++ program that finds pairs (a, b) of integers such that $3^a + 7^b$ is a perfect square[1]. Since $3^a + 7^b$ can be a very big integer, C++ built-in types (e.g., long long) cannot handle such big integers. Fortunately, we can use the GNU Multiple Precision Arithmetic Library (GMP) to deal with big integers. For GMP installation instructions, readers are referred to Appendix C.4.

Listing 3.1: Pair-searching Program.

```
// pair.cpp
#include<iostream>
#include<sstream>

#include"gmp.h"

using namespace std;

bool checkPair(unsigned int a, unsigned int b, mpz_t mpzd){
    mpz_t mpzs, mpzd2, mpz3, mpz7, mpz3a, mpz7b, mpzdiff;
    mpz_inits(mpzs,mpzd2,mpz3,mpz7,mpz3a,mpz7b,mpzdiff,NULL);
    mpz_set_ui(mpz3, 3);
```

[1]This is one of the Canadian Mathematical Olympiad (CMO) problems in 2009. Although this example has nothing to do with data clustering, it is a good example to illustrate how C++ programs are compiled, linked, and executed.

```
13      mpz_set_ui(mpz7, 7);
14      mpz_pow_ui(mpz3a, mpz3, a);
15      mpz_pow_ui(mpz7b, mpz7, b);
16      mpz_add(mpzd, mpz3a, mpz7b);
17      mpz_sqrt(mpzs, mpzd);
18      mpz_pow_ui(mpzd2, mpzs, 2);
19      mpz_sub(mpzdiff, mpzd, mpzd2);
20      bool ret;
21      if (mpz_get_si(mpzdiff) == 0) {
22          ret = true;
23      } else {
24          ret = false;
25      }
26      mpz_clears(mpzs,mpzd2,mpz3,mpz7,mpz3a,mpz7b,mpzdiff,NULL);
27      return ret;
28  }
29
30  int main(int argc, const char* argv[]){
31      unsigned int n = 100;
32      if (argc==2) {
33          istringstream is(argv[1]);
34          if ( !(is >> n) ) {
35              cout<<"The argument must be an integer"<<endl;
36              return 1;
37          }
38      } else if (argc > 2 ) {
39          cout<<"Too many arguments."<<endl;
40          return 1;
41      }
42      mpz_t mpzd;
43      mpz_init(mpzd);
44      for(unsigned int i=0; i<=n; ++i) {
45          for(unsigned int j=0; j<=n; ++j) {
46              bool b = checkPair(i,j, mpzd);
47              if ((i==j && j % 10 ==0) || b) {
48                  cout<<" ("<<(int)i<<","<<(int)j<<"): "
49                      <<mpz_get_str(NULL, 10, mpzd);
50                  if (b) {
51                      cout<<" is a perfect square";
52                  }
53                  cout<<endl;
54              }
55          }
56      }
57
58      mpz_clear(mpzd);
59      return 0;
60  }
```

The C++ program shown in Listing 3.1 contains only 60 lines of code. In Phase 1, we use some text editor to type the program and save it to a file say **pair.cpp**. Before we go through Phase 2 to Phase 6, let us give a brief description of the code. For detail description of C++ syntax, readers are referred to Lippman et al. (2005), Deitel and Deitel (2006), and Deitel and Deitel (2009).

Line 1 begins with "//", indicating that the remainder of the line is a comment. Lines 2, 3, and 5 are preprocessor directives. These lines are processed by the C++ preprocessor before the program is compiled. Line 7 specifies that members of namespace **std** can be used in the file without preceding each member with **std** and the scope resolution operator (::).

Line 9 to Line 28 is a function with three arguments. The function takes

a and b as inputs and checks whether $3^a + 7^b$ is a perfect square or not. If the number is a perfect square, the function returns `true`; otherwise, the function returns `false`. Note that when a variable of type `mpz_t` is used as a function parameter, it is effectively a call-by-reference.

Line 30 to Line 60 is the `main` function, which is a special function that takes two parameters. Line 32 to Line 41 checks for the input arguments. Line 44 to Line 56 contains two loops, which loop through and test all the pairs of integers.

To compile the program, we go to the directory of the program and type the following command:

```
g++ -c -I/usr/local/include pair.cpp
```

After the command, the program will go through Phase 2 and Phase 3. Then we type the following command:

```
g++ pair.o -L/usr/local/lib -lgmp -o pair
```

to link the program with external functions. After this command, the program will go through Phase 4. Now we type the following command:

```
./pair.exe 50
```

to execute the program. The program will go through Phase 5 and Phase 6 and produce the following output:

```
(0,0): 2
(1,0): 4 is a perfect square
(2,1): 16 is a perfect square
(10,10): 282534298
(20,20): 79792269784396402
(30,30): 22539340290898149219957898
(40,40): 6366805760909040143406894196152802
(50,50): 1798465042647412147338178328261501938021498
```

As an object-oriented programming language, C++ provides the three features presented in OO languages: encapsulation, inheritance, and polymorphism, which are also called the three pillars of OO programming. In the following sections, we introduce these three essential features and present some examples in C++.

3.3 Encapsulation

Encapsulation is the process of bundling data and methods into a single unit called class. Encapsulation allows one to break down a program into a number of small and independent objects, each of which is self-contained and does its job independently of others. Once an object is encapsulated, the object can be viewed as a black box.

An encapsulated object has an interface that lists the methods provided by the object. The methods in an object's interface have three levels of access: public, protected, and private. Public methods can be accessed by all objects; protected methods can be accessed by the object and objects of any subclasses; private methods can be accessed by the object only.

Encapsulated objects have the following valuable benefits:

- An encapsulated object is pluggable. In order to use an encapsulated object, one needs to understand only how to use the object's public interface;

- An encapsulated object can be changed without affecting other objects as long as its interface is untouched; and

- An encapsulated object can be used without causing unexpected side effects between the object and the rest of the program, since the interaction with the rest of the program is limited to its interface.

A well-encapsulated object should have all the following three characteristics (Sintes, 2001):

- Abstraction;

- Implementation hiding; and

- Division of responsibilities.

Abstraction is the process of identifying a solution to a general problem rather than a specific problem. For example, we can design a generic container to hold different types of data (e.g., integers and floats) in a sorting algorithm. In this way, we can use the sorting algorithm to sort integers and floats. Hence, abstraction helps you achieve reuse.

Implementation hiding, as its name implies, is to hide the implementation behind the interface. As long as we do not change the interface of an object, we can change the implementation of the object without affecting other parts of the program. Obviously, implementation hiding leads to flexible design.

Division of responsibilities means that each object has one or a small number of responsibilities and should do it well.

C++ supports encapsulation by providing the ability to define classes. To illustrate the concepts of encapsulation in C++, we consider the `Cluster` class. In data clustering, a cluster is a group of objects or items that are similar to each other. Listing 3.2 shows the C++ implementation of the `Cluster` class.

Listing 3.2: The **Cluster** class.

```cpp
// cluster.cpp
#include<iostream>
#include<set>

using namespace std;

class Cluster {
public:
    Cluster(size_t id): _id(id) {
    }

    void add_item(size_t item) {
        _items.insert(item);
    }

    void remove_item(size_t item) {
        _items.erase(item);
    }

    size_t size() const {
        return _items.size();
    }

    const std::set<size_t>& data() const {
        return _items;
    }
private:
    double _id;
    std::set<size_t> _items;
};

int main(int argc, char* argv[]) {
    Cluster c1(0);
    Cluster c2(1);

    for(size_t i=0; i<5; ++i) {
        c1.add_item(i);
        c2.add_item(i+5);
    }
    c2.remove_item(7);

    std::set<size_t> data1, data2;
    data1 = c1.data();
    data2 = c2.data();
    std::set<size_t>::const_iterator it;
    cout<<" Cluster 1 has "<<c1.size()<<" items:"<<endl;
    for(it = data1.begin(); it!=data1.end(); ++it) {
        cout<<*it<<" ";
    }
    cout<<endl;
    cout<<" Cluster 2 has "<<c2.size()<<" items:"<<endl;
    for(it = data2.begin(); it!=data2.end(); ++it) {
        cout<<*it<<" ";
    }
    cout<<endl;

    return 0;
}
```

Class **Cluster** is defined in lines 7–31. The class has two private data members _id and _items. The class also has five public member functions, including one constructor (lines 9–10). A constructor is a special member function that must be defined with the same name as the anme of the class.

The constructor takes one parameter, which is used to initialize the member _id.

We used the STL (Standard Template Library) container `set` to hold the items included in a cluster, since we do not want the cluster to contain the same item twice. In Section 5.1, we will see more STL containers.

The member functions `size` and `data` are accessors. The member functions `add_item` and `remove_item` are mutators. To compile the program in Listing 3.2, we can use the following simple command:

```
g++ -o cluster cluster.cpp
```

Executing the program gives the following output:

```
Cluster 1 has 5 items:
0 1 2 3 4
Cluster 2 has 4 items:
5 6 8 9
```

The cluster example shown above is very simple. In Chapter 8, we will implement a more complicate cluster class and several derived cluster clusters.

3.4 Inheritance

Inheritance is the process of creating a new class based on a pre-existing class. When one class inherits from a pre-existing class, the new class inherits all of the data and methods present in the pre-existing class. The new class is called the child class or subclass and the pre-existing class is called the base class, parent class, or superclass.

Inheritance not only allows the child class to inherit the public interface and the implementation of the parent class, but also allows the child class to redefine any methods of the parent class that it does not like through overriding, which is the process of a child class rewriting a method that appears in the parent class in order to change the method's behavior. The beautiful thing about overriding is that it allows one to change the way an object works without touching the definition of the parent class.

C++ supports single inheritance as well as multiple inheritance. In single inheritance, a child class inherits from only one base class. In multiple inheritance, a child class inherits from multiple base classes. In what follows, we illustrate the concept of inheritance in C++ using the `Distance` class and the `EuclideanDistance` class.

Listing 3.3: The `Distance` class.

```cpp
// distance.cpp
#include<iostream>
#include<string>
#include<vector>
#include<cmath>

using namespace std;

class Distance {
public:
    Distance(const string& name)
        : _name(name) {
    };

    ~Distance() {}

    virtual double operator()(const vector<double> &x,
             const vector<double> &y) const = 0;

    const string& name() {
        return _name;
    }

private:
    string _name;
};

class EuclideanDistance : public Distance {
public:
    EuclideanDistance(): Distance("Euclidean_Distance") {
    }

    double operator()(const vector<double> &x,
             const vector<double> &y) const {
        if (x.size() != y.size()) {
            return -1.0;
        }
        double dsum = 0.0;
        for(size_t i=0; i<x.size(); ++i) {
            dsum += (x[i] - y[i]) * (x[i] - y[i]);
        }

        return sqrt(dsum);
    }
};

int main() {
    vector<double> x(2), y(2);
    x[0] = 0; x[1] = 0;
    y[0] = 1; y[1] = 1;

    EuclideanDistance ed;
    cout<<"The_"<<ed.name()<<"_between_(0,_0)_and_(1,1)_is_:"
        <<ed(x,y)<<endl;

    return 0;
}
```

Listing 3.3 shows the C++ implementation of class `Distance` and the derived class `EuclideanDistance`. The `Distance` class defined in lines 9–26 has one private data member and four public member functions, which include a constructor and a virtual destructor. In C++, a class that is intended to be used as a base class should have a virtual destructor.

Line 28 shows that the inheritance is a public inheritance. C++ also supports private and protected inheritances. Table 3.1 shows the access levels of the base class's members in the derived class under different inheritances. In protected inheritance, for example, both public and protected members of the base class become protected members of the derived class.

Base	Inheritance		
	Public	Protected	Private
Public	Public	Protected	Private
Protected	Protected	Protected	Private
Private	Private	Private	Private

TABLE 3.1: Access rules of base-class members in the derived class.

The derived class `EuclideanDistance` defined in lines 28–45 redefines the operator "( )" by providing a concrete implementation. We note that the derived class used the base class's constructor to initialize the data members of the base class in its initialization list (lines 30–31). To compile the program in Listing 3.3, we can use the following command:

`g++ -o distance distance.cpp`

Executing the program gives the following output:

`The Euclidean Distance between (0, 0) and (1,1) is :1.41421`

3.5 Polymorphism

Polymorphism is the state of one having many forms, that is, a single name can be used to represent different code selected by some automatic mechanism. For example, the term "do job" can be applied to many different objects such as managers and employees, each of which interprets the term in its own way.

There are four types of polymorphism: inclusion polymorphism, parametric polymorphism, overriding, and overloading. Inclusion polymorphism, also called pure polymorphism, allows one to treat related objects generically. Parametric polymorphism allows one to write generic methods and types. Overriding allows one to change the behaviors of methods in the parent class. Overloading allows one to use the same method name for many different methods, each of which differs only in the number and type of its parameters.

C++ supports all four types of polymorphism. When we talk about polymorphism in C++, we usually refer to inclusion polymorphism and parametric polymorphism. In C++, inclusion polymorphism is also called dynamic polymorphism and parametric polymorphism is called static polymorphism. Dynamic polymorphism is achieved by virtual functions and static polymorphism is achieved by templates. In terms of flexibility, dynamic polymorphism is better than static polymorphism. In terms of performance, static polymorphism is better than dynamic polymorphism.

3.5.1 Dynamic Polymorphism

To illustrate the idea of dynamic polymorphism in C++, we consider the example of clustering algorithms. In data clustering, there are many clustering algorithm. It is a good idea to have a common base for all these clustering algorithms. The `Algorithm` class and its derived classes are shown in Listing 3.4. Class `Algorithm` is very simple and contains a pure virtual function (line 10) and a virtual destructor (line 9). We call such a class an abstract class since it contains a pure virtual function.

As we can see from Listing 3.4, dynamic polymorphism allows us to define only one `clusterize` function for class `DataMiner` to deal with different types of clustering algorithms. Without dynamic polymorphism, we would need a different `clusterize` function for each type of clustering algorithm.

Listing 3.4: Program to illustrate dynamic polymorphism.

```
1  // algorithm.cpp
2  #include<iostream>
3  #include<vector>
4
5  using namespace std;
6
7  class Algorithm {
8  public:
9      virtual ~Algorithm() {}
10     virtual void clusterize() = 0;
11 };
12
13 class Kmean: public Algorithm {
14     void clusterize() {
15         cout<<"Clustering_by_kmean"<<endl;
16     }
17 };
18
19 class Kmode: public Algorithm {
20     void clusterize() {
21         cout<<"Clustering_by_kmode"<<endl;
22     }
23 };
24
25 class DataMiner {
26 public:
27     void clusterize(Algorithm* pa) {
28         pa->clusterize();
29     }
30 };
31
32 int main() {
33     vector<Algorithm*> algos;
34     algos.push_back(new Kmean());
35     algos.push_back(new Kmean());
36     algos.push_back(new Kmode());
```

```
37      algos.push_back(new Kmean());
38
39      DataMiner dm;
40
41      for(size_t i=0; i<algos.size(); ++i){
42          dm.clusterize(algos[i]);
43      }
44
45      for(size_t i=0; i<algos.size(); ++i){
46          delete algos[i];
47      }
48
49      return 0;
50  }
```

To compile the program in Listing 3.4, we can use the following command:

`g++ -o algorithm algorithm.cpp`

Executing the program gives the following output:

```
Clustering by kmean
Clustering by kmean
Clustering by kmode
Clustering by kmean
```

The algorithm classes introduced above are very simple in that these classes do not consider different input and output of different clustering algorithms. In Chapter 10 we implement more complicate algorithm classes.

3.5.2 Static Polymorphism

To illustrate the idea of static polymorphism, we consider the classic example of clusters and partitions. As we mentioned before, a cluster is a group of records, objects, or items. A partition of a dataset is a group of clusters. Hence it is a good idea to write a container class that can hold different types of data. The program shown in Listing 3.5 gives an example of a template class **Container**.

Listing 3.5: Program to illustrate class templates.
```
 1  // container.cpp
 2  #include<iostream>
 3  #include<vector>
 4  #include<algorithm>
 5
 6  using namespace std;
 7
 8  template <typename T>
 9  class Container {
10  public:
11      void add_item(const T& item) {
12          _items.push_back(item);
13      }
14
15      void remove_item(const T& item) {
16          typename vector<T>::iterator it =
17              find(_items.begin(), _items.end(), item);
```

```
            if (it != _items.end()) {
                _items.erase(it);
            }
        }

        const vector<T>& data() const {
            return _items;
        }

        size_t size() const {
            return _items.size();
        }

    protected:
        Container() {}

        vector<T> _items;
};

class Cluster: public Container<size_t> {
};

class Partition: public Container<Cluster> {
};

int main() {
    Cluster c1, c2;
    for(size_t i=0; i<5; ++i) {
        c1.add_item(i);
    }
    for(size_t i=0; i<5; ++i) {
        c2.add_item(i+5);
    }
    c2.remove_item(8);

    Partition p;
    p.add_item(c1);
    p.add_item(c2);

    cout<<"Partition p contains "<<p.size()<<" clusters:"<<endl;
    vector<Cluster> sc = p.data();
    for(size_t i=0; i<sc.size(); ++i) {
        cout<<" Cluster "<<i<<": ";
        vector<size_t> ss = sc[i].data();
        for(size_t j=0; j<ss.size(); ++j) {
            cout<<ss[j]<<" ";
        }
        cout<<endl;
    }

    return 0;
}
```

We make the constructor of the template class **Container** protected (line 32) since we do not want the class to be used alone. When we compile and execute the above program, we see the following outputs:

```
Partition p contains 2 clusters:
Cluster 0: 0 1 2 3 4
Cluster 1: 5 6 7 9
```

The template classes illustrated above are very simple. In Section 11.1 and Section 11.2, we introduce a more complex template container class and a template double-key map class.

3.6 Exception Handling

An exception is a problem that occurs during the execution of a program. For example, an attempt to divide by zero is an exception. Dereferencing a null pointer is another example of an exception. Exception handling is a programming language feature that is designed to handle exceptions. Exception handling is very useful in that it allows programmers to create applications that can handle exceptions. In most cases, if an exception is encountered and handled, a program can continue normal execution as if no problem had occurred. Even if a program cannot continue normal execution when an exception occurs, handling the exception allows the program to notify the user of the problem and terminate in a controlled manner.

C++ is a programming language that supports exception handling. In C++, the header file `<exception>` defines the base class `exception` for all standard exceptions and the header file `<stdexcept>` defines several other exceptions derived from `exception`. Figure 3.1 shows the hierarchy of the C++ standard library exception classes.

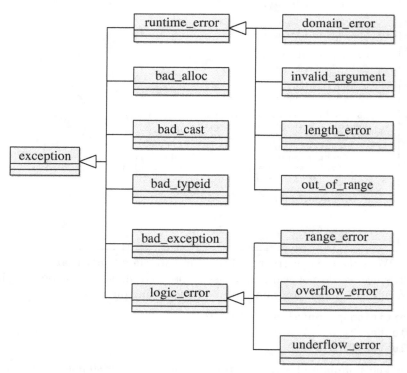

FIGURE 3.1: Hierarchy of C++ standard library exception classes.

Class **runtime_error**, class **logic_error**, and their child classes are defined in the header file **<stdexcept>**. Class **bad_exception** and class **exception** are defined in the header file **<exception>**. Class **bad_alloc** is defined in the header file **<new>**. Classes **bad_cast** and **bad_typeid** are defined in the header file **<typeinfo>**.

Now we consider a simple example illustrating C++ exception handling. The program in Listing 3.6 illustrates how we can handle the division-by-zero exception. In function **divide** (lines 7–13), we throw a **runtime_error** exception when the denominator is zero. In the main function, we use **try** and **catch** blocks to catch the exception.

Listing 3.6: Program to illustrate exception handling.

```cpp
// exception.cpp
#include<iostream>
#include<stdexcept>

using namespace std;

double divide(int x, int y) {
    if (y == 0 ) {
        throw std::runtime_error("division by zero");
    }

    return (double)x/y;
}

int main() {
    try{
        cout<<"100/3: "<<divide(100, 3)<<endl;
        cout<<"100/0: "<<divide(100, 0)<<endl;
    } catch(exception &ex) {
        cout<<"An exceptoin occurred"<<endl;
        cout<<ex.what()<<endl;
        return 1;
    } catch (...) {
        cout<<"unknown exception"<<endl;
        return 1;
    }

    return 0;
}
```

When we compile and execute the program, we see the following output:

```
100/3: 33.3333
An exceptoin occurred
division by zero
```

The program in Listing 3.6 shows the usage of an exception defined in the C++ standard library. In fact, we can define customized exception classes that derive from the base class **exception**. In Section 6.4, we shall introduce class **Error**, which is a customized exception class defined for the clustering library.

3.7 Summary

In this chapter, we introduce some basic concepts of object-oriented programming and the C++ programming language. In particular, we introduce the three pillars of object-oriented programming: encapsulation, inheritance, and polymorphism. We also illustrate the three concepts with simple C++ programs. After reading this chapter, readers should develop a sense of what object-oriented programming is and how to program in C++.

Since this book is about implementing data clustering algorithms in C++, we do not teach C++ in this book. For readers who are not familiar with the C++ programming language or how to program in C++ in an object-oriented way, some good references are Stroustrup (1994), Meyers (1997), Lippman et al. (2005), and Deitel and Deitel (2009).

Chapter 4
Design Patterns

A design pattern is a general reusable solution to a problem that occurs over and over again in software design. A design pattern has four essential elements: the pattern name, the problem, the solution, and the consequences (Gamma et al., 1994).

The pattern name is a handle we use when we talk about the design pattern. The problem describes when to apply the pattern and the solution describes how to apply the pattern. The consequences describe the results and trade-offs of applying the pattern.

Creational	Structural	Behavioral
Factory	Adapter	Interpreter
Abstract Factory	Bridge	Template Method
Builder	Composite	Chain of Responsibility
Prototype	Decorator	Command
Singleton	Facade	Iterator
	Proxy	Mediator
		Memento
		Flyweight
		Observer
		State
		Strategy
		Visitor

TABLE 4.1: Categories of design patterns.

Table 4.1 gives a list of 23 design patterns, which are classified into three categories: creational patterns, structural patterns, and behavioral patterns (Gamma et al., 1994). Creational patterns are related to the process of object creation. Structual patterns concern the composition of classes or objects. Behavioral patterns model how classes or objects interact and distribute responsibilities.

4.1 Singleton

The singleton pattern is a design pattern used to ensure that a class has exactly one instance and provides a global point of access to it. Sometime we want a class to have exactly one instance. For example, there should be only one window manager. To ensure that a class has only one instance and that the instance is easily accessible, the solution of the singleton pattern is to make the class itself responsible for keeping tracking of its sole instance and providing a way to access the instance (Gamma et al., 1994). Table 4.2 shows the four elements of the singleton pattern.

Element	Description
Pattern Name	Singleton
Problem	Only one instance of an object is allowed to exist in the system at any one time
Solution	Let the object manage its own creation and access through a class method
Consequences	Controlled access to the object instance Difficult to inherit a singleton

TABLE 4.2: The singleton pattern.

Figure 4.1 shows the UML class diagram of the singleton pattern. The singleton class contains a private static member, which is a point to the sole instance of the class. The class also contains a static method, which returns the pointer to the instance if the instance was already created or creates the instance and returns the pointer to the newly created instance. The constructor of the singleton class is protected, preventing users to instantiate the class.

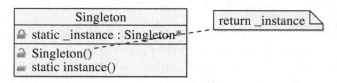

FIGURE 4.1: The singleton pattern

The singleton class shown in Figure 4.1 is not useful. To make the class useful, we need to add other functions to the class. The **Logger** class shown in Listing 4.1 is a singleton class, which we can use to write log messages to screen or a file.

Listing 4.1: Program to illustrate the singleton pattern.

```cpp
// singleton.cpp
#include<iostream>
#include<fstream>

class Logger {
public:
    static Logger* instance() {
        if(_instance == 0) {
            _instance = new Logger();
        }

        return _instance;
    }

    void setStream(std::ostream *s) {
        _s = s;
    }

    std::ostream* getStream() {
        return _s;
    }

    Logger& operator<<(const char* c) {
        if(_s && c) {
            *_s <<c;
        }

        return *this;
    }

    template <typename T>
    Logger& operator<<(const T &d) {
        if(_s) {
            *_s<<d;
        }

        return *this;
    }

    Logger& operator<<(Logger& (*manip) (Logger&)) {
        return manip(*this);
    }

    void destroy() {
        delete _instance;
        _instance = 0;
    }

protected:
    Logger(): _s(&std::cout) { }

private:
    static Logger* _instance;
    std::ostream* _s;
};

Logger* Logger::_instance = 0;

namespace std {
    Logger& endl(Logger& log) {
        std::ostream *s = log.getStream();
        *s<<endl;
        return log;
    }
}
```

```cpp
67  void funcA() {
68      Logger* logger = Logger::instance();
69      *logger << "This line is outputted to screen"
70              << std::endl
71              << std::uppercase
72              << "This line is also outputted to screen"
73              << std::endl;
74  }
75
76  void funcB() {
77      Logger* logger = Logger::instance();
78      std::ofstream fs("Log.txt");
79      logger->setStream(&fs);
80      *logger << "This line is outputted to file"
81              << std::endl
82              << "This line is also outputted to file"
83              << std::endl;
84      fs.close();
85      logger->setStream(&std::cout);
86      *logger << "This line goes to screan"
87              <<std::endl;
88  }
89
90  int main(int argc, char* argv[]) {
91      funcA();
92      funcB();
93      Logger::instance()->destroy();
94      return 0;
95  }
```

Class `Logger` has two private data members: `_instance` and `_s`. The first one is a static pointer to the sole instance of the class and is initialized to 0. The second one is a pointer to `std::ostream` and is initialized to the pointer pointing to standard output. The constructor is protected. Note that we initialize the static data member (line 57) outside the class.

We overloaded operator << in class `Logger` so that we can use class `Logger` in the same way as we use C++ stream classes. The function in lines 23–29 overloads operator << for pointer type. The function in lines 31–38 is a template function that overloads operator << for constant reference type. The function in lines 40–42 overloads operator << for function pointer type. The parameter of this function is a pointer pointing to a function, which takes a reference of `Logger` as parameter and returns a reference of `Logger`. Note that we define function `endl` in namespace `std` so that we can use this function as usual.

Function `funcA` (lines 67–74) and function `funcB` (lines 76–88) show how easily we can use class `Logger` in our program. To compile this program, we can type the following command:

`g++ -o singleton singleton.cpp`

Executing the program, we see the following screen output:

```
This line is outputted to screen
This line is also outputted to screen
This line goes to screan
```

We also see the following output in the file `Log.txt`:

```
This line is outputted to file
This line is also outputted to file
```

One more thing worth mentioning is that we write function `destroy` (lines 44–47) to delete the sole instance. We do not put the deletion in the destructor of the class because deleting the instance will call the destructor, which will delete the instance once more. If we put the deletion in the destructor, the destructor will be called infinitely many times until the computer's memory is exhausted. In our main program, we call `destroy` (line 93) to delete the instance.

4.2 Composite

The composite pattern is a design pattern used to composite objects into a tree structure to represent a part-whole hierarchy, which allows clients to treat individual objects and their containers in a uniform way. For example, the files and folders on a disk form a file-folder hierarchy. The solution of the composite pattern is to make an abstract base class that represents both components and their containers. Table 4.3 shows the four elements of the composite pattern.

Element	Description
Pattern Name	Composite
Problem	Composite objects into a tree structure so that individual objects and compositions of objects can be treated uniformly
Solution	Design an abstract class that represents both components and and their containers
Consequences	Easy to add new kinds of components Makes the design overly general

TABLE 4.3: The composite pattern.

The UML class diagram of the composite pattern is shown in Figure 4.2. The component class is an abstract base class for the leaf class and the composite class. A leaf class is used to hold an individual object. A composite class is used to hold a set of components, each of which can be an individual object or a composite. The UML class diagram in Figure 4.2 shows the basic structure of the composite pattern.

Now let us consider the file-folder example. The C++ code of the example

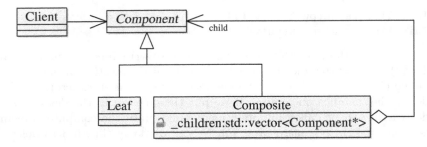

FIGURE 4.2: The composite pattern.

is shown in Listing 4.2. The program defines three classes: **Component**, **File**, and **Folder**. Class **Component** is an abstract class, which contains a virtual destructor and a pure virtual function **traverse**. A class that is used as a base class for other classes should have a virtual destructor.

Class **File** inherits from the base class **Component**. The class has a private data member and a constructor, which takes one argument. The data member is initialized by the value of the argument of the constructor. The class overrides function **traverse** by providing a concrete implementation.

Class **Folder** also inherits from the base class **Component**. The class has a data member, which is a vector of pointers. The class includes a function called **add** that can be used to add children. The class also overrides function **traverse**.

Listing 4.2: Program to illustrate the composite pattern.

```
// composite.cpp
#include<iostream>
#include<vector>
#include<string>

using namespace std;

class Component {
public:
    virtual ~Component() {}
    virtual void traverse() = 0;
};

class File: public Component {
public:
    File(const string& name): _name(name) {}
    void traverse() {
        cout<<_name<<",";
    }
private:
    string _name;
};

class Folder: public Component {
public:
    void add(Component* item) {
        _children.push_back(item);
    }
```

```cpp
            void traverse() {
                for(size_t i=0; i<_children.size(); ++i){
                    _children[i]->traverse();
                }
            }

        private:
            vector<Component*> _children;
        };

        int main() {
            File* pf1 = new File("File_1");
            File* pf2 = new File("File_2");
            File* pf3 = new File("File_3");
            Folder* pd1 = new Folder();
            Folder* pd2 = new Folder();

            pd1->add(pf1);
            pd1->add(pd2);
            pd2->add(pf2);
            pd2->add(pf3);

            cout<<"Folder_1_contains:"<<endl;
            pd1->traverse();
            cout<<endl<<"Folder_2_contains:"<<endl;
            pd2->traverse();

            delete pf1;
            delete pf2;
            delete pf3;
            delete pd1;
            delete pd2;

            return 0;
        }
```

In the main program (lines 40–64), we create three objects of class **File** and two objects of class **Folder**. We add the first file and the second folder to the first folder. We add the second file and the third file to the second folder. Then we show the files in the first folder and the second folder, respectively. Finally, we delete all objects created by **new** (lines 57–61).

To compile the program, we use the following command:

`g++ -o composite composite.cpp`

Once we execute the program, we see the following output:

```
Folder 1 contains:
File 1,File 2,File 3,
Folder 2 contains:
File 2,File 3,
```

From the output we see that function **traverse** reaches all files in a folder regardless if the file is in the folder or in subfolders of the folder.

The above example is a very simple example to illustrate the composite pattern. We can use the composite pattern in a more complex way. In Section 8.3, we use the composite pattern to design the hierarchical clustering structure.

4.3 Prototype

The prototype pattern is a creational pattern that is used when creating an object of a class is very time-consuming or complex in some way. If we want to replicate an object, we need to consider whether a shallow copy (i.e., member-wise copy) is good enough or not. If the object contains references or pointers to other objects, we need to conduct a deep copy in order to get a true copy of the object. In a deep copy, all objects referenced or pointed to are also copied. The prototype pattern is applicable when we want a deep copy or clone of an object. In such case, the **new** method provided by C++ is not good enough since it does member-wise copy.

The solution of the prototype pattern is to declare an abstract base class that specifies a pure virtual **clone** method and let the concrete derived classes override this **clone** method by implementing an operation to clone itself. Then clients can create a new object by asking a prototype to clone itself. Table 4.4 shows the four elements of the prototype pattern.

Element	Description
Pattern Name	Prototype
Problem	Need to duplicate an object for some reason but creating the object by **new** is not appropriate.
Solution	Design an abstract base class that specifies a pure virtual **clone** method
Consequences	Configure an application with classes dynamically Each subclass must implement the **clone** method

TABLE 4.4: The prototype pattern.

The UML class diagram in Figure 4.3 shows a simple structure of the prototype pattern. Class **Prototype** is an abstract class with a public member function called **clone**, which is a pure virtual function that returns a pointer pointing to itself. Another two classes are derived from the base class and override the **clone** method. We note that the **clone** method in class **ConcreteA** returns a pointer pointing to **ConcreteA** instead of **Prototype**. The **clone** method in class **ConcreteB** does similarly. In C++, this is called covariant return types.

The C++ program shown in Listing 4.3 illustrates the idea of the prototype pattern. The program defines an abstract class named **Shape** consisting of a virtual destructor and two pure virtual functions: the **draw** method and the **clone** method (lines 7–12). The virtual destructor of class **Shape** is necessary since class **Shape** is designed to be used as a base class.

Design Patterns

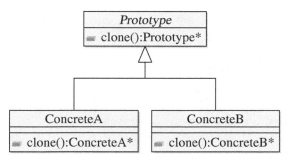

FIGURE 4.3: The prototype pattern.

Class `Triangle` (lines 14-34) and class `Square` (lines 38–58) are subclasses of class `Shape`. Class `Triangle` has a private data member and a static member. The static member is used to count the number of objects created so far. Note that the static member must be initialized outside of the class (line 36). Class `Square` is defined in a similar way.

We note that class `Triangle` declares a protected constructor with an argument (lines 27–29). In this constructor, the static member `_count` is assigned to the member `_id` and then is increased by 1. This constructor is called by the `clone` method. If we use `new` to create an object of `Triangle`, then the static member `_count` will be the same.

Listing 4.3: Program to illustrate the prototype pattern.

```cpp
// prototype.cpp
#include<iostream>
#include<vector>

using namespace std;

class Shape {
public:
    virtual ~Shape() {}
    virtual void draw() = 0;
    virtual Shape* clone() = 0;
};

class Triangle: public Shape {
public:
    void draw() {
        cout<<"I am triangle "<<_id<<endl;
    }

    Shape* clone() {
        return new Triangle(1);
    }

    Triangle() {}

protected:
    Triangle(int dummy) {
        _id = _count++;
    }

```

```
31  private:
32      int _id;
33      static int _count;
34  };
35
36  int Triangle::_count = 1;
37
38  class Square: public Shape {
39  public:
40      void draw() {
41          cout<<"I am square "<<_id<<endl;
42      }
43
44      Shape* clone() {
45          return new Square(1);
46      }
47
48      Square() {}
49
50  protected:
51      Square(int dummy) {
52          _id = _count++;
53      }
54
55  private:
56      int _id;
57      static int _count;
58  };
59
60  int Square::_count = 1;
61
62  int main() {
63      vector<Shape*> vs;
64      Triangle a;
65      Square b;
66      for(int i=0;i<2;++i) {
67          vs.push_back(a.clone());
68      }
69      for(int i=0;i<3;++i) {
70          vs.push_back(b.clone());
71      }
72
73      for(int i=0;i<vs.size();++i){
74          vs[i]->draw();
75          delete vs[i];
76      }
77
78      return 0;
79  }
```

In the main function (lines 62–79), we declare a triangle and a square. Then we clone two triangles and three squares. Finally, we draw the cloned shapes and delete them to free memory space. Once we execute the program, we see the following output:

```
I am triangle 1
I am triangle 2
I am square 1
I am square 2
I am square 3
```

More information about the prototype pattern can be found in Gamma et al. (1994) and Lasater (2007).

4.4 Strategy

The strategy pattern is a behavioral pattern that is used to encapsulate a group of algorithms inside classes to make these algorithms interchangeable so that the algorithm can vary independently from the clients that use it. For example, there are different strategies to cluster a dataset. The solution of the strategy pattern is to declare a common abstract base class for all strategies and let derived classes implement concrete strategies. Table 4.5 shows the four components of the strategy pattern.

Element	Description
Pattern Name	Strategy
Problem	Need a group of algorithmic classes that have no correlation and are not easily exchanged to be interchangeable
Solution	Design an abstract base class as an interface and let derived classes implement the details
Consequences	Eliminate conditional statements Increased number of objects

TABLE 4.5: The strategy pattern.

The class diagram in Figure 4.4 shows the basic structure of the strategy pattern. In the strategy pattern, we declare an interface common to all supported strategies. The concrete strategies classes are derived from the abstract base strategy class and implement the strategy interface. The context class maintains a pointer pointing to strategy and an interface that lets a strategy access its data.

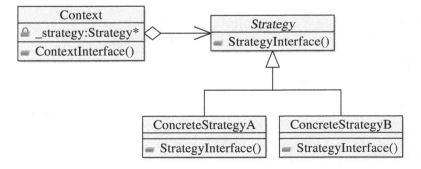

FIGURE 4.4: The strategy pattern.

The program in Listing 4.4 illustrates the strategy pattern. The program defines an abstract class **Strategy**, which has a virtual destructor and a pure virtual method named **execute**. Two concrete strategy classes **StrategyA** and **StrategyB** are subclasses of the base class **Strategy**. The two concrete classes override method **execute** by providing a concrete implementation.

Class **Context** (lines 28–44) has a private data member, which is a pointer pointing to **Strategy**, one public constructor, and two public member functions. The keyword **explicit** (line 31) in front of the constructor is used to prevent automatic conversion. For example, we cannot use the statement

```
Strategy s;
Context c = &s;
```

since the constructor is declared as **explicit**. We have to use the following explicit statement

```
Strategy s;
Context c = Context(&s);
```

Listing 4.4: Program to illustrate the strategy pattern.

```
1   // strategy.cpp
2   #include <iostream>
3   using namespace std;
4
5   class Strategy
6   {
7   public:
8       virtual ~Strategy() {}
9       virtual void execute() = 0;
10  };
11
12  class StrategyA: public Strategy
13  {
14  public:
15      void execute() {
16          cout << "Strategy_A_is_called" << endl;
17      }
18  };
19
20  class StrategyB: public Strategy
21  {
22  public:
23      void execute() {
24          cout << "Strategy_B_is_called" << endl;
25      }
26  };
27
28  class Context
29  {
30  public:
31      explicit Context(Strategy *strategy): _strategy(strategy) {
32      }
33
34      void set_strategy(Strategy *strategy)
35      {
36          _strategy = strategy;
37      }
38
```

```
        void execute() {
            _strategy->execute();
        }
private:
        Strategy * _strategy;
};

int main()
{
        StrategyA sa;
        StrategyB sb;

        Context ca(&sa);
        Context cb(&sb);

        ca.execute();
        cb.execute();

        ca.set_strategy(&sb);
        ca.execute();
        cb.set_strategy(&sa);
        cb.execute();

        return 0;
}
```

In the main function (lines 46–63), we declare two strategy objects and two context objects. We also change the strategies (lines 57 and 59) in the context objects. Once we execute the program, we see the following output:

```
Strategy A is called
Strategy B is called
Strategy B is called
Strategy A is called
```

From the above example we see that we can change strategies in context very easily. For more information about the strategy pattern, readers are referred to Gamma et al. (1994) and Lasater (2007).

4.5 Template Method

The template method pattern is a design pattern that is used to define a skeleton of an algorithm in an operation consisting of many steps and to let subclasses redefine certain steps of the algorithm without changing the algorithm's structure. For example, an algorithm might contain steps such as initialization, iteration, and post-processing and there may be many ways to initialize the algorithm. The solution of the template method pattern is to declare an abstract base class that contains a template method calling several primitive operations and let subclasses redefine certain primitive operations. Table 4.6 shows the four elements of the template method pattern.

Element	Description
Pattern Name	Template method
Problem	Need to share functionality between classes without copying functionality
Solution	Design an abstract base class that specifies the shared functionality and let subclasses define or redefine the functionality
Consequences	Reuse code by factoring out common behavior May forget to call inherited operations

TABLE 4.6: The template method pattern.

The class diagram in Figure 4.5 shows the basic structure of the template method pattern. In the template method pattern, we design an abstract base class that contains a template method, which calls other primitive operations. The subclasses `ConcreteA` and `ConcreteB` override the primitive operations.

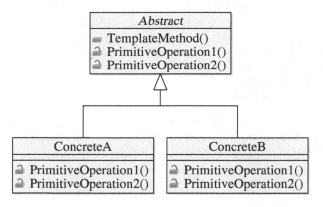

FIGURE 4.5: The template method pattern.

To illustrate the template method pattern, we consider the simple C++ program shown in Listing 4.5. The program defines three classes: `Algorithm`, `AAlgorithm`, and `BAlgorithm`. Class `Algorithm` has a virtual destructor and a public method, `calculate`, which calls another three protected methods. The three protected methods are virtual functions and have default implementation. Class `AAlgorithm` and class `BAlgorithm` are subclasses of class `Algorithm`. Class `AAlgorithm` overrides the `initialization` method. Class `BAlgorithm` overrides methods `initialization` and `postprocessing`. Note that the function `postprocessing` in class `BAlgorithm` calls the same function in the parent class by using the scope operator "::" with the parent class name (line 42).

Listing 4.5: Program to illustrate the template method pattern.

```cpp
// template.cpp
#include<iostream>

using namespace std;

class Algorithm {
public:
    virtual ~Algorithm() {}
    void calculate() {
        initialization();
        iteration();
        postprocessing();
    }
protected:
    virtual void initialization() {
        cout<<"Default initialization"<<endl;
    }

    virtual void iteration(){
        cout<<"Default iteration"<<endl;
    }

    virtual void postprocessing(){
        cout<<"Default postprocessing"<<endl;
    }
};

class AAlgorithm : public Algorithm {
private:
    void initialization() {
        cout<<"Initialization method A"<<endl;
    }
};

class BAlgorithm : public Algorithm {
private:
    void initialization() {
        cout<<"Initialization method B"<<endl;
    }

    void postprocessing(){
        Algorithm::postprocessing();
        cout<<"Output summary"<<endl;
    }
};

int main() {
    AAlgorithm aa;
    BAlgorithm ba;
    cout<<" ---- A"<<endl;
    aa.calculate();
    cout<<" ---- B"<<endl;
    ba.calculate();

    return 0;
}
```

In the main function (lines 47–56), we create an object of class **AAlgorithm** and an object of class **BAlgorithm**. Then we call the `calculate` method of the two objects. Once we execute the program, we see the following output:

```
--- A
Initialization method A
Default iteration
Default postprocessing
--- B
Initialization method B
Default iteration
Default postprocessing
Output summary
```

From the output we see that the program can find what version of the function to call. If a subclass does not override a function in the parent class, the function in the parent class is called.

From the above example we see that the template method pattern allows us to reuse code by providing a way to factor out common behavior of a system. For more information about the template method pattern, readers are referred to Gamma et al. (1994) and Lasater (2007).

4.6 Visitor

The visitor pattern is a behavioral pattern that is used to add operations on elements in a tree structure. Using the visitor pattern, we do not need to code the operations inside an element class. Instead, we pass in references to other classes that have the desired operations. The visitor pattern is a little more complex than other design patterns. The visitor pattern involves two class hierarchies: one for the elements being operated on and one for the visitors that define operations on the elements. Table 4.7 shows the four elements of the visitor pattern.

Figure 4.6 shows the basic structure of the visitor pattern. In this class diagram, we see the two class hierarchies mentioned before: the element class hierarchy and the visitor class hierarchy. The visitor class hierarchy is based on the element class hierarchy. Class **Visitor** is an abstract base class, which declares a number of visitor functions depending on the number of concrete element classes. If we add a new concrete element class, we have to add a new visitor function to every class in the visitor class hierarchy. This is a drawback of the visitor pattern.

In every class in the element class hierarchy, we define an **accept** method, which takes a reference of visitor as argument. In every class in the visitor class hierarchy, we define **VisitElementA** and **VisitElementB** since there are two concrete element classes.

The visitor pattern is usually used together with the composite pattern introduced before. The C++ program in Listing 4.6 illustrates the visitor

Element	Description
Pattern Name	Visitor
Problem	Need to perform many distinct and unrelated operations on node objects in a tree structure
Solution	Define two class hierarchies: one for the elements being operated on and one for the visitors that define operations on the elements
Consequences	Hard to add concrete element classes Easy to add new operations

TABLE 4.7: The visitor pattern.

pattern and the composite pattern. We first define the base visitor class (lines 10–15). Since the base visitor class depends on class `LeafNode` and class `InternalNode`, we forward declare the two classes (lines 7–8). Class `PrintVisitor` inherits from the base visitor class `Visitor` and provides concrete implementation of the two visit methods.

Class `Node` is an abstract base class, which contains a pure virtual `accept` method. Classes `LeafNode` and `InternalNode` inherit from class `Node` and override the `accept` method. In the `accept` method of class `InternalNode`, the visitor is applied to all children of an object of class `InternalNode` after the visitor is applied to the object.

Listing 4.6: Program to illustrate the visitor pattern.

```cpp
// visitor.cpp
#include<iostream>
#include<vector>

using namespace std;

class LeafNode;
class InternalNode;

class Visitor {
public:
    virtual ~Visitor() {}
    virtual void visit(LeafNode& lnode) = 0;
    virtual void visit(InternalNode& inode) = 0;
};

class PrintVisitor: public Visitor {
public:
    void visit(LeafNode& lnode) {
        cout<<"visit_a_leaf_node"<<endl;
    }

    void visit(InternalNode& inode) {
        cout<<"visit_an_internal_node"<<endl;
    }
};
```

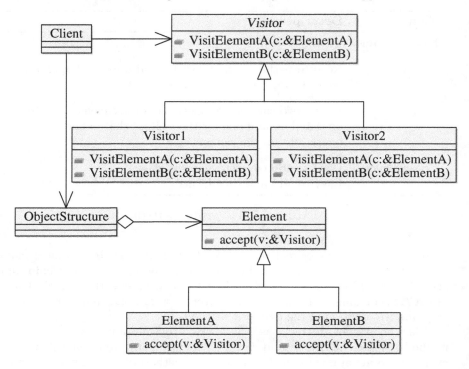

FIGURE 4.6: The visitor pattern.

```
28
29  class Node {
30  public:
31      virtual ~Node() {}
32      virtual void accept(Visitor&v) = 0;
33  };
34
35  class LeafNode: public Node {
36  public:
37      void accept(Visitor&v) {
38          v.visit(*this);
39      }
40  };
41
42  class InternalNode: public Node {
43  public:
44      void accept(Visitor&v) {
45          v.visit(*this);
46          for(size_t i=0; i<_children.size(); ++i) {
47              _children[i]->accept(v);
48          }
49      }
50
51      void add(Node* pn) {
52          _children.push_back(pn);
53      }
54
55  private:
```

```
            vector<Node*> _children;
};

int main() {
    LeafNode lna;
    LeafNode lnb;
    LeafNode lnc;
    InternalNode ina;
    InternalNode inb;
    ina.add(&lna);
    ina.add(&inb);
    inb.add(&lnb);
    inb.add(&lnc);

    PrintVisitor pv;
    ina.accept(pv);

    return 0;
}
```

In the main function (lines 59–74), we define three objects of `LeafNode` and two objects of `InternalNode` and create a node hierarchy. Then we define an object of `PrintVisitor` and apply the object to an object of `InternalNode`. Once we execute the above program, we see the following output:

```
visit an internal node
visit a leaf node
visit an internal node
visit a leaf node
visit a leaf node
```

The above example shows the usage of the visitor pattern and the composite pattern together. More information about the visitor pattern can be found in Gamma et al. (1994) and Lasater (2007).

4.7 Summary

In this chapter, we introduced several popular design patterns. Some of these patterns are used to develop our clustering library later. The examples used to illustrate these design patterns are very simple. We used simple examples without distracting complex C++ code so that readers can get the ideas of design patterns easily.

There are many other useful patterns, which can be found in Gamma et al. (1994), Shalloway and Trott (2001), Holzner (2006), and Lasater (2007). Although some books illustrate design pattern using other programming languages such as Java and C#, the ideas of design patterns are the same. Readers can also find much information about design patterns on the Internet with search engines. We did not list any web resources related to design patterns in the book since web links change over time.

Chapter 5
C++ Libraries and Tools

In this chapter, we introduce some useful C++ libraries and tools. First, we introduce the C++ standard template library (STL). Then we introduce some of the Boost C++ libraries. Finally, we introduce the GNU build system and the Cygwin software.

5.1 The Standard Template Library

The C++ Standard Template Library (STL) is a collection of template-based components that implement many common data structures and algorithms used to process those data structures. Alexander Stepanov and Meng Lee developed the STL at Hewlett-Packard based on their research in the field of generic programming, with significant contributions from David Musser (Deitel and Deitel, 2009).

STL consists of three key components: containers, iterators, and algorithms. Containers are data structures capable of storing objects of almost any data type. Iterators are similar to pointers and are used by programs to manipulate the elements in containers. Algorithms are functions that perform common data manipulations such as searching, sorting, and comparing elements.

5.1.1 Containers

STL containers can be further classified into three categories: sequence containers, associative containers and container adapters (Deitel and Deitel, 2009). Table 5.1 gives a list of STL containers grouped into the three categories.

Sequence containers implement linear data structures, such as vectors. Associative containers implement nonlinear data structures that typically can locate elements stored in the containers quickly. Stacks and queues are constrained versions of sequential containers and are implemented as container adapters that enable a program to view a sequential container in a constrained manner.

77

Class	Type	Description
`vector`	Sequence	Rapid insertions and deletions at back. Direct access to any element.
`deque`	Sequence	Rapid insertions and deletions at front or back. Direct access to any element.
`list`	Sequence	Rapid insertion and deletion anywhere.
`set`	Associative	Rapid lookup, no duplicates allowed.
`multiset`	Associative	Rapid lookup, duplicates allowed.
`map`	Associative	One-to-one mapping. No duplicates allowed. Rapid key-based lookup.
`multimap`	Associative	One-to-many mapping. Duplicates allowed. Rapid key-based lookup.
`stack`	Adapter	Last-in, first-out (LIFO).
`queue`	Adapter	First-in, first-out (FIFO).
`priority_queue`	Adapter	Highest-priority element is always the first element out.

TABLE 5.1: STL containers

STL provides three sequence containers: `vector`, `list`, and `deque`. Class `vector` is probably the most popular container and is used for applications that need frequent insertions and deletions only at back. Class `deque` is usually used for applications that require frequent insertions and deletions at both ends of a container. Class `list` is appropriate for applications that require frequent insertions and deletions in the middle and/or at the extremes of a container.

The C++ program in Listing 5.1 illustrates how to use the three sequence containers. From the program we see that STL containers are very convenient. We do not need to worry about allocating memory space for new elements. We also do not need to worry about freeing memory space at the end of the program.

In the program in Listing 5.1, we see that we can use operator [ ] to access the element of a `vector` container or `deque` container randomly. However, we have to use iterators to access the element of a `list` container.

Listing 5.1: A program that uses sequence containers.

```
// sequence.cpp
#include<iostream>
#include<vector>
#include<list>
#include<deque>
```

```cpp
int main() {
    std::vector<int> nv;
    for(int i=1;i<6;++i){
        nv.push_back(i);
    }
    std::cout<<"The size of nv is: "<<nv.size();
    std::cout<<"\nThe elements of nv are: ";
    for(size_t i=0;i<nv.size();++i){
        std::cout<<nv[i]<<" ";
    }


    std::deque<int> nd;
    for(int i=1;i<6;++i){
        nd.push_front(i);
        nd.push_back(i);
    }
    std::cout<<"\nThe size of nd is: "<<nd.size();
    std::cout<<"\nThe elements of nd are: ";
    for(size_t i=0;i<nd.size();++i){
        std::cout<<nd[i]<<" ";
    }

    std::list<int> nl;
    std::list<int>::iterator iter;
    for(int i=1;i<6;++i){
        nl.push_front(i);
        nl.push_back(i);
    }

    std::cout<<"\nThe size of nl is: "<<nl.size();
    std::cout<<"\nThe elements of nl are: ";
    for(iter=nl.begin();iter!=nl.end();++iter){
        std::cout<<*iter<<" ";
    }
    nl.sort();
    std::cout<<"\nThe elements of sorted nl are: ";
    for(iter=nl.begin();iter!=nl.end();++iter){
        std::cout<<*iter<<" ";
    }

    return 0;
}
```

When we execute the program, we get the following output:

```
The size of nv is: 5
The elements of nv are: 1 2 3 4 5
The size of nd is: 10
The elements of nd are: 5 4 3 2 1 1 2 3 4 5
The size of nl is: 10
The elements of nl are: 5 4 3 2 1 1 2 3 4 5
The elements of sorted nl are: 1 1 2 2 3 3 4 4 5 5
```

Associative containers implement the key/value data structures and provide direct access to store and retrieve elements via keys. STL provides four associative containers: set, multiset, map, and multimap. The keys of an associative container are sorted in order. Iterating through an associative container follows the sorted order of the keys of that container.

In classes **set** and **multiset**, the values are the keys—there is not a separate value associated with each key. Class **set** does not allow duplicate keys, whereas class **multiset** does allow duplicates. In classes **map** and **multimap**, keys are usually not the same as values. Class **map** does not allow duplicate keys but class **multimap** does.

Listing 5.2: A program that uses associative containers.

```cpp
// associative.cpp
#include<iostream>
#include<set>
#include<map>

typedef std::set<double> dSet;
typedef std::multiset<int> nMultiSet;
typedef std::map<int, double> ndMap;
typedef std::multimap<int, double> ndMultiMap;

int main() {
    dSet ds;
    ds.insert(1.2);
    ds.insert(3.1);
    ds.insert(0.5);

    std::cout<<"The size of ds is: "<<ds.size()<<'\n';
    std::cout<<"The elements of ds are: ";
    dSet::iterator it;
    for(it=ds.begin();it!=ds.end();++it){
        std::cout<<*it<<" ";
    }

    std::pair<dSet::const_iterator, bool> p;
    p = ds.insert(3.1);
    std::cout<<"\n"<<*(p.first)
        <<( p.second ? " was " : " was not " )
        <<"inserted"<<'\n';

    it = ds.find(0.5);
    if(it!=ds.end()){
        std::cout<<"Found value 0.5 in ds"<<'\n';
    }

    nMultiSet nms;
    nms.insert(1);
    nms.insert(-1);
    nms.insert(1);

    std::cout<<"The size of nms is: "<<nms.size()<<'\n';
    std::cout<<"The elements of nms are: ";
    nMultiSet::iterator itn;
    for(itn=nms.begin();itn!=nms.end();++itn){
        std::cout<<*itn<<" ";
    }

    ndMap ndm;
    ndm.insert(ndMap::value_type(1,3.1));
    ndm.insert(ndMap::value_type(5,-0.5));
    ndm.insert(ndMap::value_type(5,2.6));
    ndm.insert(ndMap::value_type(15,20.1));
    std::cout<<"\nndm contains: \n";
    ndMap::const_iterator itm;
    for(itm=ndm.begin(); itm!=ndm.end();++itm) {
        std::cout<<itm->first <<" --- "<<itm->second<<'\n';
    }
```

```
58      ndMultiMap ndmm;
59      ndmm.insert(ndMultiMap::value_type(15, 101.5));
60      ndmm.insert(ndMultiMap::value_type(25, -200.9));
61      ndmm.insert(ndMultiMap::value_type(35, 9.2));
62      ndmm.insert(ndMultiMap::value_type(15, -420.1));
63      std::cout<<"ndmm contains:\n";
64      ndMultiMap::const_iterator itmm;
65      for(itmm=ndmm.begin(); itmm!=ndmm.end();++itmm) {
66          std::cout<<itmm->first <<" -- "<<itmm->second<<'\n';
67      }
68
69      return 0;
70  }
```

Listing 5.2 gives a simple C++ program that shows how to use the four associative containers. Executing the program gives the following output:

```
The size of ds is: 3
The elements of ds are: 0.5 1.2 3.1
3.1 was not inserted
Found value 0.5 in ds
The size of nms is: 3
The elements of nms are: -1 1 1
ndm contains:
1 -- 3.1
5 -- -0.5
15 -- 20.1
ndmm contains:
15 -- 101.5
15 -- -420.1
25 -- -200.9
35 -- 9.2
```

From the output of the program we see that class `set` and class `map` do not allow duplicate keys.

STL provides the following three container adapters: `stack`, `queue`, and `priority_queue`. Class `stack` implements the last-in, first-out data structure. The underlying container of `stack` can be implemented with any of the sequence containers: `vector`, `list`, and `deque`. Class `queue` implements the first-in, first-out data structure. The underlying container of `queue` can be implemented with `list` or `deque`. Class `priority_queue` implements the data structure that enables insertions in sorted order into the underlying container and deletions from the front of the underlying container.

Listing 5.3: A program that uses container adapters.

```
1  // adapter.cpp
2  #include<iostream>
3  #include<stack>
4  #include<queue>
5  #include<vector>
6  #include<list>
7
8  using namespace std;
```

```
 9
10   int main() {
11       stack<int> nDequeStack;
12       stack<int, vector<int> > nVectorStack;
13
14       for(int i=-5;i<5;++i){
15           nDequeStack.push(i);
16           nVectorStack.push(i);
17       }
18
19       cout<<"nDequeStack: ";
20       while(!nDequeStack.empty()){
21           cout<<nDequeStack.top()<<" ";
22           nDequeStack.pop();
23       }
24       cout<<"\nnVectorStack: ";
25       while(!nVectorStack.empty()){
26           cout<<nVectorStack.top()<<" ";
27           nVectorStack.pop();
28       }
29
30       queue<double> dQueue;
31       dQueue.push(-2.1);
32       dQueue.push(2.5);
33       dQueue.push(-6.5);
34       cout<<"\ndQueue: ";
35       while(!dQueue.empty()){
36           cout<<dQueue.front()<<" ";
37           dQueue.pop();
38       }
39
40       priority_queue<double> dPQueue;
41       dPQueue.push(-2.1);
42       dPQueue.push(2.5);
43       dPQueue.push(-6.5);
44       cout<<"\ndPQueue: ";
45       while(!dPQueue.empty()){
46           cout<<dPQueue.top()<<" ";
47           dPQueue.pop();
48       }
49       return 0;
50   }
```

The C++ program in Listing 5.3 shows how to use the three container adapters. Executing the program gives the following output:

```
nDequeStack: 4 3 2 1 0 -1 -2 -3 -4 -5
nVectorStack: 4 3 2 1 0 -1 -2 -3 -4 -5
dQueue: -2.1 2.5 -6.5
dPQueue: 2.5 -2.1 -6.5
```

From the output we see that data in stack and priority_queue are sorted in order. Data in queue are ordered in the first-in, first-out manner.

5.1.2 Iterators

Iterators in STL are similar to pointers and are used to point to the elements of first-class containers, i.e., sequence containers and associative containers. We can use functions begin and end to get the iterators pointing to the position of the first element and the position after the last element of first-class containers, respectively.

Iterators can be classified into five categories: input, output, forward, bidirectional, and random access. An input iterator is used to read an element from a container and can move only in the forward direction one element at a time. An output iterator is used to write an element to a container. Like an input iterator, an output iterator can move only in the forward direction one element at a time. A forward iterator combines the capabilities of input and output iterators. A bidirectional iterator combines the capabilities of a forward iterator with the ability to move in the backward direction. A random access iterator has the capabilities of a bidirectional iterator and the ability to directly access any element of the container. Figure 5.1 shows the iterator hierarchy.

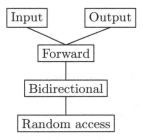

FIGURE 5.1: Iterator hierarchy.

Containers `vector` and `deque` support random access iterators. Other first-class containers (i.e., `list`, `set`, `multiset`, `map`, and `multimap`) support bidirectional iterators. Container adapters do not support iterators.

Listing 5.4: A program that illustrates iterators.

```
// iterator.cpp
#include<iostream>
#include<vector>
#include<iterator>

using namespace std;

int main() {
    vector<int> nv, nv2;
    for(int i=0;i<8;++i) {
        nv.push_back(i+1);
        nv2.push_back(-i-1);
    }
    vector<int>::const_iterator it;
    cout<<"nv: ";
    for(it = nv.begin();it!=nv.end();++it){
        cout<<*it<<" ";
    }
    it = nv.begin();
    advance(it,5);
    cout<<"\nnv: "<<*it;

    copy(nv2.begin(), nv2.end(), back_inserter(nv));
```

```
24      cout<<"\nnv_nv2_:_";
25      for(it = nv.begin();it!=nv.end();++it){
26          cout<<*it<<"_";
27      }
28
29
30      return 0;
31  }
```

The C++ program in Listing 5.4 shows the usage of iterators. This program first creates two vectors (lines 9–13) and then outputs the first vector and appends the second vector to the first vector. The function `back_inserter` is defined in the header `iterator`. The function `advance` (line 20) is also defined in the header `iterator`. Executing the program gives the following output:

nv : 1 2 3 4 5 6 7 8
nv : 6
nv nv2 : 1 2 3 4 5 6 7 8 -1 -2 -3 -4 -5 -6 -7 -8

5.1.3 Algorithms

STL provides 70 standard algorithms, including 4 numeric algorithms defined in the header file `numeric`. Tables 5.2, 5.3, 5.4, 5.5, 5.6, 5.7, 5.8, and 5.9 give the names of these algorithms. We can use these algorithms to manipulate containers.

adjacent_find	find	find_if	count
find_each	mismatch	count_if	find_end
search	equal	find_first_of	search_n

TABLE 5.2: Non-modifying sequence algorithms.

copy	remove	reverse_copy	copy_backward
remove_copy	rotate	fill	remove_copy_if
rotate_copy	fill_n	remove_if	stable_partition
generate	replace	swap	generate_n
replace_copy	swap_ranges	iter_swap	replace_copy_if
transform	partition	replace_if	unique
random_shuffle	reverse	unique_copy	

TABLE 5.3: Modifying sequence algorithms.

sort	stable_sort	partial_sort	partial_sort_copy
nth_element			

TABLE 5.4: Sorting algorithms.

lower_bound	upper_bound	equal_range	binary_search

TABLE 5.5: Binary search algorithms.

merge	inplace_merge	includes
set_union	set_intersection	set_difference
set_symmetric_difference		

TABLE 5.6: Merging algorithms.

push_heap	pop_heap	make_heap	sort_heap

TABLE 5.7: Heap algorithms.

min	max	min_element
max_element	lexicographical_compare	next_permutation
prev_permutation		

TABLE 5.8: Min/max algorithms.

accumulate	partial_sum	inner_product	adjacent_difference

TABLE 5.9: Numerical algorithms defined in the header file **numeric**.

Listing 5.5: A program that illustrates STL algorithms.

```cpp
// algorithm.cpp
#include<iostream>
#include<vector>
#include<algorithm>
#include<numeric>
#include<iterator>

using namespace std;

bool IsOdd(int x) {
    return ((x%2)==1);
}

int main() {
    ostream_iterator<int> output(cout, " ");
    vector<int>::iterator it;
    vector<int> nv(10);
    fill(nv.begin(),nv.end(),1);
    fill_n(nv.begin(),5,2);
    cout<<"nv: ";
    copy(nv.begin(), nv.end(), output);

    vector<int> nv2(10,1);
    cout<<"\nnv2: ";
    copy(nv2.begin(), nv2.end(), output);
    bool res = equal(nv.begin(),nv.end(),nv2.begin());
    cout<<"\nnv_and_nv2_are_"<<(res ? "" : "not")
        <<" equal";
```

```
29
30      for(int i=1;i<=10;++i){
31          nv[i-1] = i;
32      }
33      copy(nv.begin(), nv.end(), nv2.begin());
34      cout<<"\nBefore removing odd, nv: ";
35      copy(nv.begin(),nv.end(),output);
36      it = remove_if(nv.begin(),nv.end(),IsOdd);
37      cout<<"\nAfter removing odd, nv: ";
38      copy(nv.begin(),it,output);
39      cout<<"\nBefore replacing odd: nv: ";
40      copy(nv.begin(),nv.end(),output);
41      cout<<"\nAfter replacing odd: nv: ";
42      replace_if(nv.begin(),nv.end(),IsOdd, -1);
43      copy(nv.begin(),nv.end(),output);
44
45      cout<<"\nnv2: ";
46      copy(nv2.begin(), nv2.end(), output);
47      partial_sum(nv2.begin(),nv2.end(),nv.begin());
48      cout<<"\npartial_sum_of_nv2: ";
49      copy(nv.begin(),nv.end(),output);
50
51      return 0;
52  }
```

The C++ program in Listing 5.5 demonstrates the usage of several algorithms, such as `fill`, `copy`, `remove_if`, and `partial_sum`. Executing the program gives the following output:

```
nv: 2 2 2 2 2 1 1 1 1 1
nv2: 1 1 1 1 1 1 1 1 1 1
nv and nv2 are not equal
Before removing odd, nv: 1 2 3 4 5 6 7 8 9 10
After removing odd, nv: 2 4 6 8 10
Before replacing odd: nv: 2 4 6 8 10 6 7 8 9 10
After replacing odd: nv: 2 4 6 8 10 6 -1 8 -1 10
nv2: 1 2 3 4 5 6 7 8 9 10
partial sum of nv2: 1 3 6 10 15 21 28 36 45 55
```

Note that the old elements of the container processed by `remove_if` are still accessible. We can see this from the output.

5.2 Boost C++ Libraries

The Boost C++ libraries are a collection of free peer-reviewed portable C++ source libraries that extend the functionality of C++. Boost (Version 1.43.0) contains about 100 individual libraries, including libraries for string and text processing, containers, iterators, algorithms, concurrent programming, math and numerics, memory, and many others. For a complete list of Boost libraries, readers are referred to Boost's website at http://www.boost.org.

Most Boost libraries are header-only libraries, which consist entirely of C++ header files containing only templates and inline functions. Header-only libraries do not require separately-compiled library binaries or special treatment when linking.

5.2.1 Smart Pointers

In Boost, smart pointers are objects that store pointers to dynamically allocated objects. A smart pointer is similar to a C++ built-in pointer except that the smart pointer automatically deletes the object pointed to at the appropriate time.

Smart pointers are very useful as they ensure proper destruction of dynamically allocated objects, even in the event of an exception. In fact, smart pointers are seen as owning the object pointed to, and thus responsible for deleting the object when it is no longer needed.

The Boost smart pointer library provides six smart pointer class templates. Table 5.10 gives the descriptions of these class templates. We use `shared_ptr` extensively in the clustering library developed in this book.

Class	Description
`scoped_ptr`	Simple sole ownership of single objects. Noncopyable.
`scoped_array`	Simple sole ownership of arrays. Noncopyable.
`shared_ptr`	Object ownership shared among multiple pointers.
`shared_array`	Array ownership shared among multiple pointers.
`weak_ptr`	Non-owning observers of an object owned by `shared_ptr`.
`intrusive_ptr`	Shared ownership of objects with an embedded reference count.

TABLE 5.10: Boost smart pointer class templates.

Listing 5.6: A program that illustrates Boost `shared_ptr`.

```
1  // sharedptr.cpp
2  #include<iostream>
3  #include<boost/shared_ptr.hpp>
4
5  using namespace std;
6  using namespace boost;
7
8  class A {
9  public:
10     A() {
11         cout<<"A's constructor is called"<<endl;
12     }
13
14     ~A() {
15         cout<<"A's destructor is called"<<endl;
16     }
17
```

```cpp
18      void func () {
19          cout<<"A::func_is_called"<<endl;
20      }
21  };
22
23  int main() {
24      A* p = new A();
25      shared_ptr<A> p1(p);
26      shared_ptr<A> p2 = p1;
27      cout<<"A_use_count:_"<<p1.use_count()<<endl;
28
29      p2.reset();
30      cout<<"A_use_count:_"<<p1.use_count()<<endl;
31
32      p1->func();
33      (*p1).func();
34
35      A* q = p1.get();
36      cout<<"A_use_count:_"<<p1.use_count()<<endl;
37      q->func();
38
39      return 0;
40  }
```

The C++ program in Listing 5.6 demonstrates how to use shared_ptr. This program defines a simple class A with a constructor, a destructor, and a function func. The three member functions output different texts to screen so that we know which function is called. In the main function, we first create an object of class A (line 24) and store the pointer to the object in a shared_ptr object p1 (line 25). Object p2 is also a shared_ptr object that stores the pointer pointing to the object of A. Class shared_ptr implements operators "−>" and "∗", which are applied to the stored pointer. The function use_count returns the number of shared_ptr objects that share ownership of the stored pointer.

Since the program depends on the Boost libraries, we need to use the following command to compile the program in Cygwin or UNIX-like systems:

```
g++ -o sharedptr -I/usr/include sharedptr.cpp
```

In the compilation command, we used -I/usr/include to tell the compiler the location of the headers of the Boost libraries. In Cygwin, Boost headers are installed to /usr/include.

Executing the program gives the following output:

```
A's constructor is called
A use count: 2
A use count: 1
A::func is called
A::func is called
A use count: 1
A::func is called
A's destructor is called
```

From the output we see that the dynamically allocated object of class A is destroyed at the end as we see the destructor of class A is called.

5.2.2 Variant

The Boost **variant** class template is a safe and generic union container. Class **variant** allows programs to manipulate an object from a heterogeneous set of types in a uniform manner. Unlike **std::vector**, which stores multiple values of a single type, **variant** stores a single value of multiple types.

The small C++ program in Listing 5.7 shows the usage of **variant**. To use the Boost **variant** class template, we need to include the Boost header **boost/variant.hpp** (line 3) in the program. The variable v (line 10) is an object of class **variant** that can hold a value of type **int**, **double**, or **string**. To get the value stored in **variant** object, we can use **get** template with a type argument. If the type specified in **get** is not the same as the type of the stored value, a **bad_get** exception is raised.

Listing 5.7: A program that illustrates Boost **variant**.

```
1  // variant.cpp
2  #include<iostream>
3  #include<boost/variant.hpp>
4
5  using namespace std;
6  using namespace boost;
7
8  int main() {
9      try{
10         variant<int, double, string> v;
11         v = 1;
12         cout<<"v: "<<get<int>(v)<<endl;
13         v = 3.14;
14         cout<<"v: "<<get<double>(v)<<endl;
15         v = "a string";
16         cout<<"v: "<<get<string>(v)<<endl;
17         cout<<"v: "<<get<int>(v)<<endl;
18
19     }catch(exception &ex){
20         cout<<"exception : "<<ex.what()<<endl;
21     }catch(...){
22         cout<<"unknow error"<<endl;
23     }
24
25     return 0;
26 }
```

To compile the program, we can use the following command:

g++ -o variant -I/usr/include variant.cpp

Executing the program gives the following output:

v: 1
v: 3.14
v: a string
exception : boost::bad_get: failed value get using boost::get

5.2.3 Variant versus Any

Like `variant`, Boost `any` is another heterogeneous container. Although Boost `any` shares many of the same features of Boost `variant`, both libraries have their own advantages and disadvantages. According to the Boost library documentation, Boost `variant` has the following advantages over Boost `any`:

(a) Boost `variant` guarantees the type of its content is one of a finite, user-specified set of types;

(b) Boost `variant` provides compile-time checked visitation of its content;

(c) Boost `variant` enables generic visitation of its content;

(d) Boost `variant` avoids the overhead of dynamic allocation by offering an efficient, stack-based storage scheme.

Boost `any` has the following advantages over Boost `variant`:

(a) Boost `any` allows virtually any type for its content;

(b) Boost `any` provides the no-throw guarantee of exception safety for its swap operation;

(c) Boost `any` makes little use of template metaprogramming techniques.

Listing 5.8: A program to compare Boost variant and any.

```cpp
#include<iostream>
#include<vector>
#include<string>
#include<boost/timer.hpp>
#include<boost/variant.hpp>
#include<boost/any.hpp>

using namespace std;
using namespace boost;

int main(int argc, char* argv[]) {
    size_t N = 100000;
    if(argc >= 2) {
        N = atoi(argv[1]);
    }
    vector<variant<double, int, string> > vv;
    boost::timer t;
    t.restart();
    for(size_t i=0; i<N; ++i) {
        switch(i % 3) {
            case 0:
                vv.push_back(1.0);
                break;
            case 1:
                vv.push_back(1);
                break;
            case 2:
                vv.push_back("a string");
                break;
        }
    }
```

```cpp
32        cout<<"Creating a variant vector takes: "<<t.elapsed()
33            <<" seconds"<<endl;
34
35        t.restart();
36        for(size_t i=0; i<N; ++i) {
37            switch(i % 3) {
38                case 0:
39                    get<double>(vv[i]);
40                    break;
41                case 1:
42                    get<int>(vv[i]);
43                    break;
44                case 2:
45                    get<string>(vv[i]);
46                    break;
47            }
48        }
49        cout<<"Getting variant vector takes: "<<t.elapsed()
50            <<" seconds"<<endl;
51
52        vector<any> av;
53        t.restart();
54        for(size_t i=0; i<N; ++i) {
55            switch(i % 3) {
56                case 0:
57                    av.push_back(any(1.0));
58                    break;
59                case 1:
60                    av.push_back(any(1));
61                    break;
62                case 2:
63                    av.push_back(any(string("a string")));
64                    break;
65            }
66        }
67        cout<<"Creating any vector takes: "<<t.elapsed()
68            <<" seconds"<<endl;
69
70        t.restart();
71        for(size_t i=0; i<N; ++i) {
72            switch(i % 3) {
73                case 0:
74                    any_cast<double>(av[i]);
75                    break;
76                case 1:
77                    any_cast<int>(av[i]);
78                    break;
79                case 2:
80                    any_cast<string>(av[i]);
81                    break;
82            }
83        }
84        cout<<"Getting any vector takes: "<<t.elapsed()
85            <<" seconds"<<endl;
86
87        return 0;
88    }
```

The C++ program shown in Listing 5.8 compares Boost **variant** and **any** in terms of performance. The program creates a vector of type **variant**, populates the vector with values of types **double**, **int**, and **std::string**, and then gets the values from the vectors. The time used for this process is measured and printed to screen. The program does the same for Boost **any**.

The program also allows users to input the number of elements the vector should hold.

To compile the program, we can use the following command:

```
g++ -o compare -I/usr/include compare.cpp
```

When we execute the program using the following command (in Cygwin):

```
./compare.exe
```

we see the following output:

```
Creating a variant vector takes: 0.64 seconds
Getting variant vector takes: 0.016 seconds
Creating any vector takes: 0.625 seconds
Getting any vector takes: 0.078 seconds
```

From the output we see that creating the two vectors takes approximately the same amount of time. But getting values out of the vector of type **any** takes five times more time than getting values from the vector of type **variant**.

Now we try the program with an argument using the following command:

```
./compare.exe 1000000
```

We let the program create vectors that hold one million elements. When the above command is executed, we see the following output:

```
Creating a variant vector takes: 5.922 seconds
Getting variant vector takes: 0.109 seconds
Creating any vector takes: 5.657 seconds
Getting any vector takes: 0.75 seconds
```

From the output we have the same observations as before.

5.2.4 Tokenizer

The Boost **tokenizer** class provides a flexible and easy way to break a string into a series of tokens. For example, we can use **tokenizer** to parse a CSV (comma-separated values) file.

Listing 5.9: A program that illustrates Boost **tokenizer**.

```
1  // tokenizer.cpp
2  #include<iostream>
3  #include<boost/tokenizer.hpp>
4
5  using namespace std;
6  using namespace boost;
7
8  typedef tokenizer<char_separator<char> > Ctok;
9
10 int main() {
11     string s = "A flexible, easy tokenizer";
12
```

```
13        char_separator<char> sep(",");
14        Ctok ctok(s, sep);
15        for(Ctok::iterator it = ctok.begin(); it != ctok.end();
16               ++it) {
17           cout<<"<"<<*it<<">"<<" ";
18        }
19
20        char_separator<char> sep2(" ");
21        Ctok ctok2(s, sep2);
22        cout<<endl;
23        for(Ctok::iterator it = ctok2.begin(); it != ctok2.end();
24               ++it) {
25           cout<<"<"<<*it<<">"<<" ";
26        }
27
28        char_separator<char> sep3(", ", "", keep_empty_tokens);
29        Ctok ctok3(s, sep3);
30        cout<<endl;
31        for(Ctok::iterator it = ctok3.begin(); it != ctok3.end();
32               ++it) {
33           cout<<"<"<<*it<<">"<<" ";
34        }
35
36        return 0;
37     }
```

The C++ program in Listing 5.9 illustrates how to use the Boost **tokenizer**. The program shows three tokenizers: the first one (line 14) breaks a string by commas, the second one (line 21) breaks a string by spaces, and the last one (line 29) breaks a string by either commas or spaces and keeps the empty tokens. Executing the program gives the following output:

```
<A flexible> < easy tokenizer>
<A> <flexible,> <easy> <tokenizer>
<A> <flexible> <> <easy> <tokenizer>
```

5.2.5 Unit Test Framework

The Boost test library provides a set of components to write test programs, organize tests, and control their runtime execution. The set of components includes the execution monitor, the program execution monitor, the test tools, the unit test framework, and the minimal testing facility. The unit test framework allows us to write test cases as simple free functions or member functions and organize them into a tree of test suites. The framework also allows us to use the component of test tools to implement test cases and provides a facility to manage a log report level and a result report level.

The Boost test library can be used as a header-only library or separately-compiled library. If we want to use the library as a header-only library, we need to include the header

`boost/test/included/unit_test_framework.hpp`

as shown in line 3 of the program in Listing 5.10. The program in Listing 5.10 illustrates how to use test suites provided by the unit test framework component. We note that the main function does not appear in the program in Listing 5.10. In fact, the function init_unit_test_suite is the entry point of the program and the main function is defined in the test library.

Listing 5.10: A program that illustrates Boost unit_test_framework.

```cpp
// unittest.cpp
#include<iostream>
#include<boost/test/included/unit_test_framework.hpp>

using namespace std;
using namespace boost;

class Test {
public:
    static void test1();
    static void test2();
    static unit_test_framework::test_suite* suite();
};

void Test::test1() {
    BOOST_CHECK(10 / 2 == 5);
    BOOST_CHECK(10 / 2 == 3);
}

void Test::test2() {
    int a = 10;
    BOOST_REQUIRE(a>9);
    BOOST_REQUIRE(a>19);
}

unit_test_framework::test_suite* Test::suite(){
    unit_test_framework::test_suite* suite =
        BOOST_TEST_SUITE("A sample test suite");

    suite->add(BOOST_TEST_CASE(&Test::test1));
    suite->add(BOOST_TEST_CASE(&Test::test2));
    return suite;
}

unit_test_framework::test_suite* init_unit_test_suite(int, char*[])
{
    unit_test_framework::test_suite* test =
        BOOST_TEST_SUITE("Master test suite");
    test->add(Test::suite());
    return test;
}
```

Compiling the program is simple as we do not need to link the separately-compiled test library. We can use the following command to compile the program:

`g++ -o unittest -I/usr/include unittest.cpp`

To execute the program, we can use various log levels (see Table 5.11). For example, if we want to see all log messages, we can execute the program using the following command:

`./unittest.exe --log_level=all`

and see the following output:

`Running 2 test cases...`

```
Entering test suite "Master test suite"
Entering test suite "A sample test suite"
Entering test case "Test::test1"
unittest.cpp(16): info: check 10 / 2 == 5 passed
unittest.cpp(17): error in "Test::test1":
    check 10 / 2 == 3 failed
/usr/include/boost-1_33_1/boost/test/impl/results_collector.
    ipp(197): info: check 'Test case has less failures
    then expected' passed
Leaving test case "Test::test1"
Entering test case "Test::test2"
unittest.cpp(22): info: check a>9 passed
unittest.cpp(23): fatal error in "Test::test2":
    critical test a>19 failed
/usr/include/boost-1_33_1/boost/test/impl/results_collector.
    ipp(197): info: check 'Test case has less failures
    then expected' passed
Leaving test case "Test::test2"
Leaving test suite "A sample test suite"
Leaving test suite "Master test suite"
```

Log level	Description
all	report all log messages
success	the same as all
test_suite	show test suite messages
message	show user messages
warning	report warnings issued by user
error	report all error conditions
cpp_exception	report uncaught c++ exceptions
system_error	report system-originated nonfatal errors
fatal_error	report only user- or system- originated fatal errors
nothing	report no information

TABLE 5.11: Boost unit test log levels.

5.3 GNU Build System

Te GNU build system, also called GNU Autotools, is a set of tools produced by the GNU project. The set of tools includes **Autoconf**, **Automake**, and **Libtool**. These tools allow us to make source-code packages more portable and make life easer for users of the packages. In this section, we give a brief introduction to these tools. For detailed information about these tools, readers are referred to Vaughan et al. (2010) and Calcote (2010).

5.3.1 Autoconf

The `Autoconf` tool was designed to generate configure scripts. The generated scripts are more portable, more accurate, and more maintainable than scripts written by hand. The `Autoconf` comes with several Perl utility programs: `autoconf`, `autoheader`, `autom4te`, `autoreconf`, `autoscan`, `autoupdate`, and `ifnames` (Calcote, 2010).

The `autoheader` utility program generates a C language header file template from `configure.ac`. This template header file is usually called `config.h.in` and contains the C macros to be defined by `configure`. The `autom4te` utility program is a cache manager used internally by the Autotools and produces an `autom4te.cache` directory in the top-level project directory. The `autoreconf` utility program provides a convenient way to execute the configuration tools in the `Autoconf`, `Automake`, and `Libtool` packages as required by the project. The `autoscan` utility program is used to generate a `configure.ac` file for a new project. The `autoupdate` utility program is used to update the `configure.ac` file to the syntax of the current version of the Autotools. The `ifnames` program is used to help programmers determine what to put into the `configure.ac` file and the `Makefile.am` file for the sake of portability.

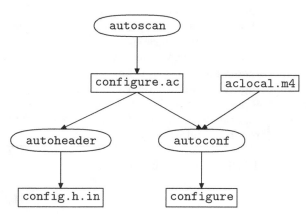

FIGURE 5.2: Flow diagram of `Autoconf`.

Figure 5.2 shows the data flow of the `Autoconf` package. One can use `autoscan` to generate a `configure.ac` file or just write the file by hand. The `autoheader` program takes the `configure.ac` file as input and generates a `config.h.in` file. The `autoconf` program takes the `configure.ac` file and the `aclocal.m4` file as inputs and generates the `configure` script.

The aclocal.m4 file is generated by the aclocal program provided by the Automake package described in the next section.

5.3.2 Automake

The Automake package is used to convert much simplified build specification files (i.e., Makefile.am files) to standard makefile templates (i.e., Makefile.in files) that provide all the standard make targets such as all, clean, dist, and install. The Automake package comes with two Perl script programs: aclocal and automake.

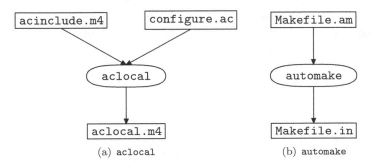

FIGURE 5.3: Flow diagram of Automake.

Figure 5.3 shows the data flow diagram of the Automake package. The aclocal program takes the acinlcude.m4 file and the configure.ac file as inputs and produces an aclocal.m4 file, which is used by autoconf to generate the configure script. The input acinclude.m4 is optional. The automake program takes the Makefile.am file as input and produces a Makefile.in file, which is a makefile template. The resulting Makefile.in files will be converted to Makefile files by the script configure.

5.3.3 Libtool

The Libtool package is used to build shared libraries on different Unix-like platforms. Different platforms use different ways to name, build, and manage shared libraries. For example, some systems name their libraries with a ".so" extension; while other systems use ".o" or ".a" as library extensions. When building shared libraries with Libtool, we do not need to consider these differences as all of these differences have been carefully considered by the authors of the Libtool package.

The Libtool package comes with libtool, libtoolize, ltdl, and ltdl.h. libtool and libtoolize are programs. ltdl is a set of static and shared libraries. ltdl.h is a header file.

5.3.4 Using GNU Autotools

The primary task of Autotools is to generate a script called **configure**, which is used by users to build a project. To produce the **configure** script, we can just execute the following command:

```
autoreconf
```

The program `autoreconf` runs `autoconf`, `autoheader`, `aclocal`, `automake`, and `libtoolize` repeatedly to update the GNU build system in the specified directory and its subdirectories. By default, the program updates only those files that are older than their sources. Sometimes we need to update all files regardless if they are older than their source or not. For example, if a new version of some tool is installed, we need to update all files. In such cases, we can execute the command with the `-f` option:

```
autoreconf -f
```

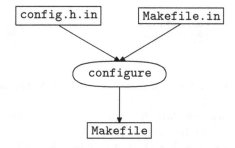

FIGURE 5.4: Flow diagram of **configure**.

Once we have the **configure** script, we can build our project. First, we run **configure** to generate **Makefile** files. Figure 5.4 shows the data flow diagram of **configure**. Then we run **make** to build the project. Appendix C.4 gives an example of how to build and install a project using the GNU build system.

5.4 Cygwin

The GNU build system was designed for Unix-type platforms. To use the GNU build system in Microsoft Windows systems, we can do it with Cygwin, which provides a virtual Unix environment and command-line interface for Microsoft Windows.

Cygwin is free open source software. Cygwin consists of two parts: a dynamic-link library (DLL) and a collection of tools. The DLL provides a substantial part of the POSIX API functionality. The collection of applications provides a Unix-like look and feel.

Cygwin comes with an extensive collection of software tools and applications. For example, we can use GCC and GNU build system in Cygwin. In fact, many Unix programs have been ported to Cygwin. One can find a complete list of Cygwin packages at Cygwin's website at http://www.cygwin.com.

To install Cygwin packages, we need to run Cygwin's setup program. The setup program downloads the necessary package files from repositories on the Internet. In addition, the setup program can install, update, and remove packages and their source code. Appendix C.3 shows how to install Cygwin and the necessary Cygwin programs in order to build the programs in this book.

5.5 Summary

In this chapter, we introduce some C++ libraries and tools. We also illustrate the usage of these libraries with simple examples. Readers should become familiar with these libraries and tools introduced in this chapter since we use them frequently to develop the clustering library. For more information about the Boost libraries, readers are referred to the book by Karlsson (2005).

Part II
A C++ Data Clustering Framework

Chapter 6
The Clustering Library

The clustering library $\mathcal{ClusLib}$ is a collection of C++ components and clustering algorithms developed in this book. $\mathcal{ClusLib}$ attempts to extract commonalities of data clustering algorithms and decompose them into components that can be reused. $\mathcal{ClusLib}$ also allows researchers to implement new clustering algorithms easily.

This chapter introduces the structure of $\mathcal{ClusLib}$. First, we introduce the organization of the source code of the library. Then, we introduce some configuration files used to compile the library. Finally, we introduce error handling and unit testing of the library.

6.1 Directory Structure and Filenames

The directory structure and filenames of the clustering library $\mathcal{ClusLib}$ are similar to that of the Boost library. Figure 6.1 shows the directory structure of the clustering library. All directories and files of $\mathcal{ClusLib}$ are contained in the highest level directory called ClusLib. Under the directory ClusLib, there are five subdirectories: cl, config, examples, m4, and test-suite. Some of the directories under ClusLib also contain subdirectories. For example, the directory cl contains subdirectories: algorithms, clusters, datasets, distances, patterns, and utilities. From the name of these directories, one can imagine the contents.

In $\mathcal{ClusLib}$, file and directory names contain only lowercase ASCII letters and numbers. Header files have a three-letter extension ".hpp" and source files have a three-letter extension ".cpp". Unlike the C++ machine learning library $\mathcal{MLC}$++ (Kohavi et al., 1998), where header files and source files reside in separate directories, $\mathcal{ClusLib}$ organizes header files and sources files according to subjects and puts header files and source files of the same subject into the same directory.

In the directory ClusLib, there are some special files: acinclude.m4, configure.ac, and Makefile.am. The file acinclude.m4 contains m4 macros used to check whether the Boost header files are available and to check the Boost version. The file configure.ac contains tests that check for conditions

103

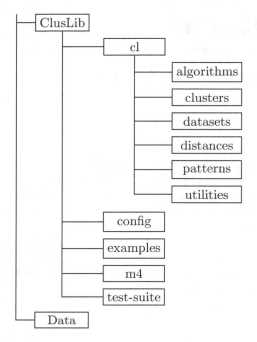

FIGURE 6.1: The directory structure of the clustering library.

that are likely to differ on different platforms. The file `configure.ac` is an input file for the program `autoconf` and is processed by the program `m4`, which produces the resulting file `configure`. The file `Makefile.am` contains variable definitions and is processed by the program `automake`, which produces the file `Makefile.in`. Every subdirectory of `ClusLib` also contains a file named `Makefile.am`.

The directory `cl` and its subdirectories contain all the C++ files of the clustering library. Each subdirectory of the directory `cl` corresponds to a subject and contains the C++ header and source files related to the subject. For example, the subdirectory `algorithms` contains the C++ header and source files related to clustering algorithms.

The directory `config` contains files automatically generated during the configuration process except the file `Makefile.am`, which is an empty file created manually.

The directory `examples` contains examples of data clustering algorithms using the clustering library $\mathcal{C}lus\mathcal{L}ib$. Each example has its own folder, which contains all the C++ files for the example. For example, the example of the k-means algorithm can be found in the subdirectory `kmean` under the directory `examples`.

The directory `m4` contains m4 macros, which are automatically generated during the configuration process. One does not have to know the contents of these macros in order to use the clustering library.

The directory `test-suite` contains some unit-test cases of the clustering library.

6.2 Specification Files

We use a set of GNU Autotools, i.e., `Autoconf`, `Automake`, and `Libtool`, to compile the clustering library. For more information about GNU Autotools, readers are referred to Calcote (2010) and Vaughan et al. (2010). In this section, we introduce some special files used by these Autotools.

6.2.1 configure.ac

The file `configure.ac` (see Listing B.1) is the configuration file for the clustering library $ClusLib$. This file contains `Autoconf` macros that test the system features the clustering library needs. One can use the program `autoscan` to create a `configure.scan` file and change it to a `configure.ac` file.

The macros AC_INIT and AC_OUTPUT are the two only macros required by the configuration script. The macro AC_INIT in the first line takes three arguments: the name of the package, the version of the package, and the bug report address. The macro AC_OUTPUT in the last line takes no arguments and generates the makefiles in the argument of the macro AC_CONFIG_FILES that are required for building the package. Note that AC_OUTPUT also creates files in the argument of the macro AC_CONFIG_HEADERS.

The macro AC_CONFIG_SRCDIR takes one argument that is a file in the package's source directory and checks for the existence of the file to make sure that the source directory does contain the file. This macro does a safety check to make sure people specify the right directory with AC_CONFIG_SRCDIR.

The macro AC_CONFIG_AUX_DIR takes one argument that is a directory and tells `configuration` to use auxiliary build tools (e.g., `install-sh`, `config.guess`, etc.) in the directory.

The macro AC_CONFIG_HEADERS takes one argument that is a header file and makes AC_OUTPUT create the file, which contains C preprocessor #define statements.

The macro AC_CONFIG_MACRO_DIR specifies the location of additional local Autoconf macros. If this macro is used, we need to set `ACLOCAL_AMFLAGS = -I dir` in the top-level `Makefile.am`, where `dir` is the location specified in the macro AC_CONFIG_MACRO_DIR.

The macro AM_INIT_AUTOMAKE has two forms: one form takes only one argument and the other form takes two or more arguments. The configure.ac file shown in List B.1 use the first form. In this form, the argument is a space-separated list of Automake options that will be applied to every Makefile.am in the tree.

The macros from line 9 to line 27 are used to generate options for configure so that users can configure the location of Boost header files and libraries. For example, one can use the following option in Cygwin:

./configure --with-boost-include=/usr/include
--with-boost-lib=/lib.

These macros are useful as the locations of Boost in different platforms are different.

The macro LT_INIT is used to add support for the --enable-shared and --disable-shared configure flags. When LT_INIT is used without arguments, it will turn on shared libraries if they are available, and also enable static libraries if the static libraries do not conflict with the shared libraries.

The macro AC_PROG_CC is used to determine a C compiler to use. The macro checks whether the environment variable CC is set. If so, it will set the output variable CC to its value. Otherwise, it will check for gcc and cc, then for other C compilers. Like AC_PROG_CC, the macro AC_PROG_CXX is used to determine a C++ compiler to use. The macro AC_PROG_CPP sets the output variable CPP to a command that runs the C preprocessor.

The macro CHECK_BOOST is a macro defined in the file acinclude.m4 (see Listing B.2). This macro checks whether the Boost headers are available and whether the Boost version meets our requirement. The macro also looks for an appropriate Boost program options library.

The macro AC_CONFIG_FILES is invoked near the end of configure.ac. The macro tells Automake which files to create.

6.2.2 Makefile.am

In addition to the configure.ac file, we need to create a special file named Makefile.am in each directory of the clustering library in order to use the GNU Autotools. Makefile.am files are processed by automake to produce standards-compliant Makefile.in files.

Listing 6.1 shows the top-level Makefile.am, which consists of only two nonempty lines. The first line defines the variable SUBDIRS, which is a list of subdirectories. The third line defines the variable ACLOCAL_AMFLAGS, which contains options that will be passed to aclocal when aclocal.m4 is to be rebuilt by make.

Listing 6.1: The top-level Makefile.am file.
```
1  SUBDIRS = cl config examples m4 test-suite
2
3  ACLOCAL_AMFLAGS = -I m4
```

Listing 6.2 shows the Makefile.am in the subdirectory cl. This file is more complex than the top-level `Makefile.am`. The first line of this file defines the variable SUBDIRS to be a list of subdirectories of `cl`. The variable AM_CPPFLAGS contains a list of arguments that will be passed to every compilation that invokes the C preprocessor.

The variable this_includedir contains the destination directory of the header files in the directory `cl` when the library is installed. The variable ${includedir} is the destination for the C++ headers of the library. The default value of ${includedir} is `/usr/local/include`.

The variable this_include_HEADERS contains a list of header files in the directory. These header files will be copied to this_includedir when the library is installed. The variable libClusLib_la_SOURCES contains a list of source files in the directory. These source files will be converted to object files by the compiler.

Listing 6.2: The `Makefile.am` file in the directory `cl`.

```
 1  SUBDIRS = algorithms clusters datasets distances patterns \
 2      utilities
 3
 4  AM_CPPFLAGS = -I${top_srcdir} -I${top_builddir}
 5
 6  this_includedir = ${includedir}/${subdir}
 7  this_include_HEADERS = \
 8      cldefines.hpp \
 9      cluslib.hpp \
10      config.hpp \
11      errors.hpp \
12      types.hpp
13
14  libClusLib_la_SOURCES = \
15      errors.cpp
16
17
18  lib_LTLIBRARIES = libClusLib.la
19  libClusLib_la_LIBADD = \
20      algorithms/libAlgorithms.la \
21      clusters/libClusters.la \
22      datasets/libDatasets.la \
23      distances/libDistances.la \
24      patterns/libPatterns.la \
25      utilities/libUtilities.la
26
27  cluslib.hpp: Makefile.am
28          echo "// This file is generated. Please do not edit!" > $@
29          echo >> $@
30          echo "#include <cl/cldefines.hpp>" >> $@
31          echo >> $@
32          for i in $(filter-out config.hpp cluslib.hpp \
33              cldefines.hpp, $(this_include_HEADERS)); do \
34              echo "#include <${subdir}/$$i>" >> $@; \
35          done
36          echo >> $@
37          subdirs='$(SUBDIRS)'; for i in $$subdirs; do \
38              echo "#include <${subdir}/$$i/all.hpp>" >> $@; \
39          done
```

The variable lib_LTLIBRARIES is the name of the Libtool library that will be produced from the object files in the directory and the libraries contained in the variable libClusLib_la_LIBADD. The code in lines 26–38 is used to generate a file called `cluslib.hpp` automatically. The header `cluslib.hpp` includes all headers in the clustering library.

In Listing 6.1, the long arrow symbol ⟶ and the symbol ␣ denote a tab and a space, respectively. We make tabs and spaces visible so that we can differentiate them. Tabs are required when we define a target in a Makefile (cf. Appendix C.1). In Makefiles, we can use a "\" to split a long line into multiple lines. Note that there are no spaces after the \.

Listing 6.3 shows the `Makefile.am` file in the directory `cl/algorithms`. Since the directory `cl/algorithms` does not contain any subdirectories, we do not need to define the variable libAlgorithms_la_LIBADD.

Listing 6.3: The `Makefile.am` file in the directory `cl/algorithms`.

```
 1  noinst_LTLIBRARIES = libAlgorithms.la
 2
 3  AM_CPPFLAGS = -I${top_srcdir} -I${top_builddir}
 4
 5  this_includedir=${includedir}/${subdir}
 6  this_include_HEADERS = \
 7      all.hpp \
 8      algorithm.hpp \
 9      average.hpp \
10      centroid.hpp \
11      cmean.hpp \
12      complete.hpp \
13      diana.hpp \
14      fsc.hpp \
15      gkmode.hpp \
16      gmc.hpp \
17      kmean.hpp \
18      kprototype.hpp \
19      lw.hpp \
20      median.hpp \
21      single.hpp \
22      ward.hpp \
23      weighted.hpp
24
25  libAlgorithms_la_SOURCES = \
26      algorithm.cpp \
27      average.cpp \
28      centroid.cpp \
29      cmean.cpp \
30      complete.cpp \
31      diana.cpp \
32      fsc.cpp \
33      gkmode.cpp \
34      gmc.cpp \
35      kmean.cpp \
36      kprototype.cpp \
37      lw.cpp \
38      median.cpp \
39      single.cpp \
40      ward.cpp \
41      weighted.cpp
42
43
44  all.hpp: Makefile.am
45      echo "// This file is generated. Please do not edit!" > $@
46      echo >> $@
47      for i in $(filter-out all.hpp, $(this_include_HEADERS)); \
48      do \
49      echo "#include <${subdir}/$$i>" >> $@; \
50      done
```

```
51                 echo >> $@
52                 subdirs='$(SUBDIRS)'; for i in $$subdirs; do \
53                 echo "#include <${subdir}/$$i/all.hpp>" >> $@; \
54                 done
```

The `Makefile.am` files in other subdirectories of `cl` are very similar to the `Makefile.am` shown in Listing 6.3. The `Makefile.am` files in the directories `config` and `m4` are empty files. The program `automake` knows what files to distribute. The `Makefile.am` files in `cl/clusters`, `cl/datasets`, `cl/distances`, `cl/patterns`, `cl/utilities`, `examples`, and `test-suite` are shown in Listings B.42, B.52, B.66, B.78, B.91, B.106, and B.126 in the Appendix, respectively.

6.3 Macros and typedef Declarations

A macro is an operation with an identifier or name that is defined by the `#define` preprocessor directive. When a macro is used in a program, its identifier will be replaced by the contents of the macro by the compiler. For more information about macros, readers are referred to Deitel and Deitel (2009).

For convenience, we defined some macros in the clustering library. For example, in the file `cldefines.hpp` (see Listing B.5), we defined the following macros

```
#define INTEGER int
#define BIGINTEGER long
#define REAL double

#define MIN_INTEGER std::numeric_limits<INTEGER>::min()
#define MAX_INTEGER std::numeric_limits<INTEGER>::max()
#define MIN_REAL -std::numeric_limits<REAL>::max()
#define MAX_REAL std::numeric_limits<REAL>::max()
#define MIN_POSITIVE_REAL std::numeric_limits<REAL>::min()
#define MIN_EPSILON std::numeric_limits<REAL>::epsilon()
#define NULL_INTEGER std::numeric_limits<INTEGER>::max()
#define NULL_REAL std::numeric_limits<REAL>::max()
```

If we use the macro REAL in a statement

$$\text{std::vector<REAL>},$$

the statement will be expanded to

$$\text{std::vector<double>}.$$

In the above list of macros, the functions **min**, **max**, and **epsilon** are member functions of the template class **numeric_limits** defined in the header **<limits>**. Listing 6.4 gives a simple program used to output various values returned by these functions.

Listing 6.4: Program to test <limits>.

```
// limits.cpp
#include<limits>
#include<iostream>

#define INTEGER int
#define BIGINTEGER long
#define REAL double

#define MIN_INTEGER std::numeric_limits<INTEGER>::min()
#define MAX_INTEGER std::numeric_limits<INTEGER>::max()
#define MIN_REAL -std::numeric_limits<REAL>::max()
#define MAX_REAL std::numeric_limits<REAL>::max()
#define MIN_POSITIVE_REAL std::numeric_limits<REAL>::min()
#define MIN_EPSILON std::numeric_limits<REAL>::epsilon()
#define NULL_INTEGER std::numeric_limits<INTEGER>::max()
#define NULL_REAL std::numeric_limits<REAL>::max()

int main() {
    std::cout<<"MIN_INTEGER: "<<MIN_INTEGER<<std::endl;
    std::cout<<"MAX_INTEGER: "<<MAX_INTEGER<<std::endl;
    std::cout<<"MIN_REAL: "<<MIN_REAL<<std::endl;
    std::cout<<"MAX_REAL: "<<MAX_REAL<<std::endl;
    std::cout<<"MIN_POSITIVE_REAL: "<<MIN_POSITIVE_REAL
        <<std::endl;
    std::cout<<"MIN_EPSILON: "<<MIN_EPSILON<<std::endl;
    std::cout<<"NULL_INTEGER: "<<NULL_INTEGER<<std::endl;
    std::cout<<"MIN_REAL: "<<MIN_REAL<<std::endl;
    std::cout<<"MIN_INTEGER: "<<MIN_INTEGER<<std::endl;

    return 0;
}
```

When the program is executed, we see the following output

```
MIN_INTEGER: -2147483648
MAX_INTEGER: 2147483647
MIN_REAL: -1.79769e+308
MAX_REAL: 1.79769e+308
MIN_POSITIVE_REAL: 2.22507e-308
MIN_EPSILON: 2.22045e-16
NULL_INTEGER: 2147483647
MIN_REAL: -1.79769e+308
MIN_INTEGER: -2147483648
```

In C++, the keyword **typedef** is used to create aliases for previously defined data types. That is, **typedef** creates a new type name but does not create a new type. For example, in the file **types.hpp** (see Listing B.6), we defined the following **typedef** declarations in the namespace **ClusLib**:

```
namespace ClusLib {
```

```
typedef INTEGER Integer;
typedef BIGINTEGER BigInteger;
typedef unsigned INTEGER Natural;
typedef unsigned BIGINTEGER BigNatural;
typedef REAL Real;
typedef std::size_t Size;
}
```

Creating new type names for the built-in data types with `typedef` is useful to make programs more portable and easier to maintain. If we want to use `float` instead of `double` in a program, for example, we have to replace every `double` in the program with `float`. With `typedef`, we only need to change one place.

6.4 Error Handling

The C++ standard exception classes (cf. Section 3.6) allow programmers to specify a message for an exception in a program. However, the message does not tell us where the exception occurred. For debugging purpose, it is very useful to know which function throws the exception and where the function is located. To this end, we define a customized exception class named `Error` in the clustering library.

Listing 6.5: Declaration of class `Error`

```
1  class Error: public std::exception {
2  private:
3      boost::shared_ptr<std::string> _msg;
4  public:
5      Error(const std::string& file,
6            long line,
7            const std::string& function,
8            const std::string& msg = "");
9      ~Error() throw() {}
10     const char* what() const throw();
11 };
```

The declaration of class `Error` is shown in Listing 6.5. The class has one private data member and three member functions. The data member is a shared pointer pointing to a message. The constructor has four arguments, the last of which has a default value. The first argument is the name of the source file where the exception occurs; the second argument is the line number of the function throwing the exception; the third argument is the name of the function; the last argument is the message. The complete header and source of class `Error` can be found in Listings B.7 and B.8.

We note that a `throw` is appended to the declarations of functions: `Error` and `what`. Since the `throw` specifier is left empty, these two functions are not allowed to throw any exceptions.

To make class `Error` useful and convenient, we defined the following two macros:

```
#define FAIL(msg) \
    std::ostringstream ss; \
    ss << msg; \
    throw ClusLib::Error(__FILE__,__LINE__, \
        BOOST_CURRENT_FUNCTION,ss.str());

#define ASSERT(condition,msg) \
    if(!(condition)) { \
        std::ostringstream ss; \
        ss << msg; \
        throw ClusLib::Error(__FILE__,__LINE__, \
            BOOST_CURRENT_FUNCTION,ss.str()); \
    }
```

The first macro has one argument. Whenever the macro is used, an exception will be thrown. The second macro has two arguments: one condition and one message. When the condition is true, no exceptions will be thrown. Otherwise, an exception will be raised.

Using the two macros, we only need to specify the message to use class `Error`. We can get the file name and line number of a function using the macros `__FILE__` and `__LINE__`, which are predefined in the C++ language. The macro `BOOST_CURRENT_FUNCTION` is defined in the Boost library and expands to a string literal containing the name of the enclosing function.

6.5 Unit Testing

Unit testing, as its name suggests, is to test individual units of source code to see if they are fit for use. A unit is the smallest testable part of a program. For more information about unit testing, readers are referred to a book by Osherove (2009). Although this book introduces unit testing in the .Net language, the ideas should apply to other languages.

In *ClusLib*, we use the Boost test library (cf. Section 5.2) to organize our test cases and suites. Section B.10 presents several unit test suites. Consider the test suite for class `DAttrInfo` and class `CAttrInfo`. For example, we first create a class with static members. The declaration of the class is shown in Listing 6.6.

Listing 6.6: Declaration of class `AttrInfoTest`

```
class AttrInfoTest {
public:
    static void testDAttrInfo();
    static void testCAttrInfo();
    static boost::unit_test_framework::test_suite* suite();
};
```

In class `AttrInfoTest`, we declared three static member functions. The first two functions contain test cases. The third function creates a test suite by including the test cases defined in the first two functions. One can see the implementation of these functions in Listing B.129.

Once we have created a test suite, we need to include the test suite in the master test suite. In Listing 6.7, we see that the test suite of `AttrInfoTest` is added to the master test suite (line 11). The complete program is shown in Listing B.127.

Listing 6.7: Function `init_unit_test_suite`

```
test_suite* init_unit_test_suite(int, char* []) {
    std::string header = "Testing ClusLib";
    std::string rule = std::string(header.length(), '=');

    BOOST_MESSAGE(rule);
    BOOST_MESSAGE(header);
    BOOST_MESSAGE(rule);

    test_suite* test = BOOST_TEST_SUITE("ClusLib_test_suite");

    test->add(AttrInfoTest::suite());
    test->add(DatasetTest::suite());
    test->add(nnMapTest::suite());
    test->add(DistanceTest::suite());
    test->add(SchemaTest::suite());

    return test;
}
```

6.6 Compilation and Installation

Once we have created the specification files, we can use GNU Autotools to build the clustering library very easily. First, we need to produce the `configure` script. To do this, we go to the top-level directory of the library `ClusLib` and execute the following command:

`autoreconf`

or the command with options:

`autoreconf -fvi`

114   *Data Clustering in C++: An Object-Oriented Approach*

The second command will reproduce all files. The program `autoreconf` will produce a `configure` script in the top-level directory of the clustering library and a `Makefile.in` file in every subdirectory.

Once we have the `configure` script, we can configure the clustering library using the following commands:

```
./configure --with-boost-include=/usr/include
--with-boost-lib=/lib
```

The first command will do many checks and will produce `Makefile` files.

Note that the option `/lib` is the location of Boost libraries in Cygwin. If we installed the Boost headers to `/usr/local/include` and the Boost libraries to `/usr/local/lib`, then we need to configure the clustering library using the following command:

```
./configure --with-boost-include=/usr/local/include
--with-boost-lib=/usr/local/lib
```

Once the configuration is finished successfully, a number of `Makefile` files will be produced. Then we can compile and install the clustering library using the following command:

```
make install
```

6.7   Summary

In this chapter, we introduced the directory structure and source file name convention of the clustering library *ClusLib*. We also introduced the configuration files used by GNU Autotools during the compilation of the library. Error handling, unit testing, and compilation of the library were also discussed.

The design of *ClusLib* borrows many ideas from QuantLib[1]. For example, organization of source files, configuration files, error handling, and unit testing were designed according to that of QuantLib.

---

[1] QuantLib is a free and open-source C++ library for quantitative finance. One can get the library from `http://quantlib.org`.

Chapter 7
Datasets

Data is an important part of a clustering algorithm and is also the purpose of developing a clustering algorithm. Hence it is important to design and implement data before designing and implementing clustering algorithms.

In this chapter, we describe how we design and implement datasets in the clustering library $ClusLib$. In the literature of data clustering, a data set is a collection of records and a record is characterized by a set of attributes (Gan et al., 2007). Hence it is natural to model attributes, records, and then datasets.

7.1 Attributes

An attribute denotes an individual component of a record. We can classify attributes into four categories: ratio, interval, ordinal, and nominal. We can use a real number to store a ratio or interval attribute value and an integer to store an ordinal or nominal attribute value.

We follow the method in MLC++ (Kohavi et al., 1998) to design and implement attributes. In MLC++, the `union` structure was used to store a real number or an integer. However, `union` is not type safe and hence using it is not a good practice in object-oriented programming. In $ClusLib$, we use `boost::variant` to store a real number or an integer.

To implement attributes, we designed four classes: `AttrValue`, `AttrInfo`, `CAttrInfo`, and `DAttrInfo`. The UML class diagram in Figure 7.1 shows the relationships among these four classes. For the sake of saving space, we do not show class methods in the class diagram. Classes `CAttrInfo` and `DAttrInfo` inherit from class `AttrInfo`. Class `AttrValue` has two friend classes, which have access to its private members.

The declarations of these four classes are shown in Listings 7.1, 7.2, 7.3, and 7.4, respectively. We describe these classes in the following subsections.

7.1.1 The Attribute Value Class

Listing 7.1 shows the declaration of class `AttrValue`. The class has a private data member, two friend classes, and a public member function. The data

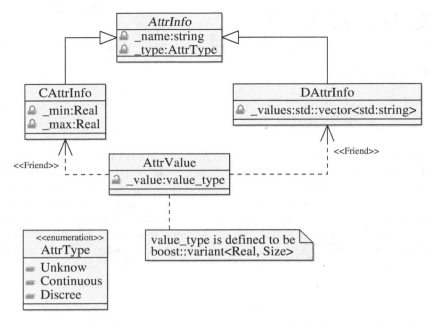

FIGURE 7.1: Class diagram of attributes.

member _value is a variant, which can store a real number or an unsigned integer. In *ClusLib*, a categorical datum is represented by an unsigned integer.

Class AttrValue is useless on its own as we cannot set or get the value it stores. The two friend classes are used to access the data member of AttrValue, since friend classes of a class can access all members of the class, including protected and private members. Since class AttrValue is capable of storing a real number or an unsigned integer, it can store any type of data. For example, a ratio or interval type can be stored as a real number. An ordinal or nominal type can be stored as an integer. To correctly interpret the data stored in AttrValue, we need to define only classes inheriting from class CAttrInfo or class DAttrInfo.

Listing 7.1: Declaration of class AttrValue.

```
class AttrValue {
public:
    friend class DAttrInfo;
    friend class CAttrInfo;

    typedef boost::variant<Real, Size> value_type;
    AttrValue();

private:
    value_type _value;
};
```

Class AttrValue has a constructor, which initializes the class's sole data member to a Null<Size>(), which is defined in the header file null.hpp (see Listing B.105).

The contents of the header file of class AttrValue is shown in Listing B.53. In the header file, the C preprocessor macros #ifndef, #define, and #endif at the begin and end of the header file are used to prevent including the header file more than once. We note that the class is defined in the namespace ClusLib. In fact, all the classes of *ClusLib* are defined in the namespace ClusLib.

The C++ keyword inline is used to tell a compiler that it should perform inline expansion on a particular function. That is, when a function is declared inline, the compiler will insert the complete body of the function in every place in the code where the function is used. The purpose of inline expansion is to eliminate the time overhead when a function is called. The inline keyword is usually required when we define functions in header files. If the inline keyword is missing, the function will be defined twice if the header file is included in two different places. The compiler will complain if a function is defined twice.

7.1.2 The Base Attribute Information Class

Listing 7.2 shows the declaration of class AttrInfo. Class AttrInfo contains many virtual function, including several pure virtual functions. Many functions are defined as inline functions in the header file shown in Listing B.54. Other functions are defined in the source file of the class shown in Listing B.55.

The constructor of class AttrInfo requires two arguments: one is the name of the attribute and the other is the type of the attribute. The type AttrType is an enum defined in the header attrinfo.hpp. The arguments of the constructor are used to initialize the values of the data members of the class.

Listing 7.2: Declaration of class AttrInfo.

```
1   enum AttrType {
2     Unknow,
3     Continuous,
4     Discrete
5   };
6
7   class DAttrInfo;
8   class CAttrInfo;
9
10  class AttrInfo {
11  public:
12    AttrInfo(const std::string &name, AttrType type);
13    virtual ~AttrInfo() {}
14
15    const std::string& name() const;
16    std::string& name();
17    AttrType type() const;
18
19    virtual bool operator==(const AttrInfo& info) const;
20    virtual bool operator!=(const AttrInfo& info) const;
21    virtual AttrInfo* clone() const = 0;
```

```
22       virtual Real distance(const AttrValue&,
23           const AttrValue&) const = 0;
24       virtual void set_d_val(AttrValue&, Size) const;
25       virtual Size get_d_val(const AttrValue&) const;
26       virtual void set_c_val(AttrValue&, Real) const;
27       virtual Real get_c_val(const AttrValue&) const;
28       virtual void set_unknown(AttrValue&) const = 0;
29       virtual bool is_unknown(const AttrValue&) const = 0;
30       virtual DAttrInfo& cast_to_d();
31       virtual const DAttrInfo& cast_to_d() const;
32       virtual CAttrInfo& cast_to_c();
33       virtual const CAttrInfo& cast_to_c() const;
34       virtual bool can_cast_to_d() const;
35       virtual bool can_cast_to_c() const;
36
37     protected:
38       bool equal_shallow(const AttrInfo&) const;
39
40     private:
41       std::string _name;
42       AttrType _type;
43     };
```

The member functions declared in lines 15–16 are used to access and modify the name of the attribute. The member function **type** (line 17) is used to get the type of the attribute. Once an attribute is created, we cannot change its type. Hence the function **type** is declared as a **const** function.

The overloaded operators "==" and "!=" (lines 19–20) are used to compare two objects of class **AttrInfo**. In this class, the operators consider only the name and the type of two attributes. If two attributes have the same name and the same type, they are deemed equal.

The member function **clone** (line 21) is a pure virtual function. In order to be useful, classes derived from class **AttrInfo** must override this function by providing concrete implementations. The function creates a deep copy of the underlying object and returns the pointer pointing to the copy.

We note that class **AttrInfo** contains a pure virtual function **distance**. This function is used to calculate the distance between two **AttrValue** objects. Usually, the distance between two records is calculated from the distances between the components of the records. This function is useful in that it provides a uniform approach to calculate the distances for different types of data.

Functions **set_d_val**, **get_d_val**, **set_c_val**, and **get_c_val** (lines 24–27) are used to set or get values in class **AttrValue**. Derived classes are supposed to implement some of these functions. For example, class **DAttrInfo** implements only **set_d_val** and **get_d_val** since the class is allowed to set or get only a discrete value (i.e., an unsigned integer). Class **CAttrInfo** implements only **set_c_val** and **get_c_val** since the class is allowed to set or get only a continuous value (i.e., a real number).

Functions **set_unknown** and **is_unknown** (lines 28–29) are used to deal with missing values as missing values are very common in data sets. We declared the two functions as pure virtual functions since the missing values of different types are represented by different values. Hence we leave the definition of these functions to derived classes.

Functions in lines 30–35 are used to do type conversions. Since we used classes `DAttrInfo` and `CAttrInfo` here, we have to declare the two classes somewhere (lines 7–8) before we use. In C++, this is called forward declaration. We cannot include the headers of classes `DAttrInfo` and `CAttrInfo`, since such inclusion will result in cyclic inclusion of headers.

The source file of class `AttrInfo` (see Listing B.55) define some member functions. Except the constructor and the function `equal_shallow`, all functions are defined in such a way that when the function is called an exception will be raised. Function `equal_shallow` will return true as long as the name and the type match. `FAIL` is a macro defined in the header file `errors.hpp` (see Listing B.7).

The destructor of class `AttrInfo` is declared virtual. Usually, the destructor of a class that is intended to be used as a base class should be virtual. This will ensure that objects of derived classes are destroyed appropriately (Meyers, 1997).

7.1.3 The Continuous Attribute Information Class

The declaration of the continuous attribute information class `CAttrInfo` is shown in Listing 7.3. The complete implementation of class `CAttrInfo` can be found in Listings B.56 and B.57.

Class `CAttrInfo` has two additional data members: `_min` and `_max`. These two data members are used to store the minimum value and the maximum value of a continuous attribute of a data set.

Class `CAttrInfo` redefines some functions (lines 5–13) in its parent class `AttrInfo`. For example, `cast_to_c` and `can_cast_to_c` are redefined. We do not override functions such as `cast_to_d`, `set_d_val` in class `CAttrInfo` because this class deals with continuous attributes only.

Functions (lines 14–17) are used to set and get the minimum and the maximum values of the attribute. Function `equal` checks whether two attributes are equal or not in terms of the attribute name and the type. This function does not consider the minimum and the maximum values. That is, as long as two attributes have the same name and type, function `equal` treats them as equal.

In the implementation of class `CAttrInfo` in Listing B.57, we note that the keyword `this` is used in the definition of function `clone`. In fact, `this` is a pointer that points to the object for which the member function is called. The `this` pointer is accessible only within nonstatic member functions of a class. Function `clone` creates a copy of the underlying object by using the default copy constructor of class `CAttrInfo` and returns the pointer pointing to the copy. In C++, every class has a default constructor if we do not provide one.

Listing 7.3: Declaration of class `CAttrInfo`.

```
1   class CAttrInfo : public AttrInfo {
2   public:
3       CAttrInfo(const std::string& name);
4
5       CAttrInfo& cast_to_c();
6       const CAttrInfo& cast_to_c() const;
7       bool can_cast_to_c() const;
8       CAttrInfo* clone() const;
9       Real distance(const AttrValue&, const AttrValue&) const;
10      void set_c_val(AttrValue&, Real) const;
11      Real get_c_val(const AttrValue&) const;
12      void set_unknown(AttrValue&) const;
13      bool is_unknown(const AttrValue&) const;
14      void set_min(Real);
15      void set_max(Real);
16      Real get_min() const;
17      Real get_max() const;
18      bool equal(const AttrInfo&) const;
19
20  protected:
21      Real _min;
22      Real _max;
23  };
```

The `distance` function is defined in the source file shown in Listing B.57. We implement this function following the method used in $\mathcal{MLC}$++ (Kohavi et al., 1998). That is, if both `AttrValue` are missing (or unknown), we set their distance to be zero; if only one of them is missing, we set their distance to be 1; otherwise, we set their distance to be the difference. The operator "^" is the exclusive or operator. The expression is true if and only if only one operand is true.

7.1.4 The Discrete Attribute Information Class

Listing 7.4 shows the declaration of class `DAttrInfo`. The complete implementation of this class can be found in Listings B.58 and B.59. Class `DAttrInfo` has a private data member: `_values`, which is a vector of strings. This member is used to hold the discrete values of an attribute. In $\mathcal{C}lus\mathcal{L}ib$, we represent discrete data by strings and store only the distinct strings in a `DAttrInfo` object. The order of these distinct strings is not important for a nominal attribute but important for an ordinal attribute. The `AttrValue` objects store unsigned integers that are indexes to the distinct strings stored in `DAttrInfo` objects.

Listing 7.4: Declaration of class `DAttrInfo`.

```
1   class DAttrInfo : public AttrInfo {
2   public:
3       DAttrInfo(const std::string& name);
4
5       Size num_values() const;
6       const std::string& int_to_str(Size i) const;
7       Size str_to_int(const std::string&) const;
8       Size add_value(const std::string&,
9           bool bAllowDuplicate = true);
10      void remove_value(const std::string&);
11      void remove_value(Size i);
12      DAttrInfo* clone() const;
```

```cpp
13        Real distance(const AttrValue&, const AttrValue&) const;
14        void set_d_val(AttrValue&, Size) const;
15        Size get_d_val(const AttrValue&) const;
16        void set_unknown(AttrValue&) const;
17        bool is_unknown(const AttrValue&) const;
18        DAttrInfo& cast_to_d();
19        const DAttrInfo& cast_to_d() const;
20        bool can_cast_to_d() const;
21        bool operator==(const AttrInfo& info) const;
22        bool operator!=(const AttrInfo& info) const;
23
24   protected:
25        typedef std::vector<std::string>::iterator iterator;
26        typedef std::vector<std::string>::const_iterator
27     const_iterator;
28
29        bool equal(const AttrInfo&) const;
30
31        std::vector<std::string> _values;
32   };
```

(a)		
Record	**Attribute**	**AttrValue**
Record 1	"A"	0
Record 2	"B"	1
Record 3	"A"	0
Record 4	"C"	2
Record 5	"B"	1

(b)	
Index	**DAttrInfo**
0	"A"
1	"B"
2	"C"

TABLE 7.1: An example of class DAttrInfo.

Table 7.1 gives an example of how we represent discrete data in $\mathcal{C}lus\mathcal{L}ib$. Table 7.0(a) shows a data set with five records described by a discrete attribute. The third column of Table 7.0(a) shows the values stored in the AttrValue objects. Table 7.0(b) shows the values stored in the DAttrInfo object.

Function num_values (line 5) is used to get the number of distinct values stored in a DAttrInfo object. Functions int_to_str (line 6) and str_to_int (line 7) are used to convert a string to its corresponding index or an index to its underlying string, respectively.

Function add_value (8–9) is used to add a value to a DAttrInfo object and returns the index of the value. This function takes two arguments: a string and a boolean. If the boolean is true and we add a value to a DAttrInfo object that already exists in the DAttrInfo object, then the function just returns the index of the value. If the boolean is false and we try to add a value to a DAttrInfo object that already exists in the DAttrInfo object, then the function throws an exception.

Overloaded functions remove_value (line 10–11) are used to delete a value from a DAttrInfo object. The two functions should be used carefully since deleting a value from a DAttrInfo object might invalidate many AttrValue objects that store the indexes to the deleted value and the values after the deleted value.

Since objects of class `DAttrInfo` are responsible to manipulate the unsigned integers stored in objects of `AttrValue`, we redefined several functions such as `set_d_val` and `get_d_val` (lines 14–15).

Functions in lines 12–22 override the corresponding functions in the base class. The protected function `equal` compares the underlying object to another object of class `AttrInfo` or its derived classes. The function returns true if the names, types, and individual stored values match.

We defined several member functions of `DAttrInfo` in the source file shown in Listing B.59. Most of these functions are long functions. We defined the `distance` function in the source. The distance is a simple matching distance (Gan et al., 2007). We also defined the `equal` function in the source.

7.2 Records

We design records following the approach in (Kohavi et al., 1998). That is, a record is a vector of `AttrValue` objects with a schema, which is a vector of shared pointers pointing to `AttrInfo` objects. Figure 7.2 shows the class diagram of a record and its dependencies. We did not show class methods in the diagram to save space.

From Figure 7.2 we see that both class `Record` and class `Schema` inherit from realizations of the template class `Container`. Class `Record` and class `Schema` have an aggregation relationship. The "1" and "*" show that a schema is associated with many records.

7.2.1 The Record Class

Listing 7.5 shows the declaration of class `Record`. The whole implementation of class `Record` can be found in Listings B.60 and B.61. Class `Record` inherits from the template class `Container` (see Listing B.92) parameterized with `AttrValue`. Class `Record` has four private data members: `_schema`, `_label`, `_data`, and `_id`. The member `_data` is inherited from class `Container`. Every object of class `Record` contains a shared pointer pointing to an object of class `Schema`. A dataset has a schema, which is shared by all the records in the dataset. Individual components of a record are stored in `_data`, which is a vector of `AttrValue` objects. A record might contain a label. A record also has an identifier.

Listing 7.5: `Record` class definition.

```
1  class Record: public Container<AttrValue> {
2  public:
3      Record(const boost::shared_ptr<Schema>& schema);
```

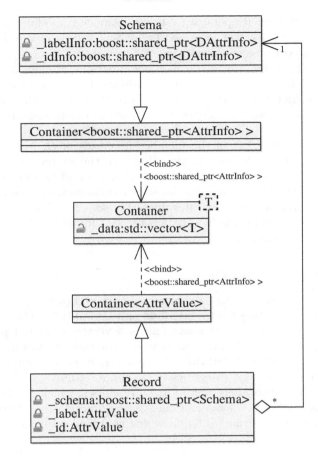

FIGURE 7.2: Class diagram of records.

```
 4
 5      const boost::shared_ptr<Schema>& schema() const;
 6      AttrValue& labelValue();
 7      const AttrValue& labelValue() const;
 8      AttrValue& idValue();
 9      const AttrValue& idValue() const;
10      Size get_id() const;
11      Size get_label() const;
12
13  private:
14      boost::shared_ptr<Schema> _schema;
15      AttrValue _label;
16      AttrValue _id;
17  };
```

Member functions of class **Record** are designed to manipulate the data member of the class. Class **Record** has one constructor, which takes a shared pointer pointing to a schema as the argument. We do not define a copy constructor or assignment operator for this class since the default ones work fine. The schema for a dataset is shared by all the records in the dataset. Hence we do not need to make a deep copy of a record.

Class **Record** inherits many member functions of class **Container** that are designed to access and modify an individual component of a record. We also defined functions that provide access to the schema, the label value, the label information, and the identifier of an object of the class. In Listing B.60, **ASSERT** is a macro defined in the header **errors.hpp**.

We implement most of the member functions of class **Record** as inline functions in the header (see Listing B.60). In the source (see Listing B.61), we implemented three functions: the constructor and two member functions of class **Schema**. The constructor initializes _schema using the argument and initializes _values using default values. The two member functions of class **Schema** are defined in this file because the two functions depend on member functions of class **Record**.

7.2.2 The Schema Class

The declaration of class **Schema** is shown in Listing 7.6. Class **Schema** has two protected data members: _data and _labelInfo. The member _data is inherited from class **Container** and is a vector of shared pointer pointing to objects of **AttrInfo**, each of which corresponds to an attribute of a dataset. The member _labelInfo is a shared pointer pointing to an object of **DAttrInfo** and contains categories of the input dataset.

Class **Schema** contains several virtual functions, which can be redefined in derived classes. Since **Schema** contains virtual functions, its destructor is also virtual. Listings B.62 and B.63 show the complete implementation of class **Schema**.

Listing 7.6: **Schema** class definition.

```
 1  class Record;
 2
 3  class Schema: public Container<boost::shared_ptr<AttrInfo> > {
 4  public:
 5      virtual ~Schema() {}
 6
 7      Schema* clone() const;
 8      boost::shared_ptr<DAttrInfo>& labelInfo();
 9      const boost::shared_ptr<DAttrInfo>& labelInfo() const;
10      boost::shared_ptr<DAttrInfo>& idInfo();
11      const boost::shared_ptr<DAttrInfo>& idInfo() const;
12      void set_label(boost::shared_ptr<Record>& r,
13                     const std::string& val);
14      void set_id(boost::shared_ptr<Record>& r,
15                  const std::string& val);
16      bool is_labelled() const;
17
18      virtual bool equal(const Schema& o) const;
19      virtual bool equal_no_label(const Schema& o) const;
20      virtual bool operator==(const Schema& o) const;
21      virtual bool operator!=(const Schema& o) const;
22      virtual bool is_member(const AttrInfo& info) const;
23
```

```
24      protected :
25          boost :: shared_ptr<DAttrInfo> _labelInfo ;
26          boost :: shared_ptr<DAttrInfo> _idInfo ;
27      };
```

We note that an object of Schema is responsible to set the label and the identifier of an object of Record. The functions set_label and set_id (lines 12–15) are designed to do this. Since the two functions depend on Record, we forward-declare class Record (line 1) before we declare class Schema. Since the two functions call member functions of class Record, the implementation of the two functions are delayed until class Record is defined. In fact, we define functions set_label and set_id in the source file of class Record (see Listing B.61).

7.3 Datasets

In data clustering, a dataset is a collection of records, which are described by the same set of attributes. Hence it is natural to design the dataset class in such a way that the class contains a collection of records and a schema. Figure 7.3 shows the UML class diagram of the Dataset class.

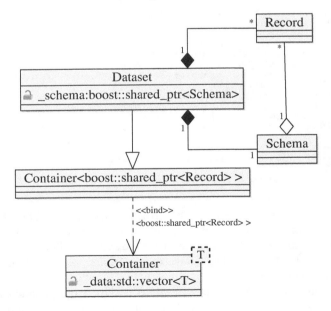

FIGURE 7.3: Class diagram of Dataset.

The declaration of class Dataset is shown in Listing 7.7. The implementation of class `Dataset` can be found in Listings B.64 and B.65. The `Dataset` class definition contains prototypes (lines 6–17) for several member functions such as `Dataset`, `num_attr`, `schema`, etc. The class includes two protected data members: `_schema` and `_data`. The member `_data` is inherited from class `Container`.

Class `Dataset` contains two constructors. The first constructor (line 6) takes one argument, which is a `const` reference to a shared pointer pointing to an object of `Schema`. The second constructor (line 7) is a copy constructor. Since we defined customized constructors, class `Dataset` does not have the default constructor. Hence we can create an object of `Dataset` either with a schema or from another object of `Dataset`.

The member function `num_attr` (line 9) is used to get the number of attributes of a dataset. The number of records of a dataset can be obtained by the function `size` defined in class `Container`. Function `schema` is a `const` member function and returns a `const` reference to the data member `_schema`. We declare this function as `const` since we do not want the schema of a dataset to be modified once the dataset is initialized.

Listing 7.7: Declaration of class `Dataset`.

```
1   class Dataset: public Container<boost::shared_ptr<Record> > {
2   public:
3       friend std::ostream& operator<<(std::ostream& os,
4               const Dataset& ds);
5
6       Dataset(const boost::shared_ptr<Schema>&);
7       Dataset(const Dataset&);
8
9       Size num_attr() const;
10      const boost::shared_ptr<Schema>& schema() const;
11      AttrValue& operator()(Size i, Size j);
12      const AttrValue& operator()(Size i, Size j) const;
13      bool is_numeric() const;
14      bool is_categorical() const;
15      void save(const std::string& filename) const;
16      std::vector<Size> get_CM() const;
17      Dataset& operator=(const Dataset&);
18
19  protected:
20      void print(std::ostream& os) const;
21
22      boost::shared_ptr<Schema> _schema;
23  };
```

We overload the operator "( )" with two arguments (lines 11–12) to access and modify an individual component of a record in a dataset. Member functions `is_numeric` (line 13) and `is_categorical` (line 14) are convenient functions used to tell whether a dataset is a numerical dataset or a categorical dataset, respectively.

Function `createRecordID` is used to assign a unique identifier to each record. We also implement the assignment operator "=". Usually, the assignment operator should be defined if the copy constructor is defined. These functions are defined in the source file (see Listing B.65).

We also note that class `Dataset` overloads the stream operator "<<" (lines

3–4) as a friend function. Usually, stream operators are overloaded as friend functions. The friend function actually calls the protected member function `print` to output the dataset. In fact, we can just put the contents of function `print` into the body of the overloaded operator since friend functions of a class can access private members of the class. In Exercise A.1, readers are asked to overload the stream operator for classes `AttrValue` and `Record`.

7.4 A Dataset Example

In this section, we give an example to illustrate the dataset classes such as class `Record`, class `Schema`, and class `Dataset`. In our example, we consider the dataset given in Table 7.2. This dataset consists of four records, which are described by four attributes. Among the four attributes, two are continuous and the other two are categorical.

RecordID	**Attr1**	**Attr2**	**Attr3**	**Attr4**	**Label**
r1	1.2	A	-0.5	X	1
r2	-2.1	B	1.5	Y	2
r3	1.5	A	-0.1	X	1
r4	-1.8	B	1.1	Y	2

TABLE 7.2: An example dataset.

The C++ program used to create and output the dataset is shown in Listing 7.8. The program first creates a schema for the dataset (lines 12–32). Then it creates a dataset and fills the dataset with the four records shown in Table 7.2 (lines 34–79). Then the program saves the dataset to a CSV file (line 81) and prints the summary of the dataset to screen (line 82). Finally, the program prints the contents of the dataset to screen (lines 83–104).

From the program we see that we can use many functions provided by the dataset classes to manipulate data. These functions allow us to deal with numeric data and categorical data in a uniform way.

Listing 7.8: Program to illustrate dataset classes.

```
1  // examples/dataset/dataset.cpp
2  #include<cl/datasets/dataset.hpp>
3  #include<iostream>
4  #include<sstream>
5  #include<iomanip>
6
7  using namespace ClusLib;
8  using namespace std;
9
10 int main(int argc, char* argv[]) {
11     try{
```

```cpp
                boost::shared_ptr<Schema> schema(new Schema());
                boost::shared_ptr<DAttrInfo> labelInfo(
                    new DAttrInfo("Label"));
                boost::shared_ptr<DAttrInfo> idInfo(
                    new DAttrInfo("Identifier"));
                schema->labelInfo() = labelInfo;
                schema->idInfo() = idInfo;
                stringstream ss;
                boost::shared_ptr<AttrInfo> ai;
                for(Size j =0; j<4; ++j) {
                    ss.str("");
                    ss<<"Attr"<<j+1;
                    if (j==0 || j==2) {
                        ai = boost::shared_ptr<CAttrInfo>(
                            new CAttrInfo(ss.str()));
                    } else {
                        ai = boost::shared_ptr<DAttrInfo>(
                            new DAttrInfo(ss.str()));
                    }
                    schema->add(ai);
                }

                boost::shared_ptr<Dataset> ds(new Dataset(schema));
                Size val;
                boost::shared_ptr<Record> r;
                r = boost::shared_ptr<Record>(new Record(schema));
                schema->set_id(r, "r1");
                schema->set_label(r, "1");
                (*schema)[0]->set_c_val((*r)[0], 1.2);
                val = (*schema)[1]->cast_to_d().add_value("A");
                (*schema)[1]->set_d_val((*r)[1], val);
                (*schema)[2]->set_c_val((*r)[2], -0.5);
                val = (*schema)[3]->cast_to_d().add_value("X");
                (*schema)[3]->set_d_val((*r)[3], val);
                ds->add(r);

                r = boost::shared_ptr<Record>(new Record(schema));
                schema->set_id(r, "r2");
                schema->set_label(r, "2");
                (*schema)[0]->set_c_val((*r)[0], -2.1);
                val = (*schema)[1]->cast_to_d().add_value("B");
                (*schema)[1]->set_d_val((*r)[1], val);
                (*schema)[2]->set_c_val((*r)[2], 1.5);
                val = (*schema)[3]->cast_to_d().add_value("Y");
                (*schema)[3]->set_d_val((*r)[3], val);
                ds->add(r);

                r = boost::shared_ptr<Record>(new Record(schema));
                schema->set_id(r, "r3");
                schema->set_label(r, "1");
                (*schema)[0]->set_c_val((*r)[0], 1.5);
                val = (*schema)[1]->cast_to_d().add_value("A");
                (*schema)[1]->set_d_val((*r)[1], val);
                (*schema)[2]->set_c_val((*r)[2], -0.1);
                val = (*schema)[3]->cast_to_d().add_value("X");
                (*schema)[3]->set_d_val((*r)[3], val);
                ds->add(r);

                r = boost::shared_ptr<Record>(new Record(schema));
                schema->set_id(r, "r4");
                schema->set_label(r, "2");
                (*schema)[0]->set_c_val((*r)[0], -1.8);
                val = (*schema)[1]->cast_to_d().add_value("B");
                (*schema)[1]->set_d_val((*r)[1], val);
                (*schema)[2]->set_c_val((*r)[2], 1.1);
                val = (*schema)[3]->cast_to_d().add_value("Y");
                (*schema)[3]->set_d_val((*r)[3], val);
```

```
79              ds->add(r);
80
81              ds->save("exampleds.csv");
82              cout<<*ds<<endl;
83              cout<<"Data:_\n";
84              cout<<setw(10)<<left<<"RecordID";
85              for(Size j=0; j<ds->num_attr(); ++j) {
86                  stringstream ss;
87                  ss<<"Attr("<<j+1<<")";
88                  cout<<setw(10)<<left<<ss.str();
89              }
90              cout<<setw(6)<<left<<"Label"<<endl;
91              for(Size i=0; i<ds->size(); ++i) {
92                  cout<<setw(10)<<left<<(*ds)[i]->get_id();
93                  for(Size j=0; j<ds->num_attr(); ++j) {
94                      if((*schema)[j]->can_cast_to_c()) {
95                          cout<<setw(10)<<left
96                              <<(*schema)[j]->get_c_val((*ds)(i,j));
97                      } else {
98                          cout<<setw(10)<<left
99                              <<(*schema)[j]->get_d_val((*ds)(i,j));
100                     }
101                 }
102                 cout<<setw(6)<<left<<(*ds)[i]->get_label()<<endl;
103             }
104
105             return 0;
106         } catch (exception& e) {
107             cout<<e.what()<<endl;
108             return 1;
109         } catch (...) {
110             cout<<"unknown_error"<<endl;
111             return 2;
112         }
113     }
```

Suppose the file `dataset.cpp` is saved in folder `examples/dataset` and the clustering library is compiled using the Autotools. To compile the program, we first go to the folder `examples/dataset` and execute the following command:

`libtool --mode=compile g++ -c -I../.. -I/usr/include dataset.cpp`

The options `-I../..` and `-I/usr/include` tell the compiler where to look for the header `dataset.hpp` and the Boost headers required by the program. In different platforms, this location might be different. Once this command is executed, an object file `dataset.o` is produced.

To produce an executable program, we need to link the program by executing the following command:

`libtool --mode=link g++ -o dataset ../../cl/libClusLib.la dataset.o`

Once this command is executed, an executable program `dataset.exe` (in Cygwin) is produced. When the program is executed, we see the following output:

```
Number of records: 4
Number of attributes: 4
Number of numerical attributes: 2
```

Number of categorical attributes: 2

```
Data:
RecordID  Attr(1)   Attr(2)   Attr(3)   Attr(4)   Label
0         1.2       0         -0.5      0         0
1         -2.1      1         1.5       1         1
2         1.5       0         -0.1      0         0
3         -1.8      1         1.1       1         1
```

The program also saves the dataset into a CSV file called **exampleds.csv** by using the function **save** (line 81). Note that the corresponding schema file called **exampleds.names** is also saved to the same folder. Listing 7.9 and Listing 7.10 show the dataset and the corresponding schema file saved by the function.

Listing 7.9: A dataset saved by the program.

```
1  r1,1.2,A,-0.5,X,1
2  r2,-2.1,B,1.5,Y,2
3  r3,1.5,A,-0.1,X,1
4  r4,-1.8,B,1.1,Y,2
```

Listing 7.10: A schema file saved by the program.

```
1  This is the schema file for dataset exampleds.csv
2  ///: schema
3  1, RecordID
4  2, Continuous
5  3, Discrete
6  4, Continuous
7  5, Discrete
8  6, Class
```

7.5 Summary

In this chapter, we introduced our design and implementation of datasets in C++. Our design and implementation follows the approaches employed in $\mathcal{MLC}$++ (Kohavi et al., 1998). In $\mathcal{ClusLib}$, however, we used a data structure that is more type-safe and that follows object-oriented design.

In addition, we used many C++ language features that were not available many years ago. For example, $\mathcal{MLC}$++ implemented its own containers to hold individual components of a record. In $\mathcal{ClusLib}$, we used class **vector** provided by the C++ standard library to hold individual components of a record.

We also gave an example to illustrate how to create datasets and manipulate data using the functions provided by the dataset classes.

Chapter 8

Clusters

In data clustering, a cluster is a collection of objects or records that share the same properties (Carmichael et al., 1968; Bock, 1989; Everitt, 1993). The result of applying a clustering algorithm to a dataset is a collection of clusters.

In this chapter, we introduce the design and implementation of clusters. First, we introduce the design of a single cluster. Then, we introduce how we design the result of a partitional clustering algorithm. Finally, we introduce the design of the result of a hierarchical clustering algorithm.

8.1 Clusters

In cluster analysis, there is no uniform definition for a cluster (Everitt, 1993). The term *cluster* is essentially used in an intuitive manner. In general, a cluster is formed by objects or records that have one or more of the following properties (Bock, 1989; Gan et al., 2007):

(a) The objects in the cluster share the same or closely related properties.

(b) The objects in the cluster show small mutual distances.

(c) The objects have relations with at least one other object in the clusters.

(d) The objects can be clearly distinguishable from objects that are not in the cluster.

Clusters come with many different forms. For example, there are one-center clusters, multi-center clusters, fuzzy clusters, subspace clusters, to just name a few. Although clusters have many forms, they share the same property: a cluster is a collection of records. Hence it is natural to design the base cluster class as a collection of records. The piece of code in Listing 8.1 shows the declaration of class `Clsuter`.

Class `Cluster` inherits from class `Container` realized with the parameter `boost::shared_ptr<Record>`. Hence class `Cluster` inherits all the methods from class `Container` that are used to access and modify an individual `Record` object. In addition, class `Cluster` has one private data member and three member functions as we can see from the declaration of the class in Listing 8.1.

Listing 8.1: Declaration of class `Cluster`

```
1  class Cluster: public Container<boost::shared_ptr<Record> > {
2  public:
3      virtual ~Cluster() {}
4
5      void set_id(Size id);
6      Size get_id() const;
7
8  protected:
9      Size _id;
10 };
```

Class `Cluster` is designed to be the base class for all other cluster classes. Hence we declared its destructor (line 3) as virtual. Since class `Cluster` does not contain any pure virtual functions, it is not an abstract class. That is, we can create objects from class `Cluster`.

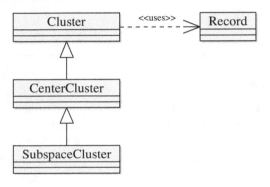

FIGURE 8.1: Hierarchy of cluster classes.

Class `SubspaceCluster` inherits class `CenterCluster`, which in turn inherits from the base class `Cluster`. Figure 8.1 shows the UML class diagram of the hierarchy of these classes. The declarations of class `CenterCluster` and class `SubspaceCluster` are shown in Listing 8.2 and Listing 8.3, respectively.

Listing 8.2: Declaration of class `CenterCluster`

```
1  class CenterCluster: public Cluster {
2  public:
3      CenterCluster() {}
4      CenterCluster(const boost::shared_ptr<Record>& center);
5      const boost::shared_ptr<Record>& center() const;
6      boost::shared_ptr<Record>& center();
7
8  protected:
9      boost::shared_ptr<Record> _center;
10 };
```

Class `CenterCluster` has a protected data member, which is used to hold a pointer pointing to the center of the cluster. Public member functions (lines

5–6) were added to access and modify the center. It is worth noting that we should create a new object of **Record** as the center when we create a new object of **CenterCluster**. If we just use the shared pointer pointing to a record in a dataset as the center, the record will be changed when we update the center during clustering.

The aforementioned class **CenterCluster** actually models a cluster with only one center. To model a multi-center cluster, we can just create a new class derived from class **Cluster** with a collection of centers.

Listing 8.3: Declaration of class **SubspaceCluster**
```
1  class SubspaceCluster : public CenterCluster {
2  public:
3      SubspaceCluster(const boost::shared_ptr<Record>& center);
4      std::vector<Real>& w();
5      const std::vector<Real>& w() const;
6      Real& w(Size i);
7      const Real& w(Size i) const;
8
9  protected:
10     std::vector<Real> _w;
11 };
```

In additional to the data members inherited from class **CenterCluster**, class **SubspaceCluster** has a protected data member, which is a vector of real numbers used to represent the relative importance of attributes for the formation of the cluster. Four public member functions (lines 4–7) were added to access and modify the protected data member and its individual components.

Complete implementation of cluster classes can be found in the listings in Section B.4.

8.2 Partitional Clustering

In this section, we describe the design and implementation of the result of a partitional clustering algorithm. For convenience, we call the result a partitional clustering. From a technical point of view, a partitional clustering is a collection of clusters. Hence it is natural to design a partitional clustering as a collection of clusters.

Listing 8.4 shows the declaration of class **PClustering**. The class inherits from class **Container** realized with parameter **boost::shared_ptr<Cluster>**. Since class **Container** implements all the functions used to access and modify an individual cluster, we do not need to implement these functions.

Class **PClustering** has several data members. Member **_bCalculated** is a boolean variable used to tell whether the member function **calculate** is called or not. Data members **_numclust** and **_numclustGiven** represent the number of clusters contained in the inherited data member **_data** and the number of clusters from the dataset, respectively. Data member _clustsize is a vector of unsigned integers which represent the sizes of clusters in _data.

Data member _clustLabel is used to store the class labels given in the dataset if such labels exist. Data members _CM and _CMGiven are the cluster membership vectors. The last data member _crosstab is a double-key map (see Section 11.2) used to store the cross tabulation of the clustering in the object and the given clustering.

Listing 8.4: Declaration of class PClustering

```
class PClustering:
    public Container<boost::shared_ptr<Cluster> > {
public:
    friend std::ostream& operator<<(std::ostream& os,
        PClustering& pc);

    PClustering();
    void removeEmptyClusters();
    void createClusterID();
    void save(const std::string& filename);

private:
    void print(std::ostream& os);
    void calculate();
    void crosstab();

    bool _bCalculated;
    Size _numclust;
    Size _numclustGiven;
    std::vector<Size> _clustsize;
    std::vector<std::string> _clustLabel;
    std::vector<Size> _CM;
    std::vector<Size> _CMGiven;
    iiiMapB _crosstab;
};
```

Member function removeEmptyClusters is used to remove empty clusters from a partitional clustering. Member function createClusterID is used to assign a unique identifier to each cluster. Member function save is used to save a copy of the dataset to disk.

Member function print is used to output a summary of the clustering to an output stream. This function is used by the friend function (lines 4–5). Member function calculate collects necessary information and calls crosstab to create the cross tabulation if the dataset contains class labels.

We note that an object of class PClustering holds a vector of shared pointers to objects of class Cluster. This is very useful in that we can use the class PClustering to hold shared pointers pointing to objects of other classes inherited from class Cluster. When we want to retrieve objects of derived classes, we can use dynamic_cast, which is a C++ operator that performs type conventions at run time. We shall see some examples of such use in later chapters.

8.3 Hierarchical Clustering

In this section, we describe the design and implementation of the result of a hierarchical clustering algorithm. For convenience, we call the result a hierarchical clustering. Similar to a partitional clustering, a hierarchical clustering is also a collection of clusters. However, a hierarchical clustering is more complicated than a partitional clustering. In fact, a hierarchical clustering is a nested tree of partitional clusterings. Hence it is natural to apply the composite design pattern to design a hierarchical clustering.

Before we introduce the design of a hierarchical clustering, let us introduce the classes used to implement the composite design pattern. The composite design pattern is implemented by three classes: **Node**, **LeafNode**, and **InternalNode**.

The declaration of class **Node** is shown in Listing 8.5. This class has three protected data members: _parent, _id, and _level. Data member _parent holds a shared pointer pointing to the parent of the node. Data member _id is the identifier of the node. Data member _level holds the level of the node in the hierarchical tree. All leaf nodes have a level of zero. Figure 8.2 shows a hierarchical tree with levels. In Figure 8.2, a circle denotes a node and the number inside a circle is the level of the circle. We see that all leaf nodes have a level of zero and internal nodes have positive levels.

Listing 8.5: Declaration of class **Node**

```
1  class Node {
2  public:
3      virtual ~Node() {}
4
5      Size get_id() const;
6      void set_id(Size id);
7      Size get_level() const;
8      void set_level(Size level);
9      boost::shared_ptr<Node> get_parent();
10     void set_parent(const boost::shared_ptr<Node>& p);
11
12     virtual void accept(NodeVisitor &v) = 0;
13     virtual Size num_children() const = 0;
14     virtual Size num_records() const = 0;
15
16 protected:
17     Node(boost::shared_ptr<Node> p, Size id)
18         : _parent(p), _id(id) {}
19
20     boost::shared_ptr<Node> _parent;
21     Size _id;
22     Size _level;
23 };
```

Class **Node** is an abstract base class as it contains pure virtual functions (lines 12–14). The function **accept** (line 12) takes one argument, which is a reference to an object of class **NodeVisitor** (see Listing B.86). Function **num_children** returns the number of children contained in the node and function **num_records** returns the number of records contained in the node and all its descendants.

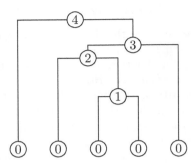

FIGURE 8.2: A hierarchical tree with levels.

The declaration of class `LeafNode` is shown in Listing 8.6, from which we see that class `LeafNode` has one private data member and five public member functions. The data member holds a shared pointer pointing to a record in a dataset. The constructor takes three arguments, two of which have default values. Member function `get_data` (line 11) returns the shared pointer stored in the class object. The other three member functions (lines 8–10) override the corresponding functions in the base class.

Some member function of class `LeafNode` were implemented as inline functions (see Listing B.83), while other member functions were implemented in the source file (see Listing B.84). In class `LeafNode`, the function **num_children** returns 0 and the function **num_records** returns 1 since its object does not have children but a record.

Listing 8.6: Declaration of class `LeafNode`

```
class LeafNode: public Node {
public:
    LeafNode(const boost::shared_ptr<Record>& r,
             Size id = 0,
             const boost::shared_ptr<Node>& p
             = boost::shared_ptr<Node>() );

    void accept(NodeVisitor &v);
    Size num_children() const;
    Size num_records() const;
    boost::shared_ptr<Record> get_data();

private:
    boost::shared_ptr<Record> _data;

};
```

The declaration of class `InternalNode` is shown in Listing 8.7, from which we see that class `InternalNode` inherits from both class `Node` and class `Container` realized with parameter `boost::shared_ptr<Node>`. This is an example of multiple inheritance supported by the C++ programming language.

Class `InternalNode` has two constructors (lines 4–10) and two private data members: `_joinValue` and `_data`. The data member `_data` is inherited from class `Container` and holds a vector of shared pointers to objects of node classes. The data member `_joinValue` is a real number that represents a threshold value when two clusters are merged or split.

Listing 8.7: Declaration of class `InternalNode`

```
1   class InternalNode: public Node,
2                       public Container<boost::shared_ptr<Node> >{
3   public:
4       InternalNode(Size id = 0,
5               const boost::shared_ptr<Node> p
6           = boost::shared_ptr<Node>() );
7       InternalNode(Real joinValue,
8               Size id = 0,
9               const boost::shared_ptr<Node> p
10          = boost::shared_ptr<Node>() );
11
12      void accept(NodeVisitor &v);
13      Size num_children() const;
14      Size num_records() const;
15      Real get_joinValue ();
16      void set_joinValue(Real joinValue);
17
18  private:
19      Real _joinValue;
20  };
```

The complete implementation of class `InternalNode` can be found in Listing B.81 and Listing B.82.

Using the node classes described above, we are now ready to introduce the design of a hierarchical clustering. The declaration of the resulting class `HClustering` is shown in Listing 8.8, from which we see that class `HClustering` has seven public member functions and one private data member. Data member `_root` is a shared pointer pointing to the root node of a hierarchical clustering tree.

Listing 8.8: Declaration of class `HClustering`

```
1   class HClustering {
2   public:
3       HClustering() {}
4       HClustering(const boost::shared_ptr<Node>& root );
5
6       boost::shared_ptr<Node> joinWith(HClustering& hc,
7               Real joinValue);
8       const boost::shared_ptr<Node>& root() const;
9       boost::shared_ptr<Node>& root();
10      PClustering get_pc(Size maxclust) const;
11      void save(const std::string &filename, Size p=100) const;
12
13  private:
14      boost::shared_ptr<Node> _root;
15  };
```

Class `HClustering` has two constructors. One constructor does not take any arguments; the other one takes one argument, which is a shared pointer pointing to a node object. Class `HClustering` also provides a member function joinWith (line 6–7) used to join another hierarchical clustering. The method joinWith returns the shared pointer pointing to the resulting parent node.

Method root is used to get the data member _root. Method get_pc is used to extract a partition from the hierarchical clustering tree. The argument specifies the maximum number of clusters to be extracted. Method save saves a dendrogram of the hierarchical clustering to an EPS file (see Section 11.6). The argument of method save specifies the maximum number of nodes to show in the dendrogram.

Listing B.46 and Listing B.47 show the complete implementation of class HClustering.

8.4 Summary

In this chapter, we introduced the design and implementation of clusters, partitional clusterings, and hierarchical clusterings. In particular, we introduced the implementation of one-center clusters and subspace clusters. We would also implement fuzzy clusters. However, fuzzy clusters might not be efficient since a fuzzy cluster may contain all the records in a dataset with different probabilities. Hence, it is better off letting clustering algorithms to handle fuzzy clusters. It is straightforward to implement multi-center clusters.

In $ClusLib$, a partitional clustering is implemented as a collection of clusters. A hierarchical clustering is more complex than a partitional clustering. We applied the composite design pattern to implement a hierarchical clustering.

Chapter 9

Dissimilarity Measures

Dissimilarity measures, or distances, play an important role in data clustering as most clustering algorithms depend on certain distance measures (Jain and Dubes, 1988; Anderberg, 1973). As a result, various distance measures have been proposed and discussed in the literature of cluster analysis (Sokal and Sneath, 1973; Anderberg, 1973; Jain and Dubes, 1988; Legendre and Legendre, 1983; Gordon, 1999; Everitt et al., 2001).

This chapter focuses on the design and implementation of some common distance measures that can be reused for different clustering algorithms.

9.1 The Distance Base Class

In data clustering, a distance measure is used to measure the dissimilarity between two records. Although we frequently measure the dissimilarity between a record and a cluster, such dissimilarity can be measured between the record and the representative records of the cluster. Hence we design the distance classes for measuring the distance between two records only.

Listing 9.1 shows the declaration of class `Distance`. We declare the destructor (line 4) of class `Distance` as virtual because the class is used as a base class. Since the class contains a pure virtual function, it is an abstract class.

Class `Distance` inherits from class `binary_function`, which is defined in the C++ standard header `<functional>`. Class `binary_function` is a template class with three arguments: the first two specify the types of the operands and the third argument specifies the type of the return value of the function. In class `Distance`, the two operands are shared pointers pointing to records of a dataset and the return value is a real number.

Listing 9.1: Declaration of class `Distance`

```
1  class Distance: std::binary_function<boost::shared_ptr<Record>,
2      boost::shared_ptr<Record>, Real> {
3  public:
4      virtual ~Distance() {}
5      Distance(const std::string &name);
6
7      const std::string& name() const;
```

```
 8          virtual Real operator()(const boost::shared_ptr<Record>&,
 9              const boost::shared_ptr<Record>& ) const = 0;
10
11      protected:
12          std::string _name;
13      };
```

Class **Distance** has one constructor (line 5), which takes one argument used to set the name of the distance measure. The class has one protected data member, which holds the name of the distance measure. The data member can be set only during construction of the object by the constructor of the class. The member function **name** (line 7) can be used to access the name only since we do not want the name to be changed once it is initialized.

Class **Distance** declares a prototype of operator "( )" with two arguments. This function is a pure virtual function and is supposed to be implemented in the derived classes of **Distance**.

9.2 Minkowski Distance

The Minkowski distance is a popular distance measure for numeric data and works well for datasets with compact or isolated clusters (Mao and Jain, 1996). Let **x** and **y** be two numeric records. Then the Minkowski distance is defined as

$$D_{min}(\mathbf{x},\mathbf{y}) = \left(\sum_{j=1}^{d} |x_j - y_j|^p\right)^{\frac{1}{p}}, \qquad (9.1)$$

where x_j is the jth component of record **x**, y_j is the jth component of record **y**, d is the number of attributes describing the record, and $p \geq 1$.

The Minkowski distance is implemented as a class **MinkowskiDistance**, which is derived from the base class **Distance**. Listing 9.2 shows the declaration of class **MinkowskiDistance**. We see that the class has one private data member and three public member functions, two of which are constructors. If the constructor is called without argument, the data member _p is set to the default value of 2.0. If the constructor is called with an argument, the data member is set to the value of the argument. One can see the complete implementation in Listing B.72 and Listing B.73.

Listing 9.2: Declaration of class **MinkowskiDistance**.

```
1   class MinkowskiDistance: public Distance {
2   public:
3       MinkowskiDistance();
4       MinkowskiDistance(Real p);
5       Real operator()(const boost::shared_ptr<Record>&,
6           const boost::shared_ptr<Record>&) const;
7
8   protected:
```

```
 9      Real _p;
10    };
```

The code in Listing 9.3 shows the implementation of the Minkowski distance between two records. This function first checks whether the schemata of the two records are equal to each other or not. If not, it will throw an exception. Once the check is passed, the function loops through all the attributes of the record and calculates the distances for all attributes. Function `distance` (line 10) is defined in class `CAttrInfo` (see Listing B.57).

Listing 9.3: Implementation of the Minkowski distance.
```
 1   Real MinkowskiDistance::operator()(
 2       const boost::shared_ptr<Record> &x,
 3       const boost::shared_ptr<Record> &y) const {
 4       boost::shared_ptr<Schema> schema = x->schema();
 5       ASSERT(*schema==*(y->schema()), "schema_does_not_match");
 6
 7       Real temp = 0.0;
 8       for(Size i=0;i<schema->size();++i){
 9           temp += std::pow(std::fabs(
10               (*schema)[i]->distance((*x)[i],(*y)[i])),_p);
11       }
12
13       return std::pow(temp,1/_p);
14   }
```

In line 10 of the code in Listing 9.3 we see that *schema is put in brackets. In fact, these brackets are necessary as operator "[ ]" has a higher precedence than operator "*".

9.3 Euclidean Distance

The Euclidean distance is a special case of the Minkowski distance when $p = 2$ in Equation (9.1). Hence it is natural to design the Euclidean distance class as a derived class of the Minkowski distance class.

Listing 9.4 shows the declaration of class `EuclideanDistance` and Listing 9.5 shows the implementation of the Euclidean distance. In our implementation shown in Listing 9.5, we just call the function in class `MinkowskiDistance` (line 4) to calculate the distance. The scope operator "::" (line 4) tells that the function `operator()` in class `MinkowskiDistance` should be called.

Listing 9.4: Declaration of class `EuclideanDistance`.
```
1   class EuclideanDistance: public MinkowskiDistance {
2   public:
3       EuclideanDistance();
4       Real operator()(const boost::shared_ptr<Record>&,
5           const boost::shared_ptr<Record>& ) const;
6   };
```

Listing 9.5: Implementation of the Euclidean distance.
```
Real EuclideanDistance::operator()(
    const boost::shared_ptr<Record> &x,
    const boost::shared_ptr<Record> &y) const {
    return MinkowskiDistance::operator()(x,y);
}
```

Complete implementation of class `EuclideanDistance` is shows in Listing B.68 and Listing B.69. In Exercise A.3, readers are asked to implement other two distance measures.

9.4 Simple Matching Distance

The simple matching distance is a distance measure for categorical data (Kaufman and Rousseeuw, 1990; Huang, 1998). Mathematically, the simple matching distance is defined as follows. Let **x** and **y** be two records described by d categorical attributes. Then the simple matching distance between **x** and **y** is calculated as

$$D_{sim} = \sum_{j=1}^{d} \delta(x_j, y_j), \qquad (9.2)$$

where x_j is the jth component of record **x**, y_j is the jth component of record **y**, and

$$\delta(x_j, y_j) = \begin{cases} 0 & \text{if } x_j = y_j, \\ 1 & \text{if } x_j \neq y_j. \end{cases} \qquad (9.3)$$

The declaration of class `SimpleMatchingDistance` is shown in Listing 9.6. The class inherits from the base distance class and has two public member functions. One member function is a constructor (line 3), which does not take any arguments. The other member function is the overloaded operator "( )".

Listing 9.6: Declaration of class `SimpleMatchingDistance`.
```
class SimpleMatchingDistance: public Distance {
public:
    SimpleMatchingDistance();
    Real operator()(const boost::shared_ptr<Record>&,
        const boost::shared_ptr<Record>&) const;
};
```

Listing 9.7 shows detail implementation of the simple matching distance. This function first checks whether the schemata of the two records are the same or not. Then it loops through all the attributes and calculates the distance from all attributes. Function `distance` (line 9) is defined in class `DAttrInfo` (see Listing B.59).

Listing 9.7: Implementation of simple matching distance.

```
Real SimpleMatchingDistance::operator()(
    const boost::shared_ptr<Record> &x,
    const boost::shared_ptr<Record> &y) const {
    boost::shared_ptr<Schema> schema = x->schema();
    ASSERT(*schema==*(y->schema()), "schema_does_not_match");

    Real temp = 0.0;
    for(Size i=0;i<schema->size();++i){
        temp += (*schema)[i]->distance((*x)[i],(*y)[i]);
    }

    return temp;
}
```

We note that the piece of code in lines 4–5 was also written for the Minkowski distance calculation (see Listing 9.3). In fact, the piece of code applies to almost all distance calculations. In Exercise A.2, readers are asked to improve this.

9.5 Mixed Distance

Mixed distance is a distance measure used for mixed-type records, which contain both numeric attributes and categorical attributes. Several distance measures for mixed-type data have been proposed and used for cluster analysis (Gan et al., 2007). For example, the general similarity coefficient (Gower, 1971; Wishart, 2002), the general distance coefficient (Gower, 1971), and the generalized Minkowski distance (Ichino, 1988; Ichino and Yaguchi, 1994) are measures for mixed-type data.

Here we implement a mixed distance that combines the squared Euclidean distance and the simple matching distance. Before we present the actual implementation of the distance, let us define the distance first. Let **x** and **y** be mixed-type records described by d attributes. Without loss of generality, we assume that the first m $(1 < m < d)$ attributes are numeric and the rest are categorical. Then the mixed distance is defined as (Huang, 1998):

$$D_{mix}(\mathbf{x},\mathbf{y}) = \sum_{j=1}^{m}(x_j - y_j)^2 + \beta \sum_{j=m+1}^{d} \delta(x_j,y_j), \qquad (9.4)$$

where β is a balance weight used to avoid favoring either type of attribute and $\delta(\cdot,\cdot)$ is defined in Equation (9.3).

The declaration of the mixed distance class `MixedDistance` is shown in Listing 9.8. The class has a protected data member `_beta`, which holds the balance weight, and two constructors. When the constructor is called without argument, the data member `_beta` is set to 1.0 (see Listing B.74). When the constructor is called with an argument, `_beta` is set to the value of the argument.

Listing 9.8: Declaration of class `MixedDistance`.

```
class MixedDistance: public Distance {
public:
    MixedDistance();
    MixedDistance(Real beta);
    Real operator()(const boost::shared_ptr<Record>&,
        const boost::shared_ptr<Record>& ) const;

protected:
    Real _beta;
};
```

Listing 9.9 shows the implementation of the mixed distance. The function first checks for schemata of the two records. Then it loops through all the attributes and calculates the distance according to the type of the attributes. In our implementation, we do not assume that the first m attributes are numeric and the rest are categorical. In fact, numeric and categorical attributes can be arranged arbitrarily. We use the function `can_cast_to_c` defined in class `CAttrInfo` and class `AttrInfo` to determine the type of an individual attribute.

Listing 9.9: Implementation of mixed distance.

```
Real MixedDistance::operator()(
    const boost::shared_ptr<Record> &x,
    const boost::shared_ptr<Record> &y) const {
    boost::shared_ptr<Schema> schema = x->schema();
    ASSERT(*schema==*(y->schema()), "schema_does_not_match");

    Real d1 = 0.0;
    Real d2 = 0.0;
    for(Size i=0;i<schema->size();++i){
        if((*schema)[i]->can_cast_to_c()) {
            d1 += std::pow(std::fabs(
                (*schema)[i]->distance((*x)[i],(*y)[i])), 2.0);
        } else {
            d2 += (*schema)[i]->distance((*x)[i],(*y)[i]);
        }
    }

    return d1 + _beta*d2;
}
```

The complete implementation of class `MixedDistance` can be found in Listing B.74 and Listing B.75.

9.6 Mahalanobis Distance

Mahalanobis distance (Jain and Dubes, 1988; Mao and Jain, 1996) is a distance measure for numeric data. This distance can be used to alleviate the distance distortion caused by linear combinations of attributes.

Mathematically, Mahalanobis distance between two numeric records **x** and

y is defined as
$$D_{mah}(\mathbf{x}, \mathbf{y}) = \sqrt{(\mathbf{x} - \mathbf{y})^T \Sigma^{-1} (\mathbf{x} - \mathbf{y})}, \qquad (9.5)$$
where Σ is a covariance matrix and Σ^{-1} denotes the inverse of Σ. Here we assume that **x** and **y** are column vectors. That is,
$$\mathbf{x} = \begin{pmatrix} x_1 \\ x_2 \\ \vdots \\ x_d \end{pmatrix}, \quad \mathbf{y} = \begin{pmatrix} y_1 \\ y_2 \\ \vdots \\ y_d \end{pmatrix}.$$

The covariance matrix Σ used in the Mahalanobis distance can be calculated from a dataset. Suppose a dataset has n records: $\mathbf{x}_1, \mathbf{x}_2, \cdots, \mathbf{x}_n$, and the dataset is described by d numeric attributes. Then the covariance between the rth attribute and the sth attribute is calculated as (Rummel, 1970)
$$c_{rs} = \frac{1}{n-1} \sum_{i=1}^{n} (x_{ir} - \mu_r)(x_{is} - \mu_s), \qquad (9.6)$$
where x_{ir} and x_{is} are the rth component and the sth component of record $\mathbf{x}_i$, respectively, and μ_r and μ_s are the rth component and the sth component of the mean $\boldsymbol{\mu}$ of the dataset, respectively. The mean $\boldsymbol{\mu} = (\mu_1, \mu_2, \cdots, \mu_d)^T$ is calculated as
$$\mu_j = \frac{1}{n} \sum_{i=1}^{n} x_{ij}, \quad j = 1, 2, \cdots, d.$$
Then the covariance matrix of the dataset is simply defined as
$$\Sigma = \begin{pmatrix} c_{11} & c_{12} & \cdots & c_{1d} \\ c_{21} & c_{22} & \cdots & c_{2d} \\ \vdots & \vdots & \ddots & \vdots \\ c_{d1} & c_{d2} & \cdots & c_{dd} \end{pmatrix}. \qquad (9.7)$$
In matrix form, we can formulate Σ as
$$\Sigma = \frac{1}{n-1} \left(X^T X - n \boldsymbol{\mu} \boldsymbol{\mu}^T \right), \qquad (9.8)$$
where X is the data matrix
$$X = \begin{pmatrix} x_{11} & x_{12} & \cdots & x_{1d} \\ x_{21} & x_{22} & \cdots & x_{2d} \\ \vdots & \vdots & \ddots & \vdots \\ x_{n1} & x_{n2} & \cdots & x_{nd} \end{pmatrix}.$$

The mathematical formula of the Mahalanobis distance is very simple.

However, implementing the Mahalanobis distance is not straightforward in C++ since the distance involves the inverse of the covariance matrix. Computing the inverse of Σ directly is computationally expensive. Since Σ is a covariance matrix, it is at least semi-positive definite. If all the diagonal elements of Σ are positive, then Σ is positive definite. Hence we can calculate the Cholesky decomposition (Press et al., 1992) of Σ first and then calculate the inverse of the Choleksy decomposition, which is a triangular matrix.

Let L be the Cholesky decomposition of Σ. That is, L is a lower triangular matrix such that
$$\Sigma = LL^T.$$

There is an efficient algorithm to calculate L (Press et al., 1992). Listing B.103 and Listing B.104 show the implementation of Cholesky decomposition and the inverse of a triangular matrix. For these matrix calculations, we use Boost uBLAS, which is a C++ template class library that provides BLAS (Basic Linear Algebra Subprograms) functionality for matrices.

Let A be the inverse of L. Then A is also a lower triangular matrix. The inverse of Σ can be calculated as
$$\Sigma^{-1} = (LL^T)^{-1} = (L^{-1})^T L^{-1} = A^T A.$$

Then the Mahalanobis distance in Equation (9.5) can be formulated as
$$D_{mah}(\mathbf{x}, \mathbf{y}) = \sqrt{[A(\mathbf{x}-\mathbf{y})]^T [A(\mathbf{x}-\mathbf{y})]}. \qquad (9.9)$$

The declaration of class `MahalanobisDistance` is shown in Listing 9.10, from which we see that the class has a protected data member and a constructor with an argument. Classes `symmetric_matrix` (line 4) and `triangular_matrix` (line 9) are defined in the Boost uBLAS library. The namespace `ublas` refers to `boost::numeric::ublas` (see `matrix.hpp` in Listing B.103).

The data member `_A` is the inverse of the Cholesky decomposition of the matrix `sigma` in the constructor (see Listing B.103). We calculate the Cholesky decomposition and its inverse only once when the distance object is constructed.

Listing 9.10: Declaration of class `MahalanobisDistance`.

```
class MahalanobisDistance: public Distance {
public:
    MahalanobisDistance(const
        ublas::symmetric_matrix<Real> &sigma);
    Real operator()(const boost::shared_ptr<Record>&,
        const boost::shared_ptr<Record>&) const;

protected:
    ublas::triangular_matrix<Real> _A;
};
```

Listing 9.11 shows the calculation of the Mahalanobis distance between two records. The function first checks whether the schemata of the two records match each other or not and checks whether the dimension of the records and the dimension of the stored matrix match or not. Then the function creates a vector that holds the differences of all components of the input records (lines 9–12). Using Boost uBLAS function `axpy_prod`, the function calculates the part $A(\mathbf{x} - \mathbf{y})$ in Equation (9.9) of the squared Mahalanobis distance (line 16). Finally, the function returns the Mahalanobis distance (line 18).

Listing 9.11: Implementation of Mahalanobis Distance.

```
1   Real MahalanobisDistance::operator()(
2       const boost::shared_ptr<Record> &x,
3       const boost::shared_ptr<Record> &y) const {
4       boost::shared_ptr<Schema> schema = x->schema();
5       ASSERT(*schema==*(y->schema()), "schema_does_not_match");
6       ASSERT(schema->size() == _A.size1(),
7               "record_and_matrix_dimensions_do_not_match");
8
9       ublas::vector<Real> v(schema->size());
10      for(Size i=0; i<schema->size(); ++i) {
11          v(i) = (*schema)[i]->distance((*x)[i],(*y)[i]);
12      }
13
14      ublas::vector<Real> w;
15      w.resize(v.size());
16      ublas::axpy_prod(_A, v, w, true);
17
18      return std::sqrt(ublas::inner_prod(w,w));
19  }
```

9.7 Summary

In this chapter, we introduced the design and implementation of several distance measures. In particular, we introduced how to implement the Minkowski distance, the simple matching distance, a distance measure for mixed-type data, and the Mahalanobis distance.

In the literature of cluster analysis, many different distance and similarity measures have been designed and used for different purposes. For example, the Manhattan segmental distance (Aggarwal et al., 1999), the generalized Mahalanobis distance (Morrison, 1967), the average Euclidean distance (Legendre and Legendre, 1983), the Chord distance (Orlóci, 1967), the cosine similarity measure (Salton and McGill, 1983), and the link-based similarity measure (Guha et al., 2000), to just name a few. For more information about distance and similarity measures, readers are referred to Gan et al. (2007) and references therein.

Chapter 10

Clustering Algorithms

A clustering algorithm is an algorithm that assigns data points or records into groups in a certain way. Over the past 50 years, many clustering algorithms have been developed and applied to various areas such as gene expression data analysis (Yeung et al., 2003) and psychology (Clatworthy et al., 2005). Although there are many different kinds of clustering algorithms, they have at least three commonalities. First, all clustering algorithms require some arguments or parameters. Second, all clustering algorithms produce clustering results. Third, all clustering algorithms perform data clustering.

In this chapter, we introduce the design and implementation of the base class for all clustering algorithms based on the aforementioned three commonalities.

10.1   Arguments

Almost all clustering algorithms require some arguments in order to perform data clustering. In addition, different clustering algorithms might require different types of arguments. For example, the k-means algorithm (Macqueen, 1967; Jain and Dubes, 1988) requires the number of clusters. The c-means algorithm (Bezdek, 1981a; Bezdek et al., 1992) requires the number of clusters as well as the fuzziness.

In order to deal with the arguments of any clustering algorithm, we designed class **Arguments** that is able to hold any arguments provided by users. The declaration of class **Arguments** is shown in Listing 10.1.

Class **Arguments** inherits from class **Additional** and has two public data members. One data member is a shared pointer pointing to a dataset and the other data member is a shared pointer pointing to a distance measure. In our design, we treat the dataset and the distance measure as arguments. If a clustering algorithm does not require any distance measure or implements its own distance measure, we can just leave the shared pointer pointing to the distance measure empty.

Listing 10.1: Declaration of class **Arguments**.

```
class Arguments : public Additional {
public:
    boost::shared_ptr<Dataset> ds;
    boost::shared_ptr<Distance> distance;
};
```

Class **Additional** is designed to hold any additional arguments. The declaration of the class is shown in Listing 10.2. This class has a public data member (line 7), a protected constructor (line 10), and two convenient member functions (lines 3–5). The data member is a map with **std::string** as the type of keys and **boost::any** as the type of values. The key is usually the name of an argument stored in a **boost::any** object. Since **boost::any** can store almost everything, the class is able to store any arguments.

The constructor of class **Additional** is declared protected. In this way, users cannot create instances of class **Additional** directly. Only derived classes of **Additional** can call the protected constructor.

Function **get** is used to retrieve an argument value from an object of class **Additional** or a class derived from class **Additional**. The argument of function **get** is the key of the argument to be retrieved. If the key is not found in the map, the function throws an exception. The complete implementation of the function can be found in Listing B.10.

Listing 10.2: Declaration of class **Additional**.

```
class Additional {
public:
    const boost::any& get(const std::string &name) const;
    void insert(const std::string &name,
                const boost::any &val);

    std::map<std::string, boost::any> additional;

protected:
    Additional() {}
};
```

Function **insert** is used to insert a value to the map. The function takes two arguments: the name of the argument of a clustering algorithm and the **boost::any** object that holds the corresponding value.

10.2 Results

Similar to arguments, which might be different for different clustering algorithms, results of different clustering algorithms might also be different. For example, the k-means algorithm produces hard clustering results. But the c-mean clustering algorithm produces fuzzy clustering results. Hence the class used to store clustering results is designed similarly to class **Arguments** described in the previous section.

The declaration of class **Results** is shown in Listing 10.3, from which we see that class **Results** also inherits from class **Additional** (see Listing 10.2). Class **Results** has one public data member CM (line 5) and one public member function reset (line 3).

The data member stores the clustering memberships. The clustering membership is a vector whose ith component is the index of the cluster to which the ith record belongs. Usually, the index of a cluster can be 0 to $k-1$, where k the number of clusters. Table 10.1 gives an example of the cluster membership of a partition of a dataset with 5 records. In the table, $\mathbf{x}_0, \mathbf{x}_1, \mathbf{x}_2, \mathbf{x}_3$, and $\mathbf{x}_4$ denote 5 records in the dataset. The 5 records are grouped into 3 clusters as shown in Table 10.0(a). The corresponding cluster membership vector is shown in Table 10.1.

(a)		(b)
Cluster Index	Members	Cluster Membership
0	$\mathbf{x}_0, \mathbf{x}_2$	$CM[0] = 0$
1	$\mathbf{x}_4$	$CM[1] = 2$
2	$\mathbf{x}_1, \mathbf{x}_3$	$CM[2] = 0$
		$CM[3] = 2$
		$CM[4] = 1$

TABLE 10.1: Cluster membership of a partition of a dataset with 5 records.

The member function reset is used to clear all clustering results. As we can see from its implementation in Listing B.11, the function clears the cluster membership vector CM and the additional results.

Listing 10.3: Declaration of class **Results**.

```
class Results : public Additional {
public:
    void reset();

    std::vector<Size> CM;
};
```

10.3 Algorithms

In this section, we introduce the base class **Algorithm** for all clustering algorithms. Listing 10.4 shows the declaration of class **Algorithm**. This class has three protected data members, five public member functions, and three protected member functions.

One of the data members of class **Algorithm** is a shared pointer pointing to a dataset. Declaring a shared pointer pointing to a dataset as a data member of the base class **Algorithm** is appropriate as almost all clustering algorithms require a data set. Some clustering algorithms also accept proxy matrices as input (Kaufman and Rousseeuw, 1990). In this case, we can just leave the shared pointer pointing to the dataset empty and put the proxy matrix to the object of **Additional**.

The other two data members of class **Algorithm** are objects of **Arguments** and **Results**, which were described in the previous two sections. Figure 10.1 shows the UML class diagram of class **Algorithm** and its dependencies. The diagram shows that class **Arguments** and class **Results** are subclasses of class **Additional**. The diagram also shows that class **Algorithm** is a composition of class **Arguments** and **Results**. That is, class **Arguments** and **Results** are part of class **Algorithm** and their instances will be destroyed if the **Algorithm** object containing them is destroyed.

We note that the C++ keyword **mutable** (line 15) is placed in front of the declaration of _results. This keyword is used to modify a data member that is intended to be changed by **const** functions. For example, we can use the member function **reset**, which is a **const** function, to clear the contents of _results.

The destructor of class **Algorithm** is declared virtual. In fact, all other member functions of class **Algorithm** are declared as virtual functions. In this way, we allow the classes derived from class **Algorithm** to redefine the behavior of these functions.

Method **getArguments** returns a reference to the member _arguments. Through this function, we can transfer parameters required by a clustering algorithm into the clustering algorithm. Since we want to change the data member _arguments, we do not return a **const** reference.

Function **getResults** is a **const** member function that returns a **const** reference to the data member _results. We declared this function as a **const** function since we do not want to change the clustering results outside the algorithm.

Listing 10.4: Declaration of class **Algorithm**.

```
1  class Algorithm {
2  public:
3      virtual ~Algorithm() {}
4      virtual Arguments& getArguments();
5      virtual const Results& getResults() const;
6      virtual void reset() const;
7      virtual void clusterize();
8
9  protected:
10     virtual void setupArguments();
11     virtual void performClustering() const = 0;
12     virtual void fetchResults() const = 0;
13
14     boost::shared_ptr<Dataset> _ds;
15     mutable Results _results;
16     Arguments _arguments;
17 };
```

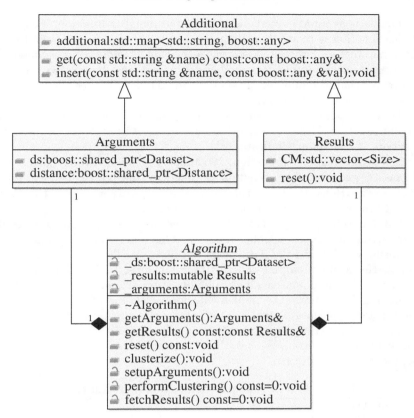

FIGURE 10.1: Class diagram of algorithm classes.

Function `reset` is a `const` function that does not return anything. This function is used to reset the clustering results. That is, this function will clear all clustering results produced by the algorithm. This function is useful, since we might want to run a clustering algorithm multiple times and do not want the old clustering results affected by the new ones. Since this function is declared as a `const` function, it is not allowed to change any data members of class `Algorithm` other than the `mutable` ones. For example, function `reset` can change the data member `_results`, which is mutable, but cannot change the data member `_arguments`, which is not mutable.

Function `clusterize` is the member function that starts the actual clustering process. Once everything is ready, we can just call this function to do clustering. In the implementation of class `Algorithm` in Listing B.11, we see that this function actually calls four other member functions. Of course, users can change the behavior of this function in derived classes since this function is virtual.

Function `setupArguments` is a protected member function. The purpose of this function is to validate the parameters provided by users and transfer these parameters into the clustering algorithm. Almost all parameters have constraints. For example, the number of clusters must be positive and cannot be greater than the number of points in the dataset. We see the usage of this function later in this section.

Function `performClustering` is a `const` member function and a pure virtual function. Every class derived from class `Algorithm` must define this function. As its name indicates, this function actually does the clustering. Since this function is protected, it cannot be called directly from outside of class `Algorithm`. This function is called by the public member function `clusterize`. However, derived classes of class `Algorithm` can call this function directly since it is not private to class `Algorithm`.

Similar to function `performClustering`, function `fetchResults` is also a pure virtual and `const` member function. This function collects and transfers clustering results into the data member _results. This function is also protected. Hence users cannot call this function directly. We call this function indirectly via the public member function `clusterize`.

Listing B.11 shows the implementation of all the aforementioned member functions of class `Algorithm` except for the two pure virtual functions: `perform- Clustering` and `fetchResults`. Listing 10.5 shows the implementation of this function. We see that function `clusterize` calls four other member functions in turn.

Listing 10.5: Implementation of function `clusterize`.

```
void Algorithm::clusterize() {
    setupArguments();
    performClustering();
    reset();
    fetchResults();
}
```

10.4 A Dummy Clustering Algorithm

In this section, we show how to using the base class `Algorithm` to create new clustering algorithms. To do this, we consider a dummy clustering algorithm that does not do anything. However, this example shows how to input parameters and output clustering results. This example also shows how to use the clustering library $\mathcal{C}lus\mathcal{L}ib$.

The C++ code of the dummy clustering algorithm is shown in Listing 10.6. Now let us take a look at the code. This first line is a comment, which shows the name and the path of the source file. We put the source file `dummy.cpp` in a folder called `dummy` under the folder `examples` (see Figure 6.1).

Listing 10.6: A dummy clustering algorithm.

```cpp
// examples/dummy/dummy.cpp
#include<cl/cluslib.hpp>

#include<iostream>
#include<cstdlib>

using namespace ClusLib;
using namespace boost;

class DummyAlgorithm: public Algorithm {
private:
    void setupArguments() {
        Algorithm::setupArguments();
        _numclust = any_cast<Size>(
            _arguments.get("numclust"));
        ASSERT(_numclust>0 && _numclust<5,
            "numclust must in range (0, 5)");
    }

    void performClustering() const {
        for(Size i=0; i<4; ++i){
            _CM.push_back(i);
        }
    }

    void fetchResults() const {
        _results.CM = _CM;
        _results.insert("msg",
            any(std::string("A dummy clustering algorithm")));
    }

    mutable std::vector<Size> _CM;
    Size _numclust;
};

int main(int argc, char* argv[]){
    try{
        Size numclust = 3;
        if (argc>1) {
            numclust = atoi(argv[1]);
        }

        boost::shared_ptr<Schema> schema(new Schema());
        boost::shared_ptr<Dataset> ds(new Dataset(schema));

        DummyAlgorithm ca;
        Arguments &Arg = ca.getArguments();
        Arg.ds = ds;
        Arg.insert("numclust", any(numclust));

        ca.clusterize();

        const Results& Res = ca.getResults();

        std::string msg =
            boost::any_cast<std::string>(Res.get("msg"));
        std::cout<<"Cluster Membership Vector: \n";
        std::cout<<"    ";
        for(Size i=0; i<Res.CM.size(); ++i) {
            std::cout<<Res.CM[i]<<" ";
        }
        std::cout<<"\nMessage: \n";
        std::cout<<msg<<'\n';

        return 0;
```

```
67      } catch (std::exception& e) {
68          std::cout<<e.what()<<std::endl;
69          return 1;
70      } catch (...) {
71          std::cout<<"unknown error"<<std::endl;
72          return 1;
73      }
74  }
```

Line 2 is a preprocessor directory used to include other C++ headers. Here we include the headers of the clustering library. The header cluslib.hpp includes all headers of the clustering library. Hence we need to include only cluslib.hpp here. Line 4 and line 5 include two standard C++ libraries. The header <iostream> defines some input and output functions. The header <cstdlib> defines the function atoi, which will be used in the program to convert an input string to an integer.

Line 7 and line 8 are statements of using directives. The statement in line 7 specifies that members of namespace ClusLib can be used in the file without preceding each member with ClusLib and the scope resolution operator "::". Similarly, the statement in line 8 specifies that members of namespace boost can be used in the file without preceding each member with boost and the scope resolution operator "::".

Lines 10–34 define the derived class DummyAlgorithm. Line 10 shows that class DummyAlgorithm is derived publicly from class Algorithm. To indicate the inheritance, we put a colon ":" right after the name of the class. To indicate that the inheritance should be public, we put the keyword public after the colon and before the name of the parent class.

Lines 12–18 redefine the setupArguments function. The function first calls the same function defined in the base class (line 13). We put the class name and the scope resolution operator before the name of the function. Then the function tries to retrieve a parameter from the data member _arguments inherited from the base class. If the parameter is obtained successfully, the function then validates the parameter (lines 16–17).

The pure virtual function performClustering of the base class is defined in lines 20–24. This function does nothing except populate the cluster membership vector. In an actual clustering algorithm, this function is implemented to do the clustering.

Lines 26–30 define the fetchResults function, which is a pure virtual function in the base class. This function first transfers the clustering membership vector into the data member _results (line 27) inherited from the base class. Then the function puts an additional result into the data member.

Line 32 declares a mutable data member of class DummyAlgorithm. Since the data member is mutable, it can be changed by const functions such as performClustering and fetchResults. In fact, this data member is changed in function performClustering described above.

Line 33 declares another data member. Since this data member is not mutable, it cannot be changed by const functions. For example, we cannot assign value to _numclust in function performClustering. Actually, we assign a value to _numclust in function setupArguments, which is not a const function.

The main function is defined in lines 37–73. In the main function, we first define variable numclust of type Size. If the program is called with at least one argument, we set the value of numclust to the value of the argument. Note that in the C++ main function argv[0] stores the name of the program. If the program is called with one argument, the argument is stored in argv[1].

Line 44 defines a shared pointer pointing to an object of Schema and initializes the shared pointer. Line 45 defines a shared pointer pointing to an object of Dataset and initializes the shared pointer. Later the shared pointer pointing to the dataset is transfered into an instance of DummyAlgorithm.

Line 47 declares an instance ca of class DummyAlgorithm. The first thing we do is to get a reference to the Arguments object in ca (line 48). Note that we put the symbol "&" before the variable Arg. This symbol is important. If this symbol is missing, the parameters cannot be transferred into the algorithm even the compiler does not complain. In lines 49 and 50, we transfer two parameters into the algorithm.

In line 52, function clusterize is called to do the clustering. As we mentioned before, this function will call another four member functions of class DummyAlgorithm. Once the function call is finished, the clustering results are retrieved through the function getResults (line 54). Lines 56–64 output the clustering results.

We note that all the code is put in the try and catch block. In this way, we can catch all exceptions and output the message. Usually, the try and catch block is put in the main function since it is too expensive to put a try and catch block inside an internal function.

In the first catch block, we see that variable e (line 67) is a reference to an object of std::exception. In Exercise A.5, readers are asked to modify the program by removing the reference symbol and see what happens when an exception occurs.

We also note that we return 0 at the end of the try block (line 66) and return 1 at end of catch blocks. In many platforms such as Unix and Windows, returning 0 means that the program is executed successfully and returning a non-zero value indicates that the program encountered some problem during execution.

Now compile the program. Suppose the clustering library is compiled using the Autotools described in Section 6.6. To compile the program, we first enter into the folder examples/dummy and execute the following command:

```
libtool --mode=compile g++ -c -I../.. -I/usr/include/ dummy.cpp
```

The option -I../.. tells the compiler the location of the header cluslib.hpp required by the program. The option -I/usr/include/ tells the compiler the location of boost headers. In different platforms, this location might be different. Once this command is executed, an object file dummy.o is produced.

To produce an executable program, we need to link the program by executing the following command:

```
libtool --mode=link g++ -o dummy ../../cl/libClusLib.la dummy.o
```

To link the program, we need to provide the location and the name of the clustering library `libClusLib.la`, which includes many function objects required by the program. Once this command is executed, an executable program is produced. In Cygwin, the name of the executable is `dummy.exe`. In Linux such as Ubuntu, the name of the executable is `dummy`.

To execute the program, we can issue the following command in Cygwin:

```
./dummy.exe 3
```

After the program is executed, we see the following output:

```
Cluster Membership Vector:
    0 1 2 3
Message:
A dummy clustering algorithm
```

Since the program is called with an argument value of 3, the parameter is in the range $(0, 5)$ and hence is valid.

Now we try the program with an invalid argument. To do this, we issue the following command:

```
./dummy.exe 5
```

In this case, we see the following output:

```
virtual void DummyAlgorithm::setupArguments():
  dummy.cpp(17):
numclust must in range (0, 5)
```

Since we input an invalid argument to the program, the program threw an exception, which is caught by a `catch` block.

We note that we used command line commands to compile and link the dummy clustering algorithm. In Exercise A.4, readers are asked to write a `Makefile.am` file to compile the dummy clustering algorithm as part of the clustering library.

10.5 Summary

In this chapter, we introduced the architecture of clustering algorithms. We first introduced class `Arguments` and class `Results` that are used to deal with arguments and results of clustering algorithms. We designed these classes in such a way that they can handle any kind of argument and result.

Then we introduced the design and implementation of class `Algorithm`, the base class from which all clustering algorithms inherit. Class `Algorithm` is designed to be an abstract class.

Finally, we gave an example of how to use these classes to develop a clustering algorithm. We illustrated the usage of class `Arguments`, `Results`, and `Algorithm` using a dummy clustering algorithm, which provides a model or template to implement new clustering algorithms. In fact, all the clustering algorithms presented in later chapters are implemented in this way.

Chapter 11
Utility Classes

In this chapter, we introduce some utility classes that are used by the clustering library $ClusLib$. In particular, we introduce a container class, a double-key map class, dataset adapter classes, and node visitor classes.

11.1 The Container Class

In previous chapters, we introduced the design of the record class, the schema class, the dataset class, and the paritional clustering class. Objects of all these classes are some containers that contain other individual items. For example, an object of the record class `Record` contains a set of objects of class `AttrValue`; an object of the schema class `Schema` contains a set of shared pointers pointing to objects of class `AttrInfo`; and so forth. Hence it is a good idea to have a template container class that can be used as a base class for all these classes. Then all these classes inherit all the public and protected data members and member functions of the base class. In this way, we can avoid writing the same kind of code many times.

However, we cannot just derive our classes mentioned above from a STL container class such as `std::vector`. There are at least two reasons why we should not use STL container classes as base classes. First, the STL container classes are not designed to be used as base classes. For example, the STL container classes do not have virtual destructor. That is, we can not extend STL types and use them dynamically. Second, a class that inherits publicly from an STL container will expose all the public member functions of the STL container to users.

Therefore, we create a container class by wrapping it around the STL container class `std::vector`. The declaration of the resulting class is shown in Listing 11.1. Class `Container` is a template class. Hence it can be used to hold many types of items.

Listing 11.1: Declaration of class `Container`.

```
template<typename T>
class Container {
public:
    typedef typename std::vector<T>::iterator iterator;
```

```
 5      typedef typename std::vector<T>::const_iterator
 6          const_iterator;
 7
 8      iterator begin();
 9      const_iterator begin() const;
10      iterator end();
11      const_iterator end() const;
12      Size size() const;
13      bool empty() const;
14      void clear();
15
16      const std::vector<T>& data() const;
17      std::vector<T>& data();
18      const T& operator[](Size i) const;
19      T& operator[](Size i);
20      void erase(const T& val);
21      void add(const T&val);
22
23  protected:
24      ~Container() {}
25
26      std::vector<T> _data;
27  };
```

Class **Container** has one protected data member (line 26) and several convenient member functions (lines 8–21). Since we want the container class to perform efficiently, we do not declare any virtual functions for this class, including its destructor. To prevent users from deleting a **Container** pointer that actually points to an object of a derived class, we make the destructor of class **Container** protected. That is, we cannot call the destructor of class **Container** directly and hence cannot use this class directly in programs. For example, the program shown in Listing 11.2 does not compile since the destructor of class **Container** is protected. However, classes derived from class **Container** can call this destructor. For example, the program shown in Listing 11.3 works.

Listing 11.2: A program to illustrate class **Container**.

```
 1  // examples/container/container1.cpp
 2
 3  #include<cl/utilities/container.hpp>
 4  #include<iostream>
 5
 6  using namespace ClusLib;
 7  using namespace std;
 8
 9  class C: public Container<int> {
10  };
11
12  int main() {
13      Container<int> *p;
14      p = new C;
15      delete p;
16
17      return 0;
18  }
```

Listing 11.3: Another program to illustrate class `Container`.

```
1   // examples/container/container2.cpp
2
3   #include<cl/utilities/container.hpp>
4   #include<iostream>
5
6   using namespace ClusLib;
7   using namespace std;
8
9   class C: public Container<int> {
10  };
11
12  int main() {
13      C *p;
14      p = new C;
15      delete p;
16
17      return 0;
18  }
```

Line 4 and line 5 define two types of aliases. Note that we used the keyword `typename` after the key word `typedef`. The keyword `typename` is required. Otherwise, a complier such as `g++` will complain. Lines 8–11 define functions that provide access to the items contained in the object of the class via iterators. Function `size` (line 12) returns the number of items contained in the object of the class. Function `empty` (line 13) returns a boolean value that tells whether the container is empty or not. Function `clear` (line 14) is used to clear all the items in the container.

Functions declared in lines 16–21 are used to access and modify the items in the container. In particular, we overload the operator "[ ]" for the container class so that we can access and modify an individual item in the container.

All the functions are defined as `inline` functions in the header as shown in Listing B.92. In fact, we cannot separate the implementation of class `Container` from its header since class `Container` is a template class. That is, we cannot put the implementation of these functions in a source file such as `container.cpp`.

In addition, it is tricky to define `inline` functions outside the class body. In the implementation of function `begin` shown in Listing 11.4, for example, we have to put the keyword `typename` after the keyword `inline`. The keyword `typename` is required since it tells the compiler that `Container<T>::iterator` is a type rather than a value.

Listing 11.4: Implementation of function `begin`.

```
1   template<typename T>
2   inline typename Container<T>::iterator Container<T>::begin() {
3       return _data.begin();
4   }
```

The complete implementation of class `Container` is shown in Listing B.92.

11.2 The Double-Key Map Class

Sometimes it very convenient to have a double-key map class that can be used to associate a value with a key, which has two components. Such a class is very useful to develop the clustering library. For example, in hierarchical clustering algorithms we need to calculate pair-wise distance between two entities. In this case, we can use a double-key map to store the pair-wise distances. Another example is to calculate the cross tabulation between two partitions of a dataset, we need to calculate the number of records contained in any two clusters from different partitions. In this case, we can use a double-key map to store the number of records common to two clusters.

In addition, the order of the two components of a key does not matter in some cases but is important in other cases. For example, suppose $D(i,j)$ is the distance between two entities i and j, where (i,j) is a key with two components i and j. Since most distance measures are symmetric, we have $D(j,i) = D(i,j)$. That is, key (i,j) and key (j,i) should associate with the same value. In cross tabulation of two partitions of a dataset, however, the order of the two components of a key matters. Suppose $N(i,j)$ denotes the number of records contained both in cluster i from the first partition and in cluster j from the second partition. Then $N(i,j)$ might not be equal to $N(j,i)$ since i and j refer to clusters from different partitions.

In this section, we introduce a template double-key map class. The declaration of the resulting class **nnMap** is shown in Listing 11.5 (lines 42–64). Before we declare class **nnMap**, we defined several comparison classes that can be used to compare the keys of the double-key map.

Listing 11.5: Declaration of class **nnMap**.

```
typedef std::pair<Size,Size> nnPair;

class compare_a {
public:
    bool operator() (const nnPair &a, const nnPair &b) const {
        Size amin = std::min(a.first, a.second);
        Size amax = std::max(a.first, a.second);
        Size bmin = std::min(b.first, b.second);
        Size bmax = std::max(b.first, b.second);

        if(amin < bmin) {
            return true;
        } else if (amin == bmin ){
            if (amax < bmax) {
                return true;
            } else {
                return false;
            }
        } else {
            return false;
        }
    }
};

class compare_b {
```

```
26      public:
27          bool operator() (const nnPair &a, const nnPair &b) const {
28              if(a.first < b.first) {
29                  return true;
30              } else if (a.first == b.first ){
31                  if (a.second < b.second) {
32                      return true;
33                  } else {
34                      return false;
35                  }
36              } else {
37                  return false;
38              }
39          }
40      };
41
42      template<class T, class C>
43      class nnMap {
44      public:
45          typedef typename std::map<nnPair, T, C>::value_type
46              value_type;
47          typedef typename std::map<nnPair, T, C>::iterator
48              iterator;
49          typedef typename std::map<nnPair, T, C>::const_iterator
50              const_iterator;
51
52          void add_item(Size i, Size j, T item);
53          bool contain_key(Size i, Size j) const;
54          T& operator()(Size i, Size j);
55          const T& operator()(Size i, Size j) const;
56          void clear();
57
58          iterator begin();
59          iterator end();
60          const_iterator begin() const;
61          const_iterator end() const;
62      private:
63          std::map<nnPair, T, C> _map;
64      };
65
66      typedef nnMap<Real, compare_a> iirMapA;
67      typedef nnMap<Size, compare_b> iiiMapB;
```

Class compare_a (lines 3–23) does not care about the order of the two components of a key. For example, class compare_a will tell us that key $(1, 2)$ and key $(2, 1)$ are the same. Unlike class compare_a, class compare_b (lines 25–40) cares about the order. The program in Listing 11.6 shows the difference of the two comparison classes.

Listing 11.6: Program to illustrate class nnMap.

```
1   // examples/nnmap/nnmap.cpp
2   #include<cl/utilities/nnmap.hpp>
3   #include<iostream>
4
5   using namespace ClusLib;
6   using namespace std;
7
8   typedef nnMap<Size, compare_a> MapA;
9   typedef nnMap<Size, compare_b> MapB;
10
11  int main() {
12      MapA mA;
13      mA.add_item(0, 1, 10);
14      mA.add_item(0, 2, 20);
```

```cpp
15      pair<MapA::iterator, bool> reta = mA.add_item(2, 0, 40);
16      cout<<"Test MapA\n    ";
17      if(reta.second == false) {
18          cout<<"element (2, 0) already existed with "
19              <<"a value of "<<reta.first->second<<endl;
20      } else {
21          cout<<" value 40 is added to the map"<<endl;
22      }
23      MapA::iterator ita;
24      cout<<"mA contains:\n";
25      for(ita = mA.begin(); ita!=mA.end(); ++ita) {
26          cout<<"("<<ita->first.first<<","
27              <<ita->first.second<<"): "<<ita->second<<endl;
28      }
29
30      MapB mB;
31      mB.add_item(0, 1, 10);
32      mB.add_item(0, 2, 20);
33      pair<MapB::iterator, bool> retb = mB.add_item(2, 0, 40);
34      cout<<"Test MapB\n    ";
35      if(retb.second == false) {
36          cout<<"element (2, 0) already existed with "
37              <<"a value of "<<retb.first->second<<endl;
38      } else {
39          cout<<" value 40 is added to the map"<<endl;
40      }
41      MapB::iterator itb;
42      cout<<"mB contains:\n";
43      for(itb = mB.begin(); itb!=mB.end(); ++itb) {
44          cout<<"("<<itb->first.first<<","
45              <<itb->first.second<<"): "<<itb->second<<endl;
46      }
47
48      return 0;
49  }
```

Since class **nnMap** is a header-only class, we can compile the program easily using the following command:

`g++ -o nnmap -I../.. nnmap.cpp`

When the program is executed, we see the following output:

```
Test MapA
    element (2, 0) already existed with a value of 20
mA contains:
(0,1): 10
(0,2): 20
Test MapB
    value 40 is added to the map
mB contains:
(0,1): 10
(0,2): 20
(2,0): 40
```

From the output of the program we see that the two comparison classes work as expected.

The complete implementation of class **nnMap** is shown in Listing B.102.

Like class **Container** described in the previous section, class **nnMap** is a good example of C++ template programming.

11.3 The Dataset Adapters

A dataset to be clusterized can come from various sources. For example, a dataset might come from a text file, a database, or some data generator. In order to have a uniform interface to interact with all data sources, we design a dataset adapter class, which is an abstract base class.

The declaration of the dataset adapter class is shown in Listing 11.7. Class **DataAdapter** is very simple in that it contains only two member functions and no data members. One member function is a virtual destructor and another one is a pure virtual function.

Listing 11.7: Declaration of class **DataAdapter**.
```
1  class DataAdapter {
2  public:
3      virtual ~DataAdapter() {}
4      virtual void fill(boost::shared_ptr<Dataset> &ds) = 0;
5  };
```

Any classes derived from class **DataAdapter** are supposed to provide a concrete implementation of the function **fill**. In the next three subsections, we introduce three classes derived from class **DataAdapter**.

11.3.1 A CSV Dataset Reader

Most of the time we read datasets from CSV (Comma Separated Values) files. In a CSV data file, each row represents a record and each column represents an attribute, a record identifier, or a class label. We want a dataset reader class that can read a dataset from a CSV file and differentiate among an attribute, a record identifier, and a class label.

To design such a class, we follow the approach used in $\mathcal{MLC}$++ (Kohavi et al., 1998). That is, the class requires two files: the data file and the schema file. The data file stores the actual dataset and the schema file stores the column information. For example, suppose the Iris dataset (Fisher, 1936; Duda and Hart, 1973) is saved in a CSV file called **bezdekIris.data**. Then we also need to prepare a schema file called **bezdekIris.names** in order to use the CSV dataset reader.

The declaration of the resulting class **DatasetReader** is shown in Listing 11.8. This class has two public member functions, four private member functions, and six private data members. The constructor takes one argument, which is the filename of the data file. Another public member function **fill** overrides the pure virtual function of the base class with a concrete implementation of the function.

Listing 11.8: Declaration of class `DatasetReader`.
```
1   class DatasetReader : public DataAdapter {
2   public:
3       DatasetReader(const std::string& fileName);
4       void fill(boost::shared_ptr<Dataset>& ds);
5
6   private:
7       void createSchema();
8       void fillData();
9       boost::shared_ptr<Record> createRecord(
10          const std::vector<std::string>& val);
11
12      std::vector<std::string> split(const std::string&);
13
14      std::string _fileName;
15      Size _labelColumn;
16      Size _idColumn;
17      Size _numColumn;
18
19      boost::shared_ptr<Schema> _schema;
20      boost::shared_ptr<Dataset> _ds;
21  };
```

As we can see from the implementation of class `DatasetReader` in Listing B.99, function `fill` first calls function `createSchema` and then calls function `fillData`, which in turn calls function `createRecord` many times until all rows of the data file are read. Function `split` is used to split a string by commas.

Data member `_fileName` holds the name of the data file, including the path. Data member `_labelColumn` stores the index of the column representing class labels if such a column exists. Data member `_idColumn` stores the index of the column representing record identifiers if such a column exists. Data member `_numColumn` is the number of columns in the data file, including the label column and the identifier column. This data member is used to ensure that each row of the data file has the same number of columns.

Now we consider a C++ program that shows how to read a CSV data file shown in Listing 11.9. The corresponding schema file of the CSV data file is shown in Listing 11.10. In the schema file, we see a special symbol "///:" (line 2). In fact, this symbol is used to indicate the beginning of attribute information.

Listing 11.9: A CSV data file.
```
1   a, 1.2,3,1
2   b,-0.8,3,2
3   c, 3.0,4,1
4   d,1e-2,2,2
```

Listing 11.10: A schema file.
```
1   4 points
2   ///: schema
3   1, RecordID
```

```
4  2, Continuous
5  3, Discrete
6  4, Class
```

The program used to read the CSV data file is shown in Listing 11.11. The program first creates a new instance of class **DatasetReader** (line 12) and then fills the dataset. The code in lines 17–39 just outputs information about the dataset and the contents of the dataset.

Listing 11.11: Program to illustrate class **DatasetReader**.
```
1   // examples/datasetreader/datasetreader.cpp
2   #include<cl/utilities/datasetreader.hpp>
3   #include<iostream>
4   #include<sstream>
5   #include<iomanip>
6
7   using namespace ClusLib;
8   using namespace std;
9
10  int main() {
11      try {
12          DataAdapter *p = new DatasetReader("./4points.data");
13          boost::shared_ptr<Dataset> ds;
14          p->fill(ds);
15          delete p;
16
17          cout<<*ds<<endl;
18          cout<<"Data:\n";
19          cout<<setw(10)<<left<<"RecordID";
20          for(Size j=0; j<ds->num_attr(); ++j) {
21              stringstream ss;
22              ss<<"Attribute("<<j+1<<")";
23              cout<<setw(14)<<left<<ss.str();
24          }
25          cout<<setw(10)<<left<<"Label"<<endl;
26          for(Size i=0; i<ds->size(); ++i) {
27              cout<<setw(10)<<left<<(*ds)[i]->get_id();
28              for(Size j=0; j<ds->num_attr(); ++j) {
29                  boost::shared_ptr<Schema> schema = ds->schema();
30                  if((*schema)[j]->can_cast_to_c()) {
31                      cout<<setw(14)<<left
32                          <<(*schema)[j]->get_c_val((*ds)(i,j));
33                  } else {
34                      cout<<setw(14)<<left
35                          <<(*schema)[j]->get_d_val((*ds)(i,j));
36                  }
37              }
38              cout<<setw(10)<<left<<(*ds)[i]->get_label()<<endl;
39          }
40
41          return 0;
42      } catch (exception& e) {
43          cout<<e.what()<<endl;
44          return 1;
45      } catch (...) {
46          cout<<"unknown error"<<endl;
47          return 2;
48      }
49  }
```

To compile the program, we use the following command:

```
libtool --mode=compile g++ -c -I../.. -I/usr/include
datasetreader.cpp
```

To link the program, we use the following command:

```
libtool --mode=link g++ -o datasetreader ../../cl/libClusLib.la
datasetreader.o
```

When the program is executed, we see the following output:

```
Number of records: 4
Number of attributes: 2
Number of numerical attributes: 1
Number of categorical attributes: 1

Data:
RecordID   Attribute(1)   Attribute(2)   Label
0          1.2            0              0
1          -0.8           0              1
2          3              1              0
3          0.01           2              1
```

From the output we see that the record identifiers, the labels, and the values of the categorical attribute are replaced by integers. In fact, we treat record identifiers and class labels as categorical and use unsigned integers to represent all categorical values in $\mathcal{C}lus\mathcal{L}ib$. In Exercise A.7, readers are asked to output the original dataset.

Class `DatasetReader` described above was designed to read CSV files. In Exercise A.8, readers are asked to modify class `DatasetReader` to read space-delimited files.

The `DatasetReader` class is able to recognize four symbols (i.e., "RecordID", "Continuous", "Discrete", and "Class") in a schema file. In Exercise A.6, readers are asked to modify the class so that it can recognize an additional symbol "Ignore" and skip the column when reading a data file.

11.3.2 A Dataset Generator

Sometimes we need to test our clustering algorithms with synthetic datasets, which are not real-world data but generated by computer programs. In this section, we introduce a dataset generator that simulates random data points from a mixture of multivariate Gaussian distributions.

The declaration of the dataset generator class `DatasetGenerator` is shown in Listing 11.12. Class `DatasetGenerator` is derived from class `DataAdapter`. The class has six protected data members, two public member functions, and one private member function.

We use Boost random library to generate random numbers. The type of the variate generator is defined in lines 10–11. The random number generator is declared as a class member (line 21). Since the class `gen_type` does not have a default constructor, we have to initialize the generator in the initialization list of class `DatasetGenerator` (see Listing B.95).

Data member _seed is the seed used to set the state of the random number generator. If we use the same seed, we will get the same random numbers. The value of _seed must be a positive unsigned integer.

Listing 11.12: Declaration of class `DatasetGenerator`.

```
class DatasetGenerator: public DataAdapter {
public:
    DatasetGenerator(ublas::matrix<Real> mu,
            std::vector<ublas::symmetric_matrix<Real> > sigma,
            std::vector<Size> records,
            Size seed = 1);
    void fill(boost::shared_ptr<Dataset> &ds);

protected:
    typedef boost::variate_generator<boost::minstd_rand,
        boost::normal_distribution<> > gen_type;

    void generate(Size ind);
    ublas::matrix<Real> _data;

    Size _seed;
    std::vector<Size> _records;
    ublas::matrix<Real> _mu;
    std::vector<ublas::symmetric_matrix<Real> > _sigma;

    gen_type _generator;
};
```

Member _records is a vector of unsigned integers, each of which represents the number of data points that should be generated from a component of Gaussian distributions. Member _mu stores the means of all components of Gaussian distributions. Each row of _mu represents a mean of a component of Gaussian distributions. Similarly, member _sigma stores the covariance matrices of all components.

Member function `DatasetGenerator` is a constructor with four arguments, the last of which has a default value of 1. These arguments are used to initialize the corresponding data members. The public member function `fill` overrides the pure virtual function of the base class with a concrete implementation of the function.

The program in Listing 11.13 shows how to use class `DatasetGenerator`. The program first tries to get the number of points from the argument of the program if the user executes the program with an argument (lines 12–17). Then the program initializes the means, covariances, and the number of records. Finally the program creates an object of class `DatasetGenerator` and fills the dataset. The code in lines 35–61 just outputs the dataset.

Listing 11.13: Program to illustrate class `DatasetGenerator`.

```
// examples/datasetgenerator/datasetgenerator.cpp
#include<cl/utilities/datasetgenerator.hpp>
#include<iostream>
#include<sstream>
#include<iomanip>

using namespace ClusLib;
using namespace std;

```

```
10  int main(int argc, char* argv[]) {
11     try{
12         stringstream ss;
13         Size N=60;
14         if(argc>1) {
15             ss<<argv[1];
16         }
17         ss>>N;
18         ublas::matrix<Real> mu(3,2);
19         vector<ublas::symmetric_matrix<Real> > sigma(3);
20         vector<Size> records(3,N);
21         mu(0,0) = 0.0; mu(0,1) = 0.0;
22         mu(1,0) = 10.0; mu(1,1) = 0.0;
23         mu(2,0) = 5.0; mu(2,1) = 5.0;
24         for(Size i=0; i<3; ++i) {
25             sigma[i].resize(2);
26             sigma[i](0,0) = 1.0;
27             sigma[i](1,0) = −0.5 + i*0.5;
28             sigma[i](1,1) = 1.0;
29         }
30         DataAdapter *p = new DatasetGenerator(mu, sigma, records);
31         boost::shared_ptr<Dataset> ds;
32         p->fill(ds);
33         delete p;
34
35         ss.clear();
36         ss.str("");
37         ss<<3*N<<"points.csv";
38         cout<<ss.str()<<endl;
39         ds->save(ss.str());
40         cout<<*ds<<endl;
41         cout<<"Data:_\n";
42         cout<<setw(10)<<left<<"RecordID";
43         for(Size j=0; j<ds->num_attr(); ++j) {
44             stringstream ss;
45             ss<<"Attribute("<<j+1<<")";
46             cout<<setw(14)<<left<<ss.str();
47         }
48         cout<<setw(10)<<left<<"Label"<<endl;
49         for(Size i=0; i<ds->size(); ++i) {
50             cout<<setw(10)<<left<<(*ds)[i]->get_id();
51             for(Size j=0; j<ds->num_attr(); ++j) {
52                 boost::shared_ptr<Schema> schema = ds->schema();
53                 if((*schema)[j]->can_cast_to_c()) {
54                     cout<<setw(14)<<left
55                         <<(*schema)[j]->get_c_val((*ds)(i,j));
56                 } else {
57                     cout<<setw(14)<<left
58                         <<(*schema)[j]->get_d_val((*ds)(i,j));
59                 }
60             }
61             cout<<setw(10)<<left<<(*ds)[i]->get_label()<<endl;
62         }
63
64         return 0;
65     } catch (exception& e) {
66         cout<<e.what()<<endl;
67         return 1;
68     } catch (...) {
69         cout<<"unknown_error"<<endl;
70         return 2;
71     }
72  }
```

We can compile and link the program in the same way as we did for the dataset reader program introduced in the previous subsection. When we execute the program using the following command (in Cygwin):

./datasetgenerator.exe 3

we will see the following output:

```
9points.csv
Number of records: 9
Number of attributes: 2
Number of numerical attributes: 2
Number of categorical attributes: 0

Data:
RecordID   Attribute(1)   Attribute(2)    Label
0          0.421584       -0.210741       0
1          -1.69489       -0.238166       0
2          0.635494       -0.430077       0
3          8.99697        -0.0946602      1
4          9.86637        1.64423         1
5          11.0844        0.683895        1
6          3.41649        3.42911         2
7          3.99683        3.93227         2
8          7.32906        4.1885          2
```

We executed the program with an argument of 3. The program then generated three points from each of the three components. Figure 11.1 shows the nine points. When we execute the program with an argument of 200, the program will generate 200 points from each of the three components. Figure 1.1 (in Section 1.1) shows the 600 points generated by the program.

11.3.3 A Dataset Normalizer

In this subsection, we introduce a dataset normalizer class, which is used to standardize numeric datasets or the numeric part of mixed-type datasets. The dataset normalizer class implements a very simple data standardization method. Let $X = \{\mathbf{x}_1, \mathbf{x}_2, \cdots, \mathbf{x}_n\}$ be a dataset containing n records, each of which is described by d attributes. Then each numeric attribute j of X will be normalized to range $[0, 1]$ according to the following formula:

$$x'_{ij} = \frac{x_{ij} - R_{min}(j)}{R_{max}(j) - R_{min}(j)}, \quad i = 1, 2, \cdots, n,$$

where $R_{min}(j)$ and $R_{max}(j)$ are the minimum and maximum values of the jth attribute, respectively, and x_{ij} is the jth component of $\mathbf{x}_i$.

The declaration of the dataset normalizer class is shown in Listing 11.14. The class also inherits from the base dataset adapter class. The constructor

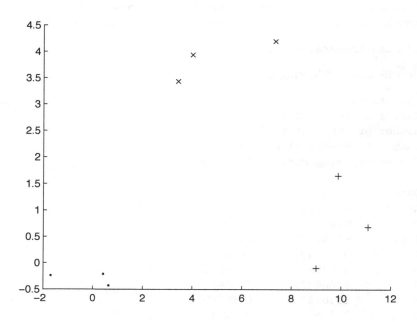

FIGURE 11.1: A generated dataset with 9 points.

of the class (line 3) takes one argument, which is a shared pointer pointing to a dataset to be normalized.

Class `DatasetNormalizer` has three protected data members and one protected member function. Data members `_dvMin` and `_dvMax` are used to hold the minimum and maximum values of numeric attributes of the dataset. Data member `_ods` holds a shared pointer pointing to the original dataset, i.e., the dataset to be normalized.

Method `get_minmax` is used only to find the minimum and maximum values of all numeric attributes. Method `fill` creates a new dataset that contains normalized values of the original dataset. All categorical attributes of the original dataset are untouched during the normalization process. The complete implementation of these functions can be found in Listing B.97.

Listing 11.14: Declaration of class `DatasetNormalizer`.

```
1  class DatasetNormalizer : public DataAdapter {
2  public:
3      DatasetNormalizer(const boost::shared_ptr<Dataset> &ds);
4      void fill(boost::shared_ptr<Dataset> &ds);
5
6  protected:
7      void get_minmax();
8
9      std::vector<Real> _dvMin;
10     std::vector<Real> _dvMax;
```

```
11      boost::shared_ptr<Dataset> _ods;
12  };
```

11.4 The Node Visitors

The node visitor classes are utility classes that are used to extract information from a hierarchical clustering, which is a tree of nodes. In this section, we introduce two node visitor classes: class JoinValueVisitor and class PCVisitor.

The declaration of the base class for all visitor classes is shown in Listing 11.15. Class NodeVisitor contains only two public member functions, which are pure virtual functions. The two member functions have the same name but different types of argument. One function is used to visit a LeafNode object and the other one is used to visit an InternalNode object.

Listing 11.15: Declaration of class NodeVisitor.
```
1   class LeafNode;
2   class InternalNode;
3
4   class NodeVisitor {
5   public:
6       virtual void visit(LeafNode& node) = 0;
7       virtual void visit(InternalNode& node) = 0;
8   };
```

We also note that we declare classes LeafNode and InternalNode (lines 1–2) before we declare class NodeVisitor. In C++, this is called forward declarations. Since the header node.hpp of class Node (see Listing B.85) already includes the header nodevisitor.hpp of class NodeVisitor, we cannot include the header node.hpp of class Node in the header nodevisitor.hpp. Doing so results in a cyclic inclusion of headers. However, class NodeVisitor depends on classes LeafNode and InternalNode. To solve this problem, we declare classes LeafNode and InternalNode here.

11.4.1 The Join Value Visitor

Class JoinValueVisitor is a child class of the base class NodeVisitor and is used to extract the join values of a hierarchical clustering. The declaration of class JoinValueVisitor is shown in Listing 11.16, from which we see that class JoinValueVisitor contains a friend function, three public member functions, one private member function, and a private data member.

Data member _joinValues is a set of values, which have the same type as the values in class iirMapA (see Listing B.102). In fact, a value of type iirMapA::value_type contains the indexes of two clusters that are merged

and their join value. These values are ordered by the comparison class
compare_iir defined in the header **nnmap.hpp**.

The overloaded operator "<<" is a friend function of the class. The friend function can access all the members of the class, including private members. As we can see from the implementation of the class in Listing B.88, the friend class called the private function **print** of the class.

Listing 11.16: Declaration of class `JoinValueVisitor`.

```
class JoinValueVisitor : public NodeVisitor {
public:
    friend std::ostream& operator<<(std::ostream& os,
             const JoinValueVisitor& jv);
    void visit(LeafNode& node);
    void visit(InternalNode& node);
    const std::set<iirMapA::value_type, compare_iir>&
        get_joinValues() const;

private:
    void print(std::ostream& os) const;

    std::set<iirMapA::value_type, compare_iir> _joinValues;
};
```

11.4.2 The Partition Creation Visitor

The partition creation visitor class `PCVisitor` is also a child class of the base class `NodeVisitor` and is used to extract a partition from a hierarchical clustering. Listing 11.17 shows the declaration of the class.

Class `PCVisitor` depends on class `CVisitor`, which is used to create a cluster from a subtree of a hierarchical clustering. The class creates a cluster by including all the leaf nodes of the subtree.

Listing 11.17: Declaration of class `JoinValueVisitor`.

```
class CVisitor : public NodeVisitor {
public:
    CVisitor();
    void visit(LeafNode& node);
    void visit(InternalNode& node);
    boost::shared_ptr<Cluster> get_cluster();

private:
    boost::shared_ptr<Cluster> _cluster;
};

class PCVisitor : public NodeVisitor {
public:
    PCVisitor(PClustering &pc, Size cutlevel);
    void visit(LeafNode& node);
    void visit(InternalNode& node);

private:
    PClustering &_pc;
    Size _cutlevel;
};
```

Class `PCVisitor` has three public member functions and two private data members. Data member _pc is a reference to an object of class PClustering. Since the member is a reference, we have to initialize it in the initialization list of the class (see Listing B.90). Another data member _cutlevel is used to determine the number of clusters that should be created from a hierarchical clustering. The higher the number, the smaller the number of clusters. If we set _cutlevel to be zero, then we get clusters, each of which contains only one record. If we set _cutlevel to be a big number, we might get a single cluster containing all the records.

11.5 The Dendrogram Class

The Dendrogram class is a class used to produce EPS (Encapsulated PostScript) figures. An EPS file is a PostScript document that describes an image or drawing and can be placed within another PostScript document.

An EPS file is essentially a text file. For example, Figure 11.2 shows a drawing from an EPS file. If we open the EPS file using some text editor, then we see the contents of the file as shown in Listing 11.18. In Listing 11.18, lines 1–34 are just headers of the EPS drawing. Lines 36–43 define two dots and a line as shown in Figure 11.2. For more information about EPS file format, readers are referred to the book by Casselman (2004).

FIGURE 11.2: An EPS figure.

Listing 11.18: An EPS file.

```
1  %!PS−Adobe−2.0  EPSF−2.0
2  %%Title:  a  line  and  two  dots
3  %%Creator:  ClusLib
```

```
 4  %%CreationDate: June 23, 2010
 5  %%BoundingBox: 100 100 300 300
 6  %Magnification: 1.0000
 7  %%EndComments
 8
 9  /cp {closepath} bind def
10  /ef {eofill} bind def
11  /gr {grestore} bind def
12  /gs {gsave} bind def
13  /sa {save} bind def
14  /rs {restore} bind def
15  /l {lineto} bind def
16  /m {moveto} bind def
17  /rm {rmoveto} bind def
18  /n {newpath} bind def
19  /s {stroke} bind def
20  /sh {show} bind def
21  /slc {setlinecap} bind def
22  /slj {setlinejoin} bind def
23  /slw {setlinewidth} bind def
24  /srgb {setrgbcolor} bind def
25  /rot {rotate} bind def
26  /sc {scale} bind def
27  /sd {setdash} bind def
28  /ff {findfont} bind def
29  /sf {setfont} bind def
30  /scf {scalefont} bind def
31  /sw {stringwidth} bind def
32  /sd {setdash} bind def
33  /tr {translate} bind def
34    0.5 setlinewidth
35
36  % Dot
37  3 slw   1 slc   0 slj n 240 125.2 m 240 125.2 l 0 0 0 srgb stroke
38
39  % Dot
40  3 slw   1 slc   0 slj n 240 250.2 m 240 250.2 l 0 0 0 srgb stroke
41
42  % Line
43  0.5 slw   1 slc   0 slj n 240 125.2 m 240 250.2 l 0 0 0 srgb stroke
44
45  showpage
46  %%Trailer
47  %%EOF
```

In fact, PostScript is also a programming language. However, PostScript files are usually produced by other programs rather than manually. In this section, we introduce a C++ class that can be used to generate dendrograms in EPS files. The declaration of the resulting class is shown in Listing 11.19.

Class **Dendrogram** has five private data members and seven public member functions. Data member _ss is a string stream used to store the EPS drawing. The other four data members are real numbers, which represent the coordinates of the lower left corner and the upper right corner of the rectangular box that covers the drawing.

Member function **setbox** is used to set the values of the four private data members. Member function **drawDot**, as its name implies, is used to draw a dot at the location specified by the arguments of the function. Other member functions such as **drawCircle**, **drawLine**, and **drawText** are used to draw other shapes. Member function **save** is used to save the drawing into an EPS file. Listing B.101 shows the complete implementation of the class.

Listing 11.19: Declaration of class **Dendrogram**.

```
class Dendrogram {
public:
    Dendrogram();
    void setbox(Real x1, Real y1, Real x2, Real y2);
    void drawDot(Real x, Real y);
    void drawCircle(Real x, Real y, Real r);
    void drawLine(Real x1, Real y1, Real x2, Real y2);
    void drawText(Real x, Real y, const std::string&txt);
    void save(const std::string &filename) const;

private:
    std::stringstream _ss;
    Real _x1;
    Real _y1;
    Real _x2;
    Real _y2;
};
```

We can use PostScript to draw color images. In our program, we draw only black and white images. In RGB color code, black is represented by $(0, 0, 0)$. For example, the dendrogram shown in Figure 1.5 was generated by our program.

11.6 The Dendrogram Visitor

Class **Dendrogram** described in the previous section provides functionalities to draw dendrograms. However, it does not provide functionalities to draw dendrograms from hierarchical clusterings. To do this, we create a dendrogram visitor class that is able to draw a dendrogram based on a hierarchical clustering.

The declaration of the resulting class is shown in Listing 11.20. This class is a child class of the visitor base class **NodeVisitor**. The constructor of the class takes three arguments. The first argument hjv is the join value (or height) of the root node of the hierarchical clustering tree. The second and the third arguments are used to calculate the maximum number of nodes to show in the dendrogram.

Listing 11.20: Declaration of class **DendrogramVisitor**.

```
class DendrogramVisitor : public NodeVisitor {
public:
    DendrogramVisitor(Real hjv,
        Size llevel, Size hlevel);
    void visit(LeafNode& node);
    void visit(InternalNode& node);
    void save(const std::string &filename);

private:
    Dendrogram _dg;
    Size _cutlevel;
    Size _count;
    Real _leftMargin;
```

```
14      Real _bottomMargin;
15      Real _boxx;
16      Real _boxy;
17      Real _height;
18      Real _width;
19      Real _hjv;
20      Real _gap;
21      bool _drawLabel;
22      std::map<Size, std::pair<Size, Size> > _lines;
23      std::map<Size, std::pair<Real, Real> > _points;
24
25      Real get_x(Size id);
26      void drawLink(Size id0, Size id1);
27  };
```

In the dendrogram produced by the program, a leaf node is denoted by a circle and an internal node is denoted by a dot. If the number of nodes of the dendrogram does not exceed 60, then the program draws a label for each node. The detail implementation of class `DendrogramVisitor` is shown in Listing B.80.

Figure 11.3 shows a dendrogram with 100 nodes. This dendrogram was produced from the results of the single linkage algorithm applied to the dataset shown in Figure 1.1. The dataset contains 600 records. Figure 11.3 shows only 100 nodes.

Figure 11.4 shows a dendrogram from the same clustering but with only 50 nodes. Labels for all the nodes are displayed in the dendrogram. Since the dataset contains 600 points, the labels in range $[0, 599]$ correspond to leaf nodes of the hierarchical clustering tree. Labels in range $[600, 1198]$ correspond to internal nodes. We come back to this later in Chapter 12.

11.7 Summary

In this section, we introduced several utility classes. These classes give us very good examples of static polymorphism and dynamic polymorphism supported by the C++ programming language. The container class and the double-key map class are template classes, which can be used for many types of data. The data adapter classes and the node visitor classes are examples of dynamic polymorphism, which is achieved by using virtual functions.

Unlike the machine learning C++ library $\mathcal{MLC}$++, which uses many utility classes developed for the library, the clustering library $\mathcal{ClusLib}$ uses only a few utility classes described in this chapter since $\mathcal{ClusLib}$ uses many classes from the Boost libraries. Without the Boost libraries, we would have to develop smart pointer classes, heterogeneous container classes, and matrix classes, to name just a few.

FIGURE 11.3: A dendrogram that shows 100 nodes.

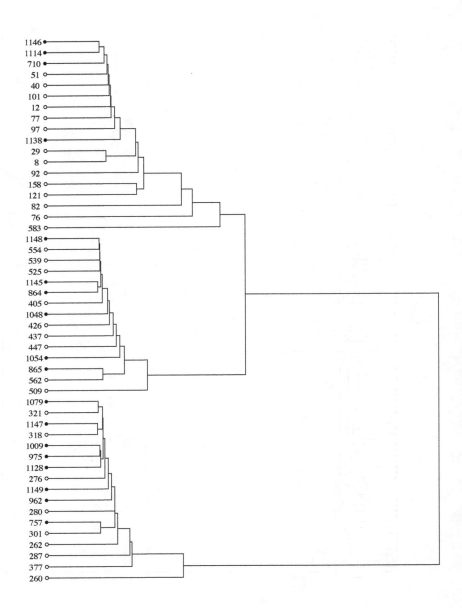

FIGURE 11.4: A dendrogram that shows 50 nodes.

Part III
Data Clustering Algorithms

Chapter 12

Agglomerative Hierarchical Algorithms

A hierarchical clustering algorithm is a clustering algorithm that divides a dataset into a sequence of nested partitions. Hierarchical clustering algorithms can be classified into two categories: agglomerative hierarchical algorithms and divisive hierarchical algorithms (Gan et al., 2007).

An agglomerative hierarchical algorithm starts with every single record in a single cluster and repeats merging the closest pair of clusters according to some similarity or dissimilarity measure until all records are in one cluster. In contrast to an agglomerative hierarchical algorithm, a divisive hierarchical algorithm starts with all records in one cluster and repeats splitting a cluster into two smaller ones until all clusters contain only a single record.

In this chapter, we introduce the implementation of several agglomerative hierarchical algorithms. In particular, we focus on agglomerative hierarchical algorithms that use the Lance-Williams recurrence formula (Lance and Williams, 1967a,b).

12.1 Description of the Algorithm

Let $\{\mathbf{x}_0, \mathbf{x}_1, \cdots, \mathbf{x}_{n-1}\}$ be a dataset with n records. An agglomerative hierarchical algorithm starts with every single record as a cluster. Let $C_i = \{\mathbf{x}_i\}$, $i = 0, 2, \cdots, n-1$ be the n clusters at the beginning. Cluster $C_i (0 \le i \le n-1)$ contains the record $\mathbf{x}_i$. At each step, two clusters that have the minimum distance are merged to form a new cluster. An agglomerative hierarchical algorithm continues merging clusters until only one cluster is left.

For convenience, we assume that at step 1, a new cluster C_n is formed by merging two clusters in the initial set of clusters $\mathcal{F}_0 = \{C_0, C_1, \cdots, C_{n-1}\}$. Then after step 1 and before step 2, we have a set of clusters $\mathcal{F}_1 = \tilde{\mathcal{F}}_0 \cup \{C_n\}$, where $\tilde{\mathcal{F}}_0$ is the set of unmerged clusters in $\mathcal{F}_0$. If C_0 and C_1 have the minimum distance among all pairs of clusters, for example, then $C_n = C_0 \cup C_1$ and $\tilde{\mathcal{F}}_0 = \{C_2, C_3, \cdots, C_{n-1}\} = \mathcal{F}_0 \setminus \{C_0, C_1\}$.

At step 2, a new cluster C_{n+1} is formed by merging two clusters in the set of clusters $\mathcal{F}_1$. Similarly, we let $\tilde{\mathcal{F}}_1$ be the set of unmerged clusters in $\mathcal{F}_1$. Then after step 2 and before step 3, we have a set of clusters $\mathcal{F}_2 = \tilde{\mathcal{F}}_1 \cup \{C_{n+1}\}$. The algorithm continues this process until at step $n-1$ when the last two clusters are merged to form the cluster C_{2n-2}. After step $n-1$, we have $\mathcal{F}_{n-1} = \{C_{2n-2}\}$, which contains only one cluster. The algorithm stops after step $n-1$.

In the above process, we have $|\mathcal{F}_0| = n$, $|\mathcal{F}_1| = n-1, \cdots$, and $|\mathcal{F}_{n-1}| = 1$, where $|\cdot|$ denotes the number of elements in the set. At each step, the algorithm merges two clusters. To decide which two clusters to merge, we need to calculate the distances between clusters. Lance and Williams (1967a) proposed a recurrence formula to compute the distance between an old cluster and a new cluster formed by two old clusters.

The Lance-Williams formula is defined as follows. Before step i ($1 \leq i < n-1$), we have a set of clusters $\mathcal{F}_{i-1}$, which contains $n-i+1$ clusters. Suppose cluster C_{i_1} and C_{i_2} have the smallest distance among all the pairs of clusters in $\mathcal{F}_{i-1}$. Then C_{i_1} and C_{i_2} will be merged to form the cluster C_{n+i-1}. The Lance-Williams formula computes the distance between an old cluster $C \in \tilde{\mathcal{F}}_{i-1} = \mathcal{F}_{i-1} \setminus \{C_{i_1}, C_{i_2}\}$ as

$$\begin{aligned} D(C, C_{n+i-1}) &= D(C, C_{i_1} \cup C_{i_2}) \\ &= \alpha_{i_1} D(C, C_{i_1}) + \alpha_{i_2} D(C, C_{i_2}) + \beta D(C_{i_1}, C_{i_2}) \\ &\quad + \gamma |D(C, C_{i_1}) - D(C, C_{i_2})|, \end{aligned} \qquad (12.1)$$

where α_{i_1}, α_{i_2}, β, and γ are parameters. DuBien and Warde (1979) investigated some properties of the Lance-Williams formula.

Algorithm	α_{i_1}	α_{i_2}	β	γ
Single linkage	$\frac{1}{2}$	$\frac{1}{2}$	0	$-\frac{1}{2}$
Complete linkage	$\frac{1}{2}$	$\frac{1}{2}$	0	$\frac{1}{2}$
Group average	$\frac{|C_{i_1}|}{|C_{i_1}|+|C_{i_2}|}$	$\frac{|C_{i_2}|}{|C_{i_1}|+|C_{i_2}|}$	0	0
Weighted group average	$\frac{1}{2}$	$\frac{1}{2}$	0	0
Centroid	$\frac{|C_{i_1}|}{|C_{i_1}|+|C_{i_2}|}$	$\frac{|C_{i_2}|}{|C_{i_1}|+|C_{i_2}|}$	$-\frac{|C_{i_1}|\cdot|C_{i_2}|}{(|C_{i_1}|+|C_{i_2}|)^2}$	0
Median	$\frac{1}{2}$	$\frac{1}{2}$	0	$-\frac{1}{4}$
Ward's method	$\frac{|C|+|C_{i_1}|}{\Sigma}$	$\frac{|C|+|C_{i_2}|}{\Sigma}$	$-\frac{|C|}{\Sigma}$	0

TABLE 12.1: Parameters for the Lance-Williams formula, where $\Sigma = |C| + |C_{i_1}| + |C_{i_2}|$.

Table 12.1 gives seven sets of parameters for the Lance-Williams formula defined in Equation (12.1). Each set of parameters results in an agglomerative hierarchical clustering algorithm. A more general recurrence formula was proposed by Jambu (1978) and discussed in (Gordon, 1996) and (Gan et al., 2007).

The first four algorithms (i.e., single linkage, complete linkage, group average, and weighted group average) are referred to as graph hierarchical methods (Murtagh, 1983). The last three algorithms (i.e., centroid, median, and Ward's method) are referred to as geometric hierarchical methods. The last three algorithms require squared Euclidean distance in the Lance-Williams formula. For geometric hierarchical algorithms, the centers of a cluster formed by merging two clusters can be calculated from the centers of the two merged clusters. In addition, the distance between two clusters can be calculated from the distance between centers of the two clusters. Table 12.2 shows the calculation of centers and distances of clusters for the geometric hierarchical algorithms.

Algorithm	$\mu(C_1 \cup C_2)$	$D(C_1, C_2)$
Centroid	$\frac{|C_1|\mu(C_1)+|C_2|\mu(C_2)}{|C_1|+|C_2|}$	$D_{euc}(\mu(C_1), \mu(C_2))^2$
Median	$\frac{\mu(C_1)+\mu(C_2)}{2}$	$D_{euc}(\mu(C_1), \mu(C_2))^2$
Ward's	$\frac{|C_1|\mu(C_1)+|C_2|\mu(C_2)}{|C_1|+|C_2|}$	$\frac{|C_1|\cdot|C_2|}{|C_1|+|C_2|}D_{euc}(\mu(C_1), \mu(C_2))^2$

TABLE 12.2: Centers of combined clusters and distances between two clusters for geometric hierarchical algorithms, where $\mu(\cdot)$ denotes a center of a cluster and $D_{euc}(\cdot, \cdot)$ is the Euclidean distance.

A hierarchical clustering algorithm is said to be monotonic if at each step we have

$$D(C, C_1 \cup C_2) \geq D(C_1, C_2), \qquad (12.2)$$

where C_1 and C_2 are the two clusters to be merged and C is an unmerged cluster. The single linkage algorithm and the complete linkage algorithm are monotonic (Johnson, 1967). However, other agglomerative hierarchical algorithms might violate the monotonic inequality (Milligan, 1979).

12.2 Implementation

To implement the agglomerative hierarchical algorithms that use the Lance-Williams formula, we use the template method design pattern. The UML class diagram of these agglomerative hierarchical algorithms is shown in Figure 12.1. Class LW is an abstract class that contains a template method (i.e., performClustering) that calls several other functions, including the pure virtual function update_dm. Each child class of class LW is a concrete class that provides concrete implementation of the function update_dm.

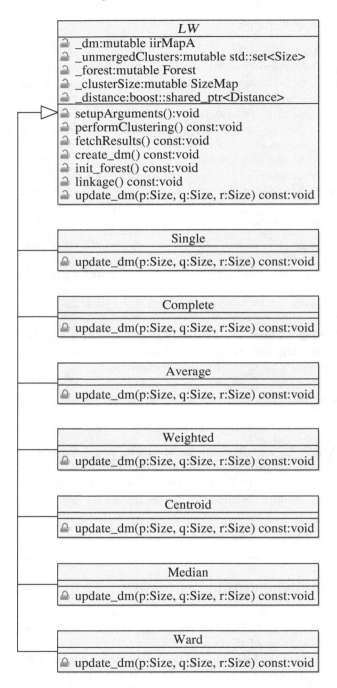

FIGURE 12.1: Class diagram of agglomerative hierarchical algorithms.

The declaration of class LW is shown in Listing 12.1. Class LW inherits from the base class Algorithm. In additional to the data members inherited from the base class, class LW declares another five data members, which are protected. Four of these data members are declared mutable and hence can be changed by the const functions.

Data member _dm is a double-key map (see Section 11.2), which is used to hold the pair-wise distances between clusters. Here the order of the two components of a key does not matter since we consider only symmetric distances.

Data member _unmergedClusters is a set of unsigned integers, which are the indexes of unmerged clusters. At the beginning, this data member contains all single-record clusters. The number of elements in the data member decreases as the algorithm proceeds. At the end, the data member becomes an empty set.

Listing 12.1: Declaration of class LW.

```
class LW: public Algorithm {
protected:
    typedef std::map<Size, boost::shared_ptr<HClustering> >
        Forest;
    typedef std::map<Size, Size> SizeMap;

    void setupArguments();
    void performClustering() const;
    void fetchResults() const;
    virtual void create_dm() const;
    virtual void init_forest() const;
    virtual void linkage() const;
    virtual void update_dm(Size p, Size q, Size r)
        const = 0;

    mutable iirMapA _dm;
    mutable std::set<Size> _unmergedClusters;
    mutable Forest _forest;
    mutable SizeMap _clusterSize;
    boost::shared_ptr<Distance> _distance;
};
```

Data member _forest is a map, which contains all the hierarchical trees generated by the algorithm. At the beginning, each cluster contains a single record and each tree contains a single cluster. When two clusters are merged, the trees that contain the two clusters are joined to form a bigger tree. At the end, a tree that contains all the clusters is produced by the algorithm.

Data member _clusterSize is a map used to keep track of the size of clusters or trees when the algorithm proceeds. This piece of information is used by the Lance-Williams formula to compute distances between clusters. By keeping track of this information, we do not have to calculate the number of elements in a cluster by traversing the hierarchical tree corresponding to the cluster.

Data member _distance is a shared pointer pointing to an object of some distance class. If the algorithm is provided with a dataset, a distance measure is required to compute the pair-wise distances between records. If the algorithm is provided with a similarity or distance matrix, a distance measure is not necessary. In our implementations, we assume that a dataset is provided to the algorithm. In Exercise A.9, we ask readers to modify the algorithm to take a similarity or distance matrix as input.

Class LW overrides several member functions (e.g., performClustering) of the base class and declares several new member functions (e.g., create_dm). The definitions of these member functions can be found in Listing B.33.

Member function setupArguments sets up the arguments (e.g., the dataset) for the algorithm. Member function performClustering calls the other three member functions: create_dm, init_forest, and linkage. Function linkage calls another member function update_dm, which is a pure virtual function to be implemented by derived classes of the base class.

The implementation of function create_dm is shown in Listing 12.2. This function calculates pair-wise distances between records in the dataset and populates the data member _dm with the distances. Since the distance measure is symmetric, we need to calculate the distance between two records only once.

Note that the distance measure and the dataset in the class are shared pointers. We need to dereference them before using them. For example, we use the operator "*" (line 6) to dereference the distance and the dataset. Also note that the operator "[ ]" has a higher precedence than the operator "*". Hence we put brackets around *_ds.

Listing 12.2: Implementation of function create_dm.

```
void LW::create_dm() const {
    Size n = _ds->size();
    for(Size i=0;i<n-1;++i){
        for(Size j=i+1;j<n;++j){
            _dm.add_item(i, j,
                (*_distance)((*_ds)[i],(*_ds)[j]));
        }
    }
}
```

Listing 12.3 shows the implementation of function init_forest. This function creates a leaf node for each single-record cluster and creates objects of class Hclustering to hold the leaf nodes. At this stage, the size of each cluster is one and all the clusters are unmerged. The identifier of a leaf node is set to the index of the record contained in the leaf node (line 5). In addition, the function assigns level zero to each leaf node (line 6).

Listing 12.3: Implementation of function init_forest.

```
void LW::init_forest() const {
    Size n = _ds->size();
    for(Size s=0;s<n;++s){
        boost::shared_ptr<Node> pln(new
            LeafNode((*_ds)[s], s));
        pln->set_level(0);
        boost::shared_ptr<HClustering> phc(new
            HClustering(pln));
        _forest.insert(Forest::value_type(s, phc));
        _clusterSize.insert(SizeMap::value_type(s,1));
        _unmergedClusters.insert(s);
    }
}
```

The definition of member function `linkage` is shown in Listing 12.4. This function loops through $n-1$ steps and merges two clusters that have the minimum distance at each step. The code in lines 7–20 is used to find the pair of clusters that have the minimum distance.

Listing 12.4: Implementation of function `linkage`.

```
void LW::linkage() const {
    Size n = _ds->size();
    std::set<Size>::iterator it;
    Real dMin, dTemp;
    Size m, s1, s2;
    for(Size s=0;s<n-1;++s){
        dMin = MAX_REAL;
        std::vector<Integer> nvTemp(_unmergedClusters.begin(),
            _unmergedClusters.end());
        m = nvTemp.size();
        for(Size i=0;i<m;++i) {
            for(Size j=i+1;j<m;++j){
                dTemp = _dm(nvTemp[i],nvTemp[j]);
                if(dTemp < dMin) {
                    dMin = dTemp;
                    s1 = nvTemp[i];
                    s2 = nvTemp[j];
                }
            }
        }
        boost::shared_ptr<Node> node =
            _forest[s1]->joinWith(*_forest[s2],dMin);
        node->set_id(n+s);
        node->set_level(s+1);
        boost::shared_ptr<HClustering> phc =
            boost::shared_ptr<HClustering>(new
            HClustering(node));
        _forest.insert(Forest::value_type(n+s, phc));
        _clusterSize.insert(SizeMap::value_type(n+s,
            _clusterSize[s1]+_clusterSize[s2]));
        _unmergedClusters.erase(s1);
        _unmergedClusters.erase(s2);
        _unmergedClusters.insert(n+s);
        update_dm(s1, s2, n+s);
    }
}
```

Once the two clusters are merged, the trees that hold the two clusters are joined to form a new tree. The function creates an object of class `HClustering` (lines 25–27) to hold the new tree and stores the tree in the data member _forest (line 28). Then the two clusters are removed from the set of unmerged clusters and the new cluster is added to the set of unmerged clusters. Finally, the function `update_dm` is called to update the distances between the new cluster and other unmerged clusters.

In class `LW`, the function `update_dm` is a pure virtual function. In following subsections, we introduce the implementations of the function `update_dm` in derived classes of class `LW`.

The function `update_dm` takes three arguments, which are unsigned integers. The first two arguments are the indexes of the two clusters to be mearged and the third argument is the index of the new cluster formed by the two clusters.

12.2.1 The Single Linkage Algorithm

According to the Lance-Williams recurrence formula, the single linkage algorithm calculates the distance between a new cluster C_{n+i-1} formed at step i and an old cluster C as

$$\begin{aligned} D(C, C_{n+i-1}) &= D(C, C_{i_1} \cup C_{i_2}) \\ &= \frac{1}{2}D(C, C_{i_1}) + \frac{1}{2}D(C, C_{i_2}) - \frac{1}{2}|D(C, C_{i_1}) - D(C, C_{i_2})| \\ &= \min\{D(C, C_{i_1}), D(C, C_{i_2})\}, \end{aligned} \quad (12.3)$$

where C_{i_1} and C_{i_2} are the two clusters merged at step i.

The single linkage algorithm is designed as a class derived from class LW. As we see from the header of class Single, the class overrides only the function update_dm, which is defined in Listing 12.5. This function calculates the distances between the unmerged clusters and the new cluster and stores the distances in the data member _dm.

Listing 12.5: Implementation of function update_dm in class Single.

```
void Single::update_dm(Size p, Size q, Size r) const {
    Real dist;
    std::set<Size>::iterator it;
    for(it = _unmergedClusters.begin();
        it != _unmergedClusters.end(); ++it) {
        if(*it == r) {
            continue;
        }

        dist = std::min(_dm(p,*it), _dm(q,*it));
        _dm.add_item(r,*it, dist);
    }
}
```

12.2.2 The Complete Linkage Algorithm

The complete linkage algorithm calculates the distance between a new cluster C_{n+i-1} formed at step i and an old cluster C as

$$\begin{aligned} D(C, C_{n+i-1}) &= D(C, C_{i_1} \cup C_{i_2}) \\ &= \frac{1}{2}D(C, C_{i_1}) + \frac{1}{2}D(C, C_{i_2}) + \frac{1}{2}|D(C, C_{i_1}) - D(C, C_{i_2})| \\ &= \max\{D(C, C_{i_1}), D(C, C_{i_2})\}, \end{aligned} \quad (12.4)$$

where C_{i_1} and C_{i_2} are the two clusters merged at step i.

The implementation of the function update_dm in class Complete is very similar to the function in class Single. Listing 12.6 shows the definition of this function in class Complete.

Listing 12.6: Implementation of function update_dm in class Complete.

```
void Complete::update_dm(Size p, Size q, Size r)
    const {
    Real dist;
    std::set<Size>::iterator it;
    for(it = _unmergedClusters.begin();
        it != _unmergedClusters.end(); ++it) {
        if(*it == r) {
            continue;
        }

        dist = std::max(_dm(p,*it), _dm(q,*it));
        _dm.add_item(r,*it,dist);
    }
}
```

12.2.3 The Group Average Algorithm

The group average algorithm calculates the distance between a new cluster C_{n+i-1} formed at step i and an old cluster C as

$$\begin{aligned}
& D(C, C_{n+i-1}) \\
= & D(C, C_{i_1} \cup C_{i_2}) \\
= & \frac{|C_{i_1}|}{|C_{i_1}| + |C_{i_2}|} D(C, C_{i_1}) + \frac{|C_{i_2}|}{|C_{i_1}| + |C_{i_2}|} D(C, C_{i_2}) \\
= & \frac{|C_{i_1}| \cdot D(C, C_{i_1}) + |C_{i_2}| \cdot D(C, C_{i_2})}{|C_{i_1}| + |C_{i_2}|},
\end{aligned} \qquad (12.5)$$

where C_{i_1} and C_{i_2} are the two clusters merged at step i and $|\cdot|$ denotes the number of elements in the underlying set.

Listing 12.7 shows the implementation of the function updated_dm in class Average. The code in lines 11–12 calculates the distance between the new cluster and an unmerged or old cluster as defined in Equation (12.5).

Listing 12.7: Implementation of function update_dm in class Average.

```
void Average::update_dm(Size p, Size q, Size r)
    const {
    Real dist;
    std::set<Size>::iterator it;
    for(it = _unmergedClusters.begin();
        it != _unmergedClusters.end(); ++it) {
        if(*it == r) {
            continue;
        }

        dist = (_clusterSize[p]*_dm(p,*it) +
                _clusterSize[q]*_dm(q,*it)) / _clusterSize[r];
        _dm.add_item(r,*it,dist);
    }
}
```

12.2.4 The Weighted Group Average Algorithm

The weighted group average algorithm is also called the weighted pair group method using arithmetic average (Jain and Dubes, 1988). According to the Lance-Williams recurrence formula, the weighted group average algorithm calculates the distance between a new cluster C_{n+i-1} formed at step i and an old cluster C as

$$\begin{aligned} D(C, C_{n+i-1}) &= D(C, C_{i_1} \cup C_{i_2}) \\ &= \frac{1}{2} D(C, C_{i_1}) + \frac{1}{2} D(C, C_{i_2}), \end{aligned} \quad (12.6)$$

where C_{i_1} and C_{i_2} are the two clusters merged at step i.

Listing 12.8 shows the implementation of the function update_dm in class Weighted.

Listing 12.8: Implementation of function update_dm in class Weighted.

```
void Weighted::update_dm(Size p, Size q, Size r)
    const {
    Real dist;
    std::set<Size>::iterator it;
    for(it = _unmergedClusters.begin();
        it != _unmergedClusters.end(); ++it) {
        if(*it == r) {
            continue;
        }

        dist = (_dm(p,*it) + _dm(q,*it)) / 2;
        _dm.add_item(r,*it,dist);
    }
}
```

12.2.5 The Centroid Algorithm

The centroid algorithm is also called the unweighted pair group method using centroids (Jain and Dubes, 1988). According to the Lance-Williams recurrence formula, the centroid algorithm calculates the distance between a new cluster C_{n+i-1} formed at step i and an old cluster C as

$$\begin{aligned} &D(C, C_{n+i-1}) \\ =\ & D(C, C_{i_1} \cup C_{i_2}) \\ =\ & \frac{|C_{i_1}|}{|C_{i_1}| + |C_{i_2}|} D(C, C_{i_1}) + \frac{|C_{i_2}|}{|C_{i_1}| + |C_{i_2}|} D(C, C_{i_2}) \\ & - \frac{|C_{i_1}| \cdot |C_{i_2}|}{(|C_{i_1}| + |C_{i_2}|)^2} D(C_{i_1}, C_{i_2}), \end{aligned} \quad (12.7)$$

where C_{i_1} and C_{i_2} are the two clusters merged at step i, and $D(\cdot, \cdot)$ is the squared Euclidean distance defined in Equation (1.2).

Listing 12.9 shows the calculation of the distances in the centroid algorithm. Since data member _dm stores the Euclidean distances, we need to square the distances before using the Lance-Williams formula. We also take square root of the calculated distance before storing it to _dm.

Listing 12.9: Implementation of function update_dm in class Centroid.
```
void Centroid::update_dm(Size p, Size q, Size r)
            const {
    Real dist;
    std::set<Size>::iterator it;
    Real sp = _clusterSize[p];
    Real sq = _clusterSize[q];
    for(it = _unmergedClusters.begin();
         it != _unmergedClusters.end(); ++it) {
        if(*it == r) {
            continue;
        }

        dist = std::pow(_dm(p,*it), 2.0)*sp/(sp+sq) +
            std::pow(_dm(q,*it), 2.0)*sq/(sp+sq) -
            std::pow(_dm(p,q), 2.0)*sp*sq/((sp+sq)*(sp+sq));
        _dm.add_item(r,*it, std::sqrt(dist));
    }
}
```

As we can see from the implementation of the function setupArguments in Listing B.15, we create a Euclidean distance object and assign it to _distance. This ensures that the centroid algorithm uses the Euclidean distance regardless the distance provided by users.

12.2.6 The Median Algorithm

The median algorithm has another name, which is called the weighted pair group method using centroids (Jain and Dubes, 1988). According to the Lance-Williams recurrence formula, the centroid algorithm calculates the distance between a new cluster C_{n+i-1} formed at step i and an old cluster C as

$$\begin{aligned} & D(C, C_{n+i-1}) \\ &= D(C, C_{i_1} \cup C_{i_2}) \\ &= \frac{1}{2} D(C, C_{i_1}) + \frac{1}{2} D(C, C_{i_2}) - \frac{1}{4} D(C_{i_1}, C_{i_2}), \end{aligned} \quad (12.8)$$

where C_{i_1} and C_{i_2} are the two clusters merged at step i, and $D(\cdot, \cdot)$ is the squared Euclidean distance defined in Equation (1.2).

The implementation of the median algorithm is shown in Listing 12.10. The implementation of the median algorithm is very similar to that of the centroid algorithm described in the previous subsection. In the median algorithm, however, the distance calculation does not depend on the size of the cluster.

Listing 12.10: Implementation of function update_dm in class Median.
```
void Median::update_dm(Size p, Size q, Size r) const {
    Real dist;
    std::set<Size>::iterator it;
    for(it = _unmergedClusters.begin();
```

```
5              it != _unmergedClusters.end(); ++it) {
6                  if(*it == r) {
7                      continue;
8                  }
9
10                 dist = 0.5*std::pow(_dm(p,*it), 2.0)+
11                     0.5*std::pow(_dm(q,*it), 2.0)-
12                     0.25*std::pow(_dm(p,q), 2.0);
13                 _dm.add_item(r,*it, std::sqrt(dist));
14             }
15         }
```

12.2.7   Ward's Algorithm

Ward's algorithm was proposed by Ward, Jr. (1963) and Ward, Jr. and Hook (1963). This algorithm aims to minimize the loss of information associated with each merging. Hence Ward's algorithm is also referred to as the "minimum variance" method.

According to the Lance-Williams recurrence formula, the centroid algorithm calculates the distance between a new cluster C_{n+i-1} formed at step i and an old cluster C as

$$\begin{aligned} &D(C, C_{n+i-1}) \\ = \; &D(C, C_{i_1} \cup C_{i_2}) \\ = \; &\frac{|C|+|C_{i_1}|}{|C|+|C_{i_1}|+|C_{i_2}|} D(C, C_{i_1}) + \frac{|C|+|C_{i_2}|}{|C|+|C_{i_1}|+|C_{i_2}|} D(C, C_{i_2}) \\ &- \frac{|C|}{|C|+|C_{i_1}|+|C_{i_2}|} D(C_{i_1}, C_{i_2}), \end{aligned} \tag{12.9}$$

where C_{i_1} and C_{i_2} are the two clusters merged at step i, and $D(\cdot,\cdot)$ is the squared Euclidean distance defined in Equation (1.2).

The implementation of Ward's algorithm is shown in Listing 12.11. The implementation is very similar to that of the centroid algorithm.

Listing 12.11: Implementation of function `update_dm` in class `Ward`.
```
1   void Ward::update_dm(Size p, Size q, Size r) const {
2       Real dist;
3       std::set<Size>::iterator it;
4       Real sp = _clusterSize[p];
5       Real sq = _clusterSize[q];
6       for(it = _unmergedClusters.begin();
7           it != _unmergedClusters.end(); ++it) {
8           if(*it == r) {
9               continue;
10          }
11
12          Real sk = _clusterSize[*it];
13          Real st = sp+sq+sk;
14          dist = std::pow(_dm(p,*it), 2.0)*(sp+sk)/st +
15              std::pow(_dm(q,*it), 2.0)*(sk+sq)/st -
16              std::pow(_dm(p,q), 2.0)*sk/st;
17          _dm.add_item(r,*it, std::sqrt(dist));
18      }
19  }
```

12.3 Examples

In this section, we present some examples of applying the agglomerative hierarchical clustering algorithms implemented in the previous section. The main program is shown in Listing B.108.

The program uses the Boost program options to parse command line arguments. The code in lines 16–57 deals with command line options. The code in lines 59–63 reads a dataset from the input file and prints the summary of the dataset to screen. The code in lines 65–83 creates an object of the Euclidean distance and an object of the clustering algorithm selected by users. The default algorithm is single linkage if users do not provide an option for clustering algorithms.

The code in lines 85–87 transfers the dataset and the distance into the clustering algorithm. The code in lines 89–93 does the clustering and measures the time used by the clustering algorithm. The code in lines 95–120 outputs clustering results. The dendgrogram, join vlaues, and the summary of the partition created from the hierarchical clustering are saved to files.

The program is compiled as part of the clustering library when we issue the make command in the top folder of the clustering library (i.e., in the folder ClusLib). The Makefile.am file for this program is shown in Listing B.107. Before compiling the clustering library, we need to specify the location of the Boost program options library by setting the LDFLAGS.

When we execute the program without arguments or with the argument --help, we see the help message:

```
Allowed options:
  --help                produce help message
  --method arg (=single) method (single, complete, gaverage, wgaverage,
                                centroid, median, ward)
  --datafile arg        the data file
  --p arg (=50)         maximum number of nodes to show in dendrogram
  --maxclust arg (=3)   maximum number of clusters
```

The command line options allow us to set the algorithm, the data file, the maximum number of nodes to show in the dendrogram, and the maximum number of clusters to extract from the hierarchical clustering tree. Except for the data file, all options have default values. For example, the default maximum number of nodes to show in a dendrogram is 50.

In the following several subsections, we test these agglomerative hierarchical clustering algorithms using two datasets. The first dataset is a two-dimensional dataset consisting of 600 records (see Figure 1.1) and is generated by the dataset generator program (see Section 11.3.2). For convenience, we call the first dataset the synthetic dataset. The second dataset is the Iris dataset, which is a four-dimensional dataset consisting of 150 records.

12.3.1 The Single Linkage Algorithm

Now we test the single linkage clustering algorithm with two datasets. Suppose the datasets are saved in folder **Data** (see Figure 6.1) and we are in the folder **ClusLib**. To cluster the Iris dataset using the single linkage algorithm, we use the following command:

```
examples/agglomerative/agglomerative.exe
--datafile=../Data/bezdekIris.data
```

When the program is executed, we see the following output on the screen:

```
Number of records: 150
Number of attributes: 4
Number of numerical attributes: 4
Number of categorical attributes: 0

completed in 0.235 seconds
Clustering Summary:
Number of clusters: 3
Size of Cluster 0: 50
Size of Cluster 1: 2
Size of Cluster 2: 98

Number of given clusters: 3
Cross Tabulation:
Cluster ID    Iris-setosa    Iris-versicolor    Iris-virginica
0             50             0                  0
1             0              0                  2
2             0              50                 48
```

From the output we see that it took the single linkage algorithm 0.235 seconds to cluster the Iris dataset. The output dendrogram is shown in Figure 12.2. The dendrogram shows only 50 nodes. From the dendrogram, we can see two clusters clearly. From the cross tabulation shown in the output we see that 48 records were misclassified.

If we want the dendrogram to show all the nodes, we can use the following command:

```
examples/agglomerative/agglomerative.exe
--datafile=../Data/bezdekIris.data --p=0
```

The resulting dendrogram is the one shown in Figure 1.5.

To cluster the synthetic dataset using the single linkage algorithm, we use the following command:

```
examples/agglomerative/agglomerative.exe
--datafile=../Data/600points.csv
```

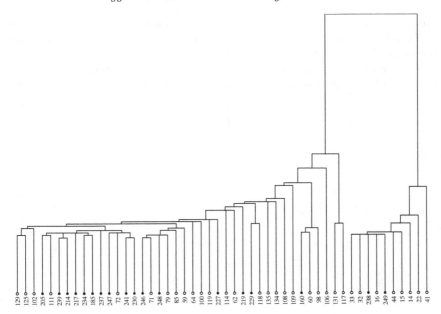

FIGURE 12.2: The dendrogram produced by applying the single linkage algorithm to the Iris dataset.

When the program is executed, we see the following output on the screen:

```
Number of records: 600
Number of attributes: 2
Number of numerical attributes: 2
Number of categorical attributes: 0

completed in 12.484 seconds
Clustering Summary:
Number of clusters: 3
Size of Cluster 0: 200
Size of Cluster 1: 199
Size of Cluster 2: 201

Number of given clusters: 3
Cross Tabulation:
Cluster ID   1    2    3
0            0    200  0
1            0    0    199
2            200  0    1
```

From the output we see that it took the single linkage algorithm 12.484 seconds to cluster the synthetic dataset. The size of the synthetic dataset is 4 times the size of the Iris dataset. But it takes the single linkage algorithm about 50 times more time to cluster the second dataset than to cluster the Iris data. From the cross tabulation shown in the output we see that 1 record was misclassified.

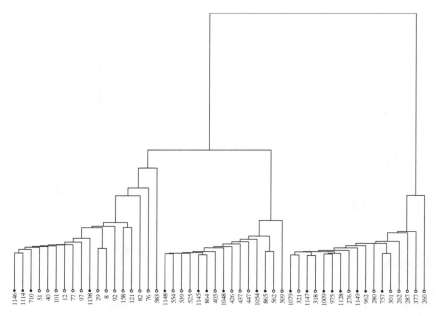

FIGURE 12.3: The dendrogram produced by applying the single linkage algorithm to the synthetic dataset.

Figure 12.3 shows the truncated dendrogram produced by the program. The dendrogram shows only 50 nodes. However, we can still see three clusters clearly.

12.3.2 The Complete Linkage Algorithm

To cluster the Iris dataset using the complete linkage algorithm, we can use the following command:

```
examples/agglomerative/agglomerative.exe
--datafile=../Data/bezdekIris.data --method=complete
```

When the program is executed, we see the following output on the screen:

```
Number of records: 150
Number of attributes: 4
Number of numerical attributes: 4
```

```
Number of categorical attributes: 0

completed in 0.234 seconds
Clustering Summary:
Number of clusters: 3
Size of Cluster 0: 72
Size of Cluster 1: 28
Size of Cluster 2: 50

Number of given clusters: 3
Cross Tabulation:
Cluster ID    Iris-setosa    Iris-versicolor    Iris-virginica
0             0              23                 49
1             0              27                 1
2             50             0                  0
```

From the cross tabulation shown in the output, we see that the complete linkage algorithm clusters 24 records incorrectly. The dendrogram produced by the complete linkage algorithm is shown in Figure 12.4.

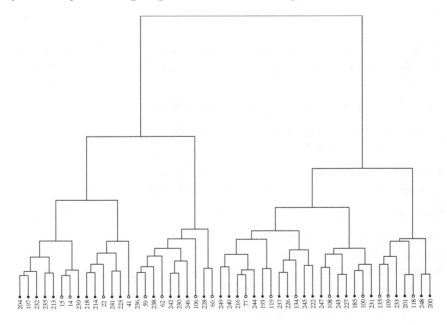

FIGURE 12.4: The dendrogram produced by applying the complete linkage algorithm to the Iris dataset.

Similarly, we can cluster the synthetic data using the complete linkage algorithm. The summary of the clustering results is

```
Number of records: 600
Number of attributes: 2
Number of numerical attributes: 2
Number of categorical attributes: 0

completed in 12.453 seconds
Clustering Summary:
Number of clusters: 3
Size of Cluster 0: 201
Size of Cluster 1: 200
Size of Cluster 2: 199

Number of given clusters: 3
Cross Tabulation:
Cluster ID    1    2    3
0            200   0    1
1             0   200   0
2             0    0   199
```

From the output we see that there is one record clustered incorrectly. Figure 12.5 shows the dendrogram produced by the complete linkage algorithm.

12.3.3 The Group Average Algorithm

To cluster the Iris dataset using the group average algorithm, we can use the following command:

```
examples/agglomerative/agglomerative.exe
--datafile=../Data/bezdekIris.data --method=gaverage
```

When the program is executed, we see the following output on the screen:

```
Number of records: 150
Number of attributes: 4
Number of numerical attributes: 4
Number of categorical attributes: 0

completed in 0.235 seconds
Clustering Summary:
Number of clusters: 3
Size of Cluster 0: 50
Size of Cluster 1: 36
Size of Cluster 2: 64
```

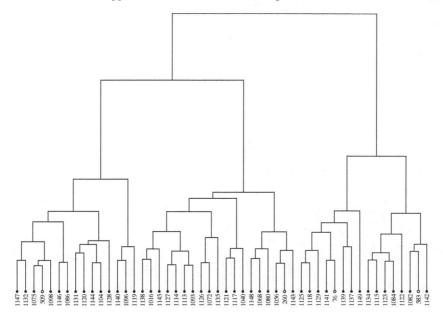

FIGURE 12.5: The dendrogram produced by applying the complete linkage algorithm to the synthetic dataset.

```
Number of given clusters: 3
Cross Tabulation:
Cluster ID    Iris-setosa    Iris-versicolor    Iris-virginica
0             50             0                  0
1             0              0                  36
2             0              50                 14
```

The algorithm clustered 14 records incorrectly. The dendrogram produced by the group average algorithm is shown in Figure 12.6.

We can use a similar command to cluster the synthetic dataset with the group average algorithm. The program produced the following output:

```
Number of records: 600
Number of attributes: 2
Number of numerical attributes: 2
Number of categorical attributes: 0

completed in 12.453 seconds
Clustering Summary:
Number of clusters: 3
```

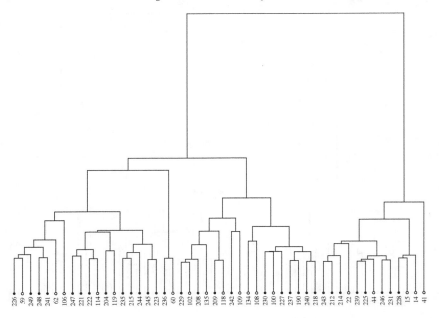

FIGURE 12.6: The dendrogram produced by applying the group average algorithm to the Iris dataset.

```
Size of Cluster 0: 201
Size of Cluster 1: 199
Size of Cluster 2: 200

Number of given clusters: 3
Cross Tabulation:
Cluster ID   1    2    3
0            200  0    1
1            0    0    199
2            0    200  0
```

We can see that one record is clustered incorrectly. The resulting dendrogram is shown in Figure 12.7.

12.3.4 The Weighted Group Average Algorithm

To cluster the Iris dataset using the weighted group average algorithm, we can use the following command:

```
examples/agglomerative/agglomerative.exe
--datafile=../Data/bezdekIris.data --method=wgaverage
```

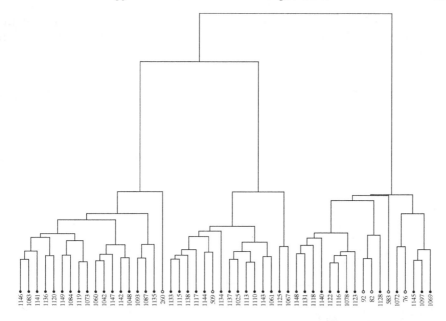

FIGURE 12.7: The dendrogram produced by applying the group average algorithm to the synthetic dataset.

When the program is executed, we see the following output on the screen:

```
Number of records: 150
Number of attributes: 4
Number of numerical attributes: 4
Number of categorical attributes: 0

completed in 0.235 seconds
Clustering Summary:
Number of clusters: 3
Size of Cluster 0: 50
Size of Cluster 1: 35
Size of Cluster 2: 65

Number of given clusters: 3
Cross Tabulation:
Cluster ID   Iris-setosa   Iris-versicolor   Iris-virginica
0            50            0                 0
1            0             0                 35
2            0             50                15
```

The weighted group average algorithm misclassified 15 records. The dendrogram produced by the weighted group average algorithm is shown in Figure 12.8.

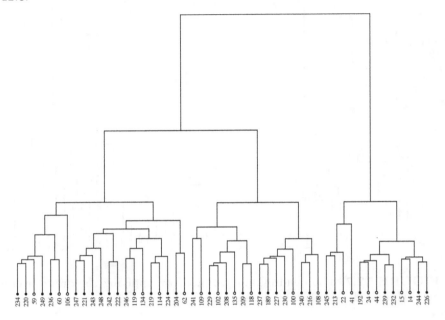

FIGURE 12.8: The dendrogram produced by applying the weighted group average algorithm to the Iris dataset.

When applying the weight group average algorithm to cluster the synthetic dataset, we get the following output:

```
Number of records: 600
Number of attributes: 2
Number of numerical attributes: 2
Number of categorical attributes: 0

completed in 12.281 seconds
Clustering Summary:
Number of clusters: 3
Size of Cluster 0: 200
Size of Cluster 1: 199
Size of Cluster 2: 201

Number of given clusters: 3
Cross Tabulation:
Cluster ID    1    2    3
```

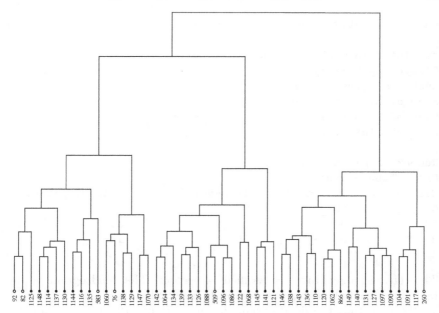

FIGURE 12.9: The dendrogram produced by applying the weighted group average algorithm to the synthetic dataset.

```
0            0    200  0
1            0    0    199
2            200  0    1
```

Again only one record was clustered incorrectly. The dendrogram produced by applying the weighted group average algorithm to the synthetic dataset is shown in Figure 12.9.

12.3.5 The Centroid Algorithm

To cluster the Iris dataset using the centroid algorithm, we can use the following command:

```
examples/agglomerative/agglomerative.exe
--datafile=../Data/bezdekIris.data --method=centroid
```

When the program is executed, we see the following output on the screen:

```
Number of records: 150
Number of attributes: 4
Number of numerical attributes: 4
```

```
Number of categorical attributes: 0

completed in 0.235 seconds
Clustering Summary:
Number of clusters: 3
Size of Cluster 0: 50
Size of Cluster 1: 36
Size of Cluster 2: 64

Number of given clusters: 3
Cross Tabulation:
Cluster ID    Iris-setosa    Iris-versicolor    Iris-virginica
0             50             0                  0
1             0              0                  36
2             0              50                 14
```

The centroid algorithm misclassified 14 records. The dendrogram produced by the centroid algorithm is shown in Figure 12.10.

We note that the dendrogram shown in Figure 12.10 is not monotonic. For example, the cluster consisting of nodes 206 and 119 has a cross link with another cluster.

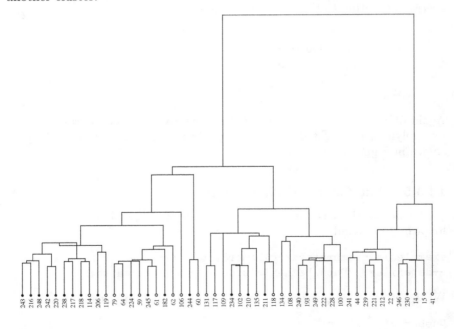

FIGURE 12.10: The dendrogram produced by applying the centroid algorithm to the Iris dataset.

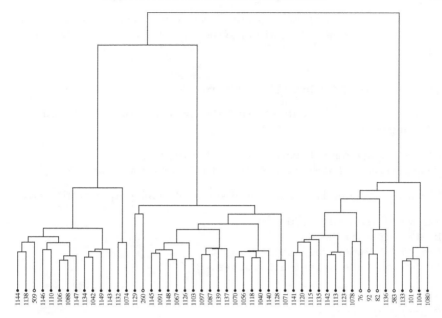

FIGURE 12.11: The dendrogram produced by applying the centroid algorithm to the synthetic dataset.

When applying the centroid algorithm to cluster the synthetic dataset, we get the following output:

```
Number of records: 600
Number of attributes: 2
Number of numerical attributes: 2
Number of categorical attributes: 0

completed in 12.422 seconds
Clustering Summary:
Number of clusters: 3
Size of Cluster 0: 201
Size of Cluster 1: 200
Size of Cluster 2: 199

Number of given clusters: 3
Cross Tabulation:
Cluster ID    1    2    3
0             200  0    1
1             0    200  0
2             0    0    199
```

Again only one record was clustered incorrectly. The dendrogram produced by applying the centroid algorithm to the synthetic dataset is shown in Figure 12.11.

12.3.6 The Median Algorithm

To cluster the Iris dataset using the median algorithm, we can use the following command:

```
examples/agglomerative/agglomerative.exe
--datafile=../Data/bezdekIris.data --method=median
```

When the program is executed, we see the following output on the screen:

```
Number of records: 150
Number of attributes: 4
Number of numerical attributes: 4
Number of categorical attributes: 0

completed in 0.235 seconds
Clustering Summary:
Number of clusters: 3
Size of Cluster 0: 13
Size of Cluster 1: 87
Size of Cluster 2: 50

Number of given clusters: 3
Cross Tabulation:
Cluster ID    Iris-setosa    Iris-versicolor    Iris-virginica
0             0              0                  13
1             0              50                 37
2             50             0                  0
```

The centroid algorithm misclassified 37 records. The dendrogram produced by the median algorithm is shown in Figure 12.12.

The dendrogram shown in Figure 12.12 is not monotonic as cross links exist. For example, the cluster consisting of nodes 249 and 236 has a cross link with another cluster.

When applying the centroid algorithm to cluster the synthetic dataset, we get the following output:

```
Number of records: 600
Number of attributes: 2
Number of numerical attributes: 2
Number of categorical attributes: 0
```

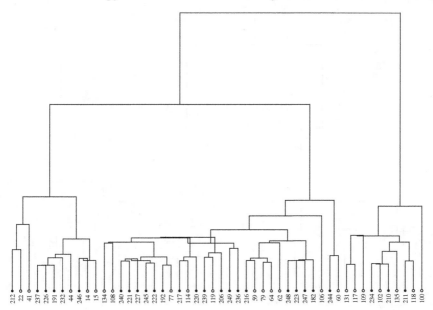

FIGURE 12.12: The dendrogram produced by applying the median algorithm to the Iris dataset.

```
completed in 12.625 seconds
Clustering Summary:
Number of clusters: 3
Size of Cluster 0: 200
Size of Cluster 1: 199
Size of Cluster 2: 201

Number of given clusters: 3
Cross Tabulation:
Cluster ID    1    2    3
0             0    200  0
1             0    0    199
2             200  0    1
```

Again only one record was clustered incorrectly. The dendrogram produced by applying the median algorithm to the synthetic dataset is shown in Figure 12.13. This dendrogram is also not monotonic as the cluster containing node 1144 has a cross link with another cluster.

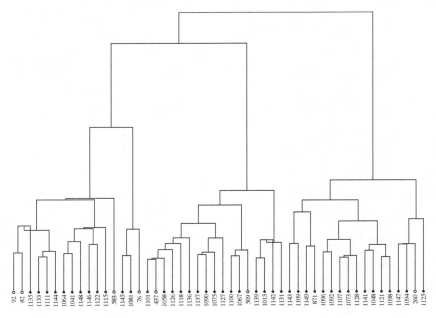

FIGURE 12.13: The dendrogram produced by applying the median algorithm to the synthetic dataset.

12.3.7 Ward's Algorithm

To cluster the Iris dataset using Ward's algorithm, we can use the following command:

```
examples/agglomerative/agglomerative.exe
--datafile=../Data/bezdekIris.data --method=ward
```

When the program is executed, we see the following output on the screen:

```
Number of records: 150
Number of attributes: 4
Number of numerical attributes: 4
Number of categorical attributes: 0

completed in 0.235 seconds
Clustering Summary:
Number of clusters: 3
Size of Cluster 0: 50
Size of Cluster 1: 36
Size of Cluster 2: 64

Number of given clusters: 3
```

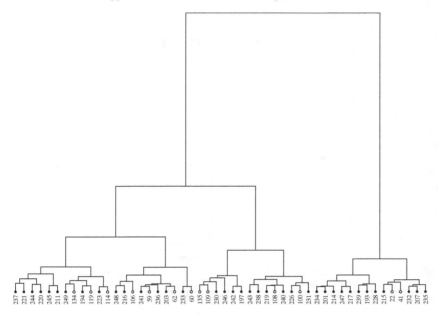

FIGURE 12.14: The dendrogram produced by applying the ward algorithm to the Iris dataset.

```
Cross Tabulation:
Cluster ID    Iris-setosa    Iris-versicolor    Iris-virginica
0             50             0                  0
1             0              1                  35
2             0              49                 15
```

The centroid algorithm misclassified 16 records. The dendrogram produced by Ward's algorithm is shown in Figure 12.14.

When applying the centroid algorithm to cluster the synthetic dataset, we get the following output:

```
Number of records: 600
Number of attributes: 2
Number of numerical attributes: 2
Number of categorical attributes: 0

completed in 12.984 seconds
Clustering Summary:
Number of clusters: 3
Size of Cluster 0: 201
Size of Cluster 1: 200
```

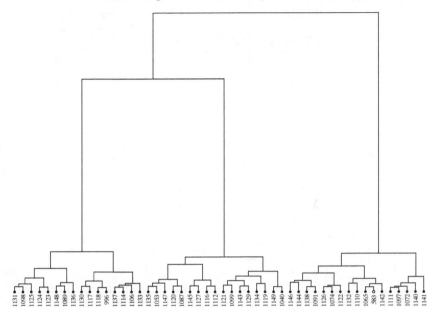

FIGURE 12.15: The dendrogram produced by applying Ward's algorithm to the synthetic dataset.

```
Size of Cluster 2: 199

Number of given clusters: 3
Cross Tabulation:
Cluster ID    1    2    3
0             200  0    1
1             0    200  0
2             0    0    199
```

Again only one record was clustered incorrectly. The dendrogram produced by applying Ward's algorithm to the synthetic dataset is shown in Figure 12.15.

12.4 Summary

In this chapter, we described the implementation of several agglomerative hierarchical clustering algorithms based on the Lance-Williams framework. We also presented examples of applying these agglomerative hierarchical algorithms to a synthetic dataset and the Iris dataset. In our examples, we tried only the Euclidean distance to calculate the pair-wise distances between records and tried to get a partition of maximum three clusters from the hierarchical clustering tree. We encourage users to try other distances and apply these algorithms to cluster other datasets.

Chapter 13

DIANA

Hierarchical clustering algorithms can be agglomerative or divisive. In the previous chapter, we implemented several agglomerative hierarchical clustering algorithms. In this chapter, we implement a divisive hierarchical clustering algorithm, DIANA, which was described in Kaufman and Rousseeuw (1990, Chapter 6).

13.1 Description of the Algorithm

DIANA (DIVisive ANAlysis) is a divisive hierarchical clustering algorithm based on the idea of Macnaughton-Smith et al. (1964). Given a dataset consisting of n records, there are $2^{n-1} - 1$ ways to divide the dataset into two nonempty groups. The DIANA algorithm does not consider all these divisions.

Specifically, let $X = \{\mathbf{x}_0, \mathbf{x}_1, \cdots, \mathbf{x}_{n-1}\}$ be a dataset with n records. At the beginning, all the n records are in one cluster. In the first step, the algorithm divides the dataset into two groups using an iterative process. To do this, the algorithm first finds the record that has the greatest average distance to the rest of the records. The average distance between record $\mathbf{x}_i$ to the rest is calculated as

$$D_i = \frac{1}{n-1} \sum_{j=0, j \neq i}^{n-1} D(\mathbf{x}_i, \mathbf{x}_j),$$

where $D(\cdot, \cdot)$ is a distance measure.

Suppose $D_0 = \max_{0 \leq i \leq n-1} D_i$; i.e., $\mathbf{x}_0$ has the greatest average distance to the rest of the records in the dataset. Then $\mathbf{x}_0$ is first split from the dataset. We have two groups now: $G_1 = \{\mathbf{x}_0\}$ and $G_2 = X \setminus G_1 = \{\mathbf{x}_1, \mathbf{x}_2, \cdots, \mathbf{x}_{n-1}\}$. Then the algorithm checks every record in G_2 to see if the record should be moved to G_1. To do this, the algorithm calculates the distance between $\mathbf{x}$ and G_1 and the distance between $\mathbf{x}$ and $G_2 \setminus \{\mathbf{x}\}$ for all $\mathbf{x} \in G_2$. The distance between $\mathbf{x}$ and G_1 is calculated as

$$D_{G_1}(\mathbf{x}) = \frac{1}{|G_1|} \sum_{\mathbf{y} \in G_1} D(\mathbf{x}, \mathbf{y}), \quad \mathbf{x} \in G_2 \qquad (13.1)$$

where $|G_1|$ denotes the number of records in G_1. The distance between $\mathbf{x}$ and

$G_2 \setminus \{\mathbf{x}\}$ is calculated as

$$D_{G_2}(\mathbf{x}) = \frac{1}{|G_2|-1} \sum_{\mathbf{y} \in G_2} D(\mathbf{x}, \mathbf{y}), \quad \mathbf{x} \in G_2. \tag{13.2}$$

If $D_{G_1}(\mathbf{x}) < D_{G_2}(\mathbf{x})$, then $\mathbf{x}$ is moved from G_2 to G_1. The algorithm continues checking all other records in G_2 until no records should be moved. At this stage, the dataset is divided into two clusters: G_1 and G_2.

In the second step, the algorithm first finds the cluster that has the largest diameter. The diameter of a cluster is defined as the maximum distance between any two records in the cluster. That is,

$$Diam(G) = \max_{\mathbf{x},\mathbf{y} \in G} D(\mathbf{x}, \mathbf{y}). \tag{13.3}$$

If $Diam(G_1) > Diam(G_2)$, then the algorithm applies the process described in the first step to divide cluster G_1 into two clusters: G_3 and G_4.

The algorithm repeats the above procedure until every cluster contains only one record. The algorithm can finish the process in $n-1$ steps.

13.2 Implementation

To implement the divisive hierarchical clustering algorithm, we need to assign a unique identifier to each cluster in the hierarchical clustering tree and a level to these clusters. To do this, we follow the same approach we used to implement the agglomerative hierarchical clustering algorithms. That is, given a dataset $X = \{\mathbf{x}_0, \mathbf{x}_1, \cdots, \mathbf{x}_{n-1}\}$, the initial cluster that contains all the records has an identifier of $2n-2$. We denote the first cluster by X_{2n-2}.

At the first step, X_{2n-2} is divided into two clusters. If both clusters have more than one record, then the two clusters are denoted as X_{2n-3} and X_{2n-4}. If one of the two clusters has only one record, then we denote the one-record cluster by X_i if the cluster contains $\mathbf{x}_i$. If the other cluster contains more than one record, then we denote the other cluster by X_{2n-3}. At the end, the clustering tree includes clusters $X_{2n-2}, X_{2n-3}, \cdots, X_{n-1}, \cdots$, and X_0. The clusters $X_0, X_1, \cdots, X_{n-1}$ are one-record clusters. All other clusters contain more than one record.

The DIANA algorithm is implemented as a class called **Diana**, whose declaration is shown in Listing 13.1. Class **Diana** contains several data members. Data member **_dm** is used to store the pair-wise distances of records in a dataset. Class **iirMapA** is a double-key map class described in Section 11.2.

Data member **_unsplittedClusters** is used to store the identifiers of unsplit clusters, i.e., clusters containing more than one record. At the beginning, this data member contains one element, $2n-2$, which is the identifier of the cluster that contains all records in the dataset.

Data member _clusterDiameter is a map used to store cluster diameters. The key of the map is the identifier of a cluster and the value of the map represents the diameter of the cluster. By storing the diameters of clusters, we do not need to recalculate them.

Data member _leaf is a map used to store the handles of all leaf nodes. There are n leaf nodes, each of which contains a record of the dataset. The key of the map is the identifier of a leaf node and the value of the map is a shared pointer pointing to the leaf node. The identifier of a leaf node is the same as that of the record contained in the leaf node.

Data member _internal is also a map used to store the handles of all internal nodes. There are $n-1$ internal nodes, each of which represents a cluster containing more than one record. The key of the map is the identifier of an internal node and the value of the map is a shared pointer pointing to the internal node. The identifier of the top internal node is $2n-2$.

Data member _clusterID is a set used to store unused identifiers. At the beginning, this set is $\{2n-3, 2n-4, \cdots, n\}$. Other identifiers $2n-2$, $n-1$, $\cdots$, and 0 are used. Once an identifier is used, it will be removed from the set.

Data member _distance is a shared pointer pointing to a distance measure, which is used to calculate the pair-wise distances _dm. Class Diana also inherits the data member _ds from the base algorithm class. The inherited data member _ds is a shared pointer pointing to a dataset.

Listing 13.1: Declaration of class Diana.

```
class Diana: public Algorithm {
protected:
    void setupArguments();
    void performClustering() const;
    void fetchResults() const;
    virtual void create_dm() const;
    virtual void initialization() const;
    virtual void division() const;
    virtual void do_split(Size ind) const;
    virtual void create_cluster(const std::set<Size> ele,
        Size ind) const;

    mutable iirMapA _dm;
    mutable std::set<Size> _unsplitClusters;
    mutable std::map<Size, Real> _clusterDiameter;
    mutable std::map<Size, boost::shared_ptr<LeafNode> > _leaf;
    mutable std::map<Size, boost::shared_ptr<InternalNode> >
        _internal;
    mutable std::set<Size> _clusterID;
    boost::shared_ptr<Distance> _distance;
};
```

Method _setupArguments is very simple in that it transfers a dataset and a distance measure into the algorithm. The method also checks the inputs to ensure the dataset and the distance measure are valid.

Method _performClsutering first calls method create_dm, then method initialization, and then method division. Method create_dm calculates the pair-wise distances of records in the dataset and stores the distances in the data member _dm.

Method `initialization` creates n leaf nodes and an internal node that contains the n leaf nodes as children. The implementation of this method is shown in Listing 13.2. As we can see from the code, data members `_leaf`, `_internal`, `_unsplitClusters`, `_clusterDiameter`, and `_clusterID` are also initialized in this function.

Listing 13.2: Implementation of method `initialization`.

```
1   void Diana::initialization() const {
2       Size n = _ds->size();
3       Size id = 2*n-2;
4       boost::shared_ptr<InternalNode> pin(new InternalNode(id));
5       for(Size s=0;s<n;++s){
6           boost::shared_ptr<LeafNode> pln(new
7               LeafNode((*_ds)[s], s));
8           pln->set_level(0);
9           pin->add(pln);
10          _leaf.insert(std::pair<Size,
11              boost::shared_ptr<LeafNode> >(s, pln));
12      }
13      _internal.insert(std::pair<Size,
14          boost::shared_ptr<InternalNode> >(id, pin));
15      _unsplitClusters.insert(id);
16
17      Real dMax = MIN_REAL;
18      for(Size i=0;i<n-1;++i){
19          for(Size j=i+1;j<n;++j){
20              if (dMax < _dm(i,j) ) {
21                  dMax = _dm(i,j);
22              }
23          }
24      }
25      _clusterDiameter.insert(std::pair<Size, Real>(id,dMax));
26
27      for(Size s=2*n-3; s>n-1; --s) {
28          _clusterID.insert(s);
29      }
30  }
```

Method `division` repeats dividing a larger cluster into two smaller ones until all clusters contain only one record. The process is done in $n-$ steps. The implementation of this method is shown in Listing 13.3. The function first finds the cluster that has the largest diameter among all unsplit clusters (lines 7–15). Then the function sets the level, the identifier, and the join value for the selected cluster. Afterwards, the function calls another method `do_split` to divide the selected cluster.

Listing 13.3: Implementation of method `division`.

```
1   void Diana::division() const {
2       Size n = _ds->size();
3       std::set<Size>::iterator it;
4       Real dMax;
5       Size ind;
6       for(Size s=2*n-2; s>n-1; --s) {
7           dMax= MIN_REAL;
8           std::vector<Size> nvTemp(_unsplitClusters.begin(),
9               _unsplitClusters.end());
10          for(Size i=0; i<nvTemp.size(); ++i) {
11              if(dMax < _clusterDiameter[nvTemp[i]]) {
12                  dMax = _clusterDiameter[nvTemp[i]];
13                  ind = nvTemp[i];
```

```
                    }
              }

              _internal[ind]->set_level(s-n+1);
              _internal[ind]->set_id(s);
              _internal[ind]->set_joinValue(dMax);
              do_split(ind);
        }
}
```

Method **do_split** is used to split a cluster into two clusters. The implementation of the method is shown in Listing 13.4. This function is longer than other functions. But the logic is very simple. The function first gets all the nodes contained in the internal node representing the cluster to be split. Then the function declares two sets: **splinter** and **remaining**. At first, set **splinter** is empty and set **remaining** is populated with all the identifiers of the child nodes (lines 9-15).

At the beginning, method **do_split** finds the record that has the greatest average distance from the rest of the records in the cluster (lines 17–33). This record is moved from set **remaining** to set **splinter**. Then the function checks for every record in set **remaining** and moves the record if the record is closer to the splinter (lines 37–68). At the end, the function removes the cluster from set **_unsplitClusters** (line 70), removes children of the internal node (line 71), and calls method **create_cluster** twice to create clusters.

Listing 13.4: Implementation of method **do_split**.

```
void Diana::do_split(Size ind) const {
      std::vector<boost::shared_ptr<Node> > data =
            _internal[ind]->data();
      Size n = data.size();

      Size ra;
      std::set<Size> splinter;
      std::set<Size> remaining;
      for(Size i=0; i<n; ++i) {
            boost::shared_ptr<LeafNode> leaf =
                  boost::static_pointer_cast<LeafNode>(
                  data[i]);
            Size id = leaf->get_data()->get_id();
            remaining.insert(id);
      }

      std::set<Size>::iterator it, it1;
      Real dMax = MIN_REAL;
      for(it = remaining.begin();
          it != remaining.end(); ++it) {
            Real dSum = 0.0;
            for(it1 = remaining.begin();
                it1 != remaining.end(); ++it1) {
                  if(*it == *it1) {
                        continue;
                  }
                  dSum += _dm(*it, *it1);
            }
            if(dMax < dSum){
                  dMax = dSum;
                  ra = *it;
            }
      }
```

```cpp
34          splinter.insert(ra);
35          remaining.erase(ra);
36
37          bool bChanged = true;
38          while(bChanged) {
39              bChanged = false;
40              for(it = remaining.begin();
41                  it != remaining.end(); ++it) {
42                  Real d1 = 0.0;
43                  for(it1 = splinter.begin();
44                      it1 != splinter.end(); ++it1) {
45                      d1 += _dm(*it, *it1);
46                  }
47                  d1 /= splinter.size();
48
49                  Real d2 = 0.0;
50                  for(it1 = remaining.begin();
51                      it1 != remaining.end(); ++it1) {
52                      if(*it == *it1) {
53                          continue;
54                      }
55                      d2 += _dm(*it, *it1);
56                  }
57                  if(remaining.size() > 1) {
58                      d2 /= (remaining.size()-1.0);
59                  }
60
61                  if(d1 < d2) {
62                      bChanged = true;
63                      splinter.insert(*it);
64                      remaining.erase(it);
65                      break;
66                  }
67              }
68          }
69
70          _unsplitClusters.erase(ind);
71          _internal[ind]->clear();
72          create_cluster(splinter, ind);
73          create_cluster(remaining, ind);
74      }
```

Method `create_cluster` takes two arguments: the first is a set of identifiers of leaf nodes and the second is the identifier of an internal node. The internal node will be the parent of the node representing the cluster formed by the set of leaf nodes. If the set contains only one element, then a leaf node is added to the internal node as a child (lines 34–35). In this case, we do not need to create a new leaf node since all leaf nodes were already created in method `initialization`.

If the set contains more than one element, the function creates a new internal node to hold all the leaf nodes and adds the new internal node to the input internal node as a child (lines 6–32). In this case, we have to create a new internal node. The identifier of the new internal node is set to the last element of set _clusterID. The last element of the set is the largest element of the set since all elements of the set are ordered. Once the identifier is used, it is removed from the set (line 16). The function then updates data members _internal, _unsplitClusters, and _clusterDiameter.

Listing 13.5: Implementation of method `create_cluster`.

```cpp
void Diana::create_cluster(const std::set<Size> ele,
    Size ind) const {
    std::set<Size>::iterator it;
    Real dMax;
    if(ele.size() > 1) {
        boost::shared_ptr<InternalNode> pin(new
            InternalNode(0, _internal[ind]));
        _internal[ind]->add(pin);
        for(it = ele.begin(); it != ele.end(); ++it) {
            pin->add(_leaf[*it]);
        }

        it = _clusterID.end();
        --it;
        Size id = *it;
        _clusterID.erase(it);

        _internal.insert(std::pair<Size,
            boost::shared_ptr<InternalNode> >(id, pin));
        _unsplitClusters.insert(id);

        dMax = MIN_REAL;
        std::vector<Size> nvTemp(ele.begin(), ele.end());
        for(Size i=0; i<nvTemp.size(); ++i) {
            for(Size j=i+1; j<nvTemp.size(); ++j) {
                if(dMax < _dm(nvTemp[i], nvTemp[j])) {
                    dMax = _dm(nvTemp[i], nvTemp[j]);
                }
            }
        }
        _clusterDiameter.insert(
            std::pair<Size, Real>(id,dMax));
    } else {
        it = ele.begin();
        _internal[ind]->add(_leaf[*it]);
    }
}
```

The complete implementation of class `Diana` can be found in Listing B.21.

13.3 Examples

In this section, we apply the DIANA algorithm implemented in the previous section to two datasets. The two datasets are used to illustrate the agglomerative hierarchical algorithms in Chapter 12.

The code in Listing 13.6 shows how to use class `Diana` and how to apply the DIANA algorithm. To use the DIANA algorithm, we first create a new instance of class `Diana`. Then we transfer a dataset and a distance measure into the instance. Then we call method `clusterize` to cluster the dataset using the DIANA algorithm. The runtime is measured and printed to screen.

Listing 13.6: C++ code to illustrate the DIANA algorithm.

```cpp
boost::shared_ptr<Algorithm> ca(new Diana());
boost::shared_ptr<Distance> dist(new EuclideanDistance());
```

```
 3
 4   Arguments &Arg = ca->getArguments();
 5   Arg.ds = ds;
 6   Arg.distance = dist;
 7
 8   boost::timer t;
 9   t.restart();
10   ca->clusterize();
11   double seconds = t.elapsed();
12   std::cout<<"completed in "<<seconds<<" seconds"<<std::endl;
```

The complete program can be found in B.110. In this program, we use the Boost program options library to parse command line options. The program is compiled as part of the clustering library. The Makefile.am for this example is shown in Listing B.109.

When we execute the program with argument --help or without any arguments, we see the following output:

```
Allowed options:
  --help                produce help message
  --datafile arg        the data file
  --p arg (=50)         maximum number of nodes to show in dendrogram
  --maxclust arg (=3)   maximum number of clusters
```

From the output message we see that the only required argument to run the program is the data file. Other options have default values. For example, the default maximum number of clusters is 3.

Now let us apply the DIANA algorithm to the synthetic dataset (see Figure 1.1). Suppose we are in the directory ClusLib (see Figure 6.1). Once we type the following command (in Cygwin):

```
examples/diana/diana.exe --datafile=../Data/600points.csv --p=60
```

we see the following output:

```
Number of records: 600
Number of attributes: 2
Number of numerical attributes: 2
Number of categorical attributes: 0

completed in 2 seconds
Clustering Summary:
Number of clusters: 3
Size of Cluster 0: 208
Size of Cluster 1: 200
Size of Cluster 2: 192

Number of given clusters: 3
Cross Tabulation:
Cluster ID    1    2    3
```

```
0                   200  0    8
1                    0  200   0
2                    0   0   192
```

From the output we see that it took the DIANA algorithm about 2 seconds to cluster the dataset and the algorithm clustered 8 records incorrectly. The dendrogram produced by the algorithm is shown in Figure 13.1. This dendrogram shows only 60 nodes.

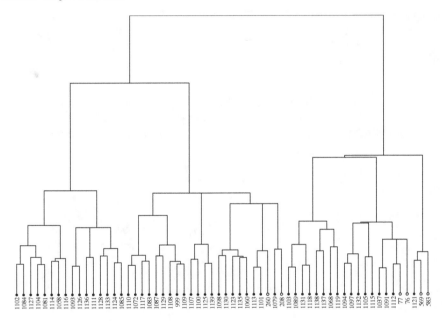

FIGURE 13.1: The dendrogram produced by applying the DIANA algorithm to the synthetic dataset.

Now let us apply the DIANA algorithm to the Iris dataset. To do this, we type the following command:

```
examples/diana/diana.exe --datafile=../Data/bezdekIris.data
--p=0
```

We see the following output:

```
Number of records: 150
Number of attributes: 4
Number of numerical attributes: 4
Number of categorical attributes: 0
```

```
completed in 0.172 seconds
Clustering Summary:
Number of clusters: 3
Size of Cluster 0: 41
Size of Cluster 1: 59
Size of Cluster 2: 50

Number of given clusters: 3
Cross Tabulation:
Cluster ID      Iris-setosa     Iris-versicolor     Iris-virginica
0               0               5                   36
1               0               45                  14
2               50              0                   0
```

From the output we see that 19 points were clustered incorrectly. The dendrogram produced by the algorithm is shown in Figure 13.2. Since we used command line argument --p=0, the dendrogram shows all the nodes.

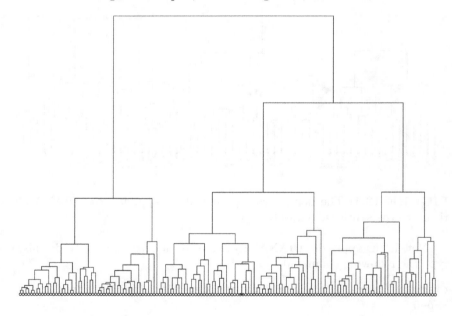

FIGURE 13.2: The dendrogram produced by applying the DIANA algorithm to the Iris dataset.

The dendrograms shown in Figure 13.1 and Figure 13.2 are monotonic. In fact, all dendrograms produced by the DIANA algorithm are monotonic.

However, as a divisive hierarchical clustering algorithm, DIANA is not the exact counterpart of an agglomerative hierarchical clustering algorithm.

13.4 Summary

In this chapter, we introduced the implementation of a divisive hierarchical clustering algorithm called DIANA. Examples to illustrate the DIANA algorithm were also presented.

The DIANA algorithm was based on the idea presented in Macnaughton-Smith et al. (1964). However, the original idea of Macnaughton-Smith et al. (1964) is to split all available clusters. Kaufman and Rousseeuw (1990) modified this original method by defining the diameter for a cluster and first splitting the cluster with the largest diameter. The modified method proposed by Kaufman and Rousseeuw (1990) produces a monotonic hierarchical clustering. One drawback of this modified method is that it is sensitive to outliers.

Chapter 14

The k-means Algorithm

The k-means algorithm is the most popular and the simplest partitional clustering algorithm (Jain, 2010). The k-means algorithm has many variations (see Section 1.5). In this chapter, we implement the standard k-means algorithm.

14.1 Description of the Algorithm

Let $X = \{\mathbf{x}_0, \mathbf{x}_1, \cdots, \mathbf{x}_{n-1}\}$ be a numeric dataset containing n records and k be an integer in $\{1, 2, \cdots n\}$. The k-means algorithm tries to divide the dataset into k clusters $C_0, C_1, \cdots,$ and C_{k-1} by minimizing the following objective function

$$E = \sum_{i=0}^{k-1} \sum_{\mathbf{x} \in C_i} D(\mathbf{x}, \boldsymbol{\mu}_i), \qquad (14.1)$$

where $D(\cdot, \cdot)$ is a distance measure and $\boldsymbol{\mu}_i$ is the mean of cluster C_i, i.e.,

$$\boldsymbol{\mu}_i = \frac{1}{|C_i|} \sum_{\mathbf{x} \in C_i} \mathbf{x}.$$

Let γ_i be the cluster membership of record $\mathbf{x}_i$ for $i = 0, 1, \cdots, n-1$. That is, $\gamma_i = j$ if $\mathbf{x}_i$ belongs to cluster C_j. Then Equation (14.1) can be rewirtten as

$$E = \sum_{i=0}^{n-1} D(\mathbf{x}_i, \boldsymbol{\mu}_{\gamma_i}). \qquad (14.2)$$

To minimize the objective function, the k-means algorithm employs an iterative process. At the beginning, the k-means algorithm selects k random records from the dataset X as initial cluster centers.

Suppose $\boldsymbol{\mu}_0^{(0)}, \boldsymbol{\mu}_1^{(0)}, \cdots,$ and $\boldsymbol{\mu}_{k-1}^{(0)}$ are the initial cluster centers. Based on these cluster centers, the k-means algorithm updates the cluster memberships $\gamma_0^{(0)}, \gamma_1^{(0)}, \cdots, \gamma_{n-1}^{(0)}$ as follows:

$$\gamma_i^{(0)} = \operatorname*{argmin}_{0 \leq j \leq k-1} D(\mathbf{x}_i, \boldsymbol{\mu}_j^{(0)}), \qquad (14.3)$$

where argmin is the argument that minimizes the distance. That is, $\gamma_i^{(0)}$ is set to the index of the cluster to which $\mathbf{x}_i$ has the smallest distance.

Based on the cluster memberships $\gamma_0^{(0)}, \gamma_1^{(0)}, \cdots, \gamma_{n-1}^{(0)}$, the k-means algorithm updates the cluster centers as follows:

$$\boldsymbol{\mu}_j^{(1)} = \frac{1}{|\{i : \gamma_i^{(0)} = j\}|} \sum_{i=0, \gamma_i^{(0)} = j}^{n-1} \mathbf{x}_i, \quad j = 0, 1, \cdots, k-1. \tag{14.4}$$

Then the k-means algorithm repeats updating the cluster memberships based on Equation (14.3) and updating the cluster centers based on Equation (14.4) until some condition is satisfied. For example, the k-means algorithm stops when the cluster memberships do not change any more.

14.2 Implementation

The k-means algorithm is implemented as a class called **Kmean**. The declaration of the class is shown in Listing 14.1, from which we see that class **Kmean** has several data members and member functions. Some of the data members are declared as **mutable**. These **mutable** data members can be modified by **const** functions. Usually, clustering results are declared as **mutable** and parameters are not declared as **mutable**. In this way, a clustering algorithm is not allowed to change its parameters since the member function **performClustering** is a **const** function.

Listing 14.1: Declaration of class **Kmean**.

```
 1  class Kmean: public Algorithm {
 2  protected:
 3      void setupArguments();
 4      void performClustering() const;
 5      void fetchResults() const;
 6      virtual void initialization() const;
 7      virtual void iteration() const;
 8      virtual void updateCenter() const;
 9
10      mutable std::vector<boost::shared_ptr<CenterCluster> >
11          _clusters;
12      mutable std::vector<Size> _CM;
13      mutable Real _error;
14      mutable Size _numiter;
15
16      Size _numclust;
17      Size _maxiter;
18      Size _seed;
19      boost::shared_ptr<Distance> _distance;
20  };
```

Data member **_clusters** is a vector of shared pointers pointing to objects of **CenterCluster**. This data member is used to hold the k clusters, each of which has a center. Data member _CM is a vector of unsigned integers, which represent the cluster memberships of records in the dataset.

Data member _error is used to hold the value of the objective function defined in Equation (14.1). Data member _numiter is the number of iterations the k-means algorithm goes through.

Data member _numclust is a parameter supplied by users. This member represents the number of clusters, i.e., k. Data member _maxiter is the maximum number of iterations the k-means algorithm is allowed to run. This data member is also supplied by users.

Data memeber _seed is a parameter supplied by users. This data member is used as a seed to generate random initial cluster centers. Parameterizing the seed used in the algorithm is useful. For example, we can run the k-means algorithm using different initial cluster centers, which are generated by different seeds.

Data member _distance and another inherited member, _ds, represent a distance measure and a dataset, respectively.

Class Kmean includes six member functions in its class declaration. Class Kmean also has other member functions that are inherited from the base class Algorithm. For example, clusterize, which is a member function defined in class Algorithm, is also a member function of Kmean. In what follows, we introduce the implementation of the member functions.

Method setupArguments is used to transfer parameters supplied by users into the k-means algorithm and validate these parameters. Listing 14.2 shows the implementation of this function. The function first calls the same function defined in the base class (line 2) and checks whether the dataset is numeric or not (line 3).

Afterwards, method setupArguments gets and validates other data members: _distance, _numclust, _maxiter, and _seed. For example, the k-means algorithm requires that the shared pointer pointing to the distance measure be nonempty. The k-means algorithm also requires that both the seed and the maximum number of iterations be positive.

Listing 14.2: Implementation of method setupArguments.

```
1   void Kmean::setupArguments() {
2       Algorithm::setupArguments();
3       ASSERT(_ds->is_numeric(), "dataset_is_not_numeric");
4
5       _distance = _arguments.distance;
6       ASSERT(_distance, "distance_is_null");
7
8       _numclust = boost::any_cast<Size>(
9           _arguments.get("numclust"));
10      ASSERT(_numclust>=2 && _numclust<=_ds->size(),
11          "invalid_numclust");
12
13      _maxiter = boost::any_cast<Size>(
14          _arguments.get("maxiter"));
15      ASSERT(_maxiter>0, "invalide_maxiter");
16
17      _seed = boost::any_cast<Size>(
```

```
18         _arguments.get("seed"));
19     ASSERT(_seed>0, "invalide_seed");
20  }
```

Method `performClustering` calls the two functions: `initialization` and `iteration`. Method `initialization` does some initialization work. The implementation of method `initialization` is shown in Listing 14.3.

Method `initialization` first allocates space for the data member `_CM` (line 4) and creates a local vector `index` to hold the indices $0, 1, \cdots, n-1$. Then the function uses a Boost uniform random number generator to select k records from the dataset as initial cluster centers (lines 9–22). Once a record is selected, it is removed from the pool. This will ensure no duplicate records are selected as initial cluster centers. Data member `_clusters` is initialized with these centers.

Note that we do not use the shared pointers pointing to the selected records as initial centers. Rather, we make copies of these records and use these copies as initial cluster centers (line 16). Since the dataset holds shared pointers pointing to records, the records will be changed if we use the shared pointers directly.

Once the initial cluster centers are selected, the function updates the cluster memberships based on the initial cluster centers (lines 24–38). Data member `_clusters` is also updated here.

Listing 14.3: Implementation of method `initialization`.

```
1   void Kmean::initialization() const {
2       Size numRecords = _ds->size();
3       std::vector<Size> index(numRecords,0);
4       _CM.resize(numRecords);
5       for(Size i=0;i<index.size();++i){
6           index[i] = i;
7       }
8
9       boost::minstd_rand generator(_seed);
10      for(Size i=0;i<_numclust;++i){
11          boost::uniform_int<> uni_dist(0,numRecords-i-1);
12          boost::variate_generator<boost::minstd_rand&,
13              boost::uniform_int<> > uni(generator, uni_dist);
14          Size r = uni();
15          boost::shared_ptr<Record> cr = boost::shared_ptr
16              <Record>(new Record(*(*_ds)[r]));
17          boost::shared_ptr<CenterCluster> c = boost::shared_ptr
18              <CenterCluster>(new CenterCluster(cr));
19          c->set_id(i);
20          _clusters.push_back(c);
21          index.erase(index.begin()+r);
22      }
23
24      Size s;
25      Real dMin, dDist;
26      for(Size i=0;i<numRecords;++i){
27          dMin = MAX_REAL;
28          for(Size j=0;j<_numclust;++j){
29              dDist = (*_distance)((*_ds)[i],
30                  _clusters[j]->center());
31              if (dDist<dMin){
32                  s = j;
33                  dMin = dDist;
```

```
34            }
35        }
36        _clusters[s]->add((*_ds)[i]);
37        _CM[i] = s;
38    }
39 }
```

After calling `initialization`, method `performClustering` calls method `iteration`. The implementation of method `iteration` is shown in Listing 14.4. The function repeats updating the cluster centers based on the cluster memberships and updating the cluster memberships based on the cluster centers until no cluster memberships change or the maximum number of iterations is reached.

Method `iteration` first updates the cluster centers since the cluster memberships were updated in method `initialization` but the cluster centers were not. Hence the number of iteration is set to 1 before the iterative process begins (line 5).

In the iterative process (lines 6–34), the function checks for every record to see if it should be moved to another cluster. If a record is closer to another cluster, then its cluster membership will be changed accordingly and it will be moved from its current cluster to another cluster (lines 22–23). Once all records are checked, the function calls the method `updateCenter` to update the cluster centers based on the updated cluster memberships. Afterward, the function checks for the number of iterations to see if it reaches the maximum number of iterations (lines 31–33). If the maximum number of iterations is reached, the iterative process stops.

Listing 14.4: Implementation of method `iteration`.

```
1  void Kmean::iteration() const {
2      bool bChanged = true;
3
4      updateCenter();
5      _numiter = 1;
6      while(bChanged) {
7          bChanged = false;
8          Size s;
9          Real dMin, dDist;
10         for(Size i=0;i<_ds->size();++i) {
11             dMin = MAX_REAL;
12             for(Size k=0;k<_clusters.size();++k) {
13                 dDist = (*_distance)((*_ds)[i],
14                     _clusters[k]->center());
15                 if (dMin > dDist) {
16                     dMin = dDist;
17                     s = k;
18                 }
19             }
20
21             if (_CM[i] != s){
22                 _clusters[_CM[i]]->erase((*_ds)[i]);
23                 _clusters[s]->add((*_ds)[i]);
24                 _CM[i] = s;
25                 bChanged = true;
26             }
27         }
28
29         updateCenter();
```

```
30        ++_numiter;
31        if (_numiter > _maxiter){
32            break;
33        }
34    }
35 }
```

Method `updateCenter` is very simple. The implementation of this method is shown in Listing 14.5. The method iterates through all clusters and all attributes to update the centers.

Listing 14.5: Implementation of method `updateCenter`.
```
1  void Kmean::updateCenter() const {
2      Real dTemp;
3      boost::shared_ptr<Schema> schema = _ds->schema();
4      for(Size k=0;k<_clusters.size();++k){
5          for(Size j=0;j<schema->size();++j){
6              dTemp = 0.0;
7              for(Size i=0; i<_clusters[k]->size();++i){
8                  boost::shared_ptr<Record> rec =
9                      (*_clusters[k])[i];
10                 dTemp += (*schema)[j]->get_c_val((*rec)[j]);
11             }
12             (*schema)[j]->set_c_val(
13                 (*_clusters[k]->center())[j],
14                 dTemp/_clusters[k]->size());
15         }
16     }
17 }
```

Once the clustering process is finished, the k-means algorithm will call method `fetchResults` to transfer some clustering results to the data member `_results`. The implementation of this method is shown in Listing 14.6. The function first creates an object of class `PClustering` and populates it with the clusters. Then the function assigns the cluster memberships to the cluster memberships of `_results`. The function also calculates the value of the objective function and inserts the value to `_results`. The number of iterations is also put into `_results`.

Listing 14.6: Implementation of method `fetchResults`.
```
1  void Kmean::fetchResults() const {
2      PClustering pc;
3      for(Size i=0;i<_clusters.size();++i){
4          pc.add(_clusters[i]);
5      }
6      _results.CM = _CM;
7      _results.insert("pc", boost::any(pc));
8
9      _error = 0.0;
10     for(Size i=0;i<_ds->size();++i) {
11         _error += (*_distance)((*_ds)[i],
12             _clusters[_CM[i]]->center());
13     }
14     _results.insert("error", boost::any(_error));
15     _results.insert("numiter", boost::any(_numiter));
16 }
```

The complete implementation of class `Kmean` can be found in Listing B.29. The complete header file of class `Kmean` can be found in Listing B.28.

14.3 Examples

In this section, we show how to use the k-means algorithm implemented in the previous section. The k-means algorithm requires some parameters such as the number of clusters and the seed used to select random cluster centers. We use the Boost program options library to allow users to supply these parameters.

The k-means algorithm is sensitive to the initial cluster centers. To alleviate this effect, we run the k-means algorithm multiple times and use a different seed for each run. Running the k-means algorithm multiple times is very easy to implement. The C++ code in Listing 14.7 shows how to run the k-means algorithm **numrun** times, where **numrun** is a parameter supplied by users.

Listing 14.7: Code to run the k-means algorithm multiple times.

```
1   Results Res;
2   Real avgiter = 0.0;
3   Real avgerror = 0.0;
4   Real dMin = MAX_REAL;
5   Real error;
6   for(Size i=1; i<=numrun; ++i) {
7       Kmean ca;
8       Arguments &Arg = ca.getArguments();
9       Arg.ds = ds;
10      Arg.distance = ed;
11      Arg.insert("numclust", numclust);
12      Arg.insert("maxiter", maxiter);
13      Arg.insert("seed", seed);
14      if (numrun == 1) {
15          Arg.additional["seed"] = seed;
16      } else {
17          Arg.additional["seed"] = i;
18      }
19
20      ca.clusterize();
21
22      const Results &tmp = ca.getResults();
23      avgiter += boost::any_cast<Size>(tmp.get("numiter"));
24      error = boost::any_cast<Real>(tmp.get("error"));
25      avgerror += error;
26      if (error < dMin) {
27          dMin = error;
28          Res = tmp;
29      }
30  }
31  avgiter /= numrun;
32  avgerror /= numrun;
```

In Listing 14.7, we see that the program defines a new object of class **Kmean** for each run and transfers the parameters to the new object. In fact, defining a new object for each run is necessary since the internal state of a **Kmean** object is changed after each run.

If the number of runs is one, then the seed supplied by users is used. Otherwise, the program uses $1, 2, \cdots$ as seeds for all the runs. That is, if the number of runs is larger than one, then the seed supplied by users is ignored. The program saves the clustering results of the run that has the lowest error so far. The average number of iterations and the average error are also calculated.

The code shown in Listing 14.7 is part of the program. The complete code can be found in Listing B.112. The program is complied as part of the clustering library. The `Makefile.am` file is shown in Listing B.111.

When we execute the program with the argument `--help` or without any arguments, we see the following output:

```
Allowed options:
  --help                 produce help message
  --datafile arg         the data file
  --k arg (=3)           number of clusters
  --seed arg (=1)        seed used to choose random initial centers
  --maxiter arg (=100)   maximum number of iterations
  --numrun arg (=1)      number of runs
```

From the help message we see that the program allows users to specify values for five parameters, four of which have default values.

Now let us apply the k-means algorithm to cluster the synthetic dataset shown in Figure 1.1. We first run the program with a seed of 5 and 3 clusters. To do this, we execute the following command in the directory `ClusLib`:

```
examples/kmean/kmean.exe --datafile=../Data/600points.csv
--seed=5
```

Once the command is executed, we see the following output:

```
Number of records: 600
Number of attributes: 2
Number of numerical attributes: 2
Number of categorical attributes: 0

completed in 0 seconds
Clustering Summary:
Number of clusters: 3
Size of Cluster 0: 201
Size of Cluster 1: 200
Size of Cluster 2: 199

Number of given clusters: 3
Cross Tabulation:
Cluster ID    1     2    3
0             200   0    1
1             0     200  0
2             0     0    199
```

```
Number of runs: 1
Average number of iterations: 4
Average error: 748.08
Best error: 748.08
```

From the output we see that the k-means algorithm is very fast. The running time is too short to display on screen. The algorithm converged in 4 iterations and clustered one record incorrectly.

Now let us run the program with a different seed by executing the following command:

```
examples/kmean/kmean.exe --datafile=../Data/600points.csv
--seed=20
```

That is, we change the seed to 20. Once the command is executed, we see the following output:

```
Number of records: 600
Number of attributes: 2
Number of numerical attributes: 2
Number of categorical attributes: 0

completed in 0.015 seconds
Clustering Summary:
Number of clusters: 3
Size of Cluster 0: 85
Size of Cluster 1: 394
Size of Cluster 2: 121

Number of given clusters: 3
Cross Tabulation:
Cluster ID   1    2    3
0            85   0    0
1            0    200  194
2            115  0    6

Number of runs: 1
Average number of iterations: 9
Average error: 1653.19
Best error: 1653.19
```

This time the k-means algorithm converged in 9 iterations and clustered many records incorrectly. The results confirmed that the k-means algorithm is sensitive to initial cluster centers.

Now let us run the program multiple times. To do this, we execute the following command:

```
examples/kmean/kmean.exe --datafile=../Data/600points.csv
--numrun=100
```

Once the command is executed, we see the following output:

```
Number of records: 600
Number of attributes: 2
Number of numerical attributes: 2
Number of categorical attributes: 0

completed in 0.953 seconds
Clustering Summary:
Number of clusters: 3
Size of Cluster 0: 201
Size of Cluster 1: 199
Size of Cluster 2: 200

Number of given clusters: 3
Cross Tabulation:
Cluster ID    1    2    3
0             200  0    1
1             0    0    199
2             0    200  0

Number of runs: 100
Average number of iterations: 5.23
Average error: 1010.68
Best error: 748.08
```

From the output we see that running the k-means algorithm with the synthetic dataset 100 times took less than 1 second. The average error is 1010.68 and the average number of iterations is 5.23. The lowest error over the 100 runs is 748.08. The run with the lowest error clustered one record incorrectly.

In the rest of this section, we apply the k-means algorithm to the Iris dataset. Let use run the program with the Iris dataset 100 times. To do this, we issue the following command:

```
examples/kmean/kmean.exe --datafile=../Data/bezdekIris.data
--numrun=100
```

Once the command is executed, we see the following output:

```
Number of records: 150
Number of attributes: 4
Number of numerical attributes: 4
Number of categorical attributes: 0
```

```
completed in 0.469 seconds
Clustering Summary:
Number of clusters: 3
Size of Cluster 0: 50
Size of Cluster 1: 38
Size of Cluster 2: 62

Number of given clusters: 3
Cross Tabulation:
Cluster ID    Iris-setosa    Iris-versicolor    Iris-virginica
0             50             0                  0
1             0              2                  36
2             0              48                 14

Number of runs: 100
Average number of iterations: 7.99
Average error: 109.005
Best error: 97.2046
```

From the output we see that it took the algorithm about half seconds to run 100 times. The average error is 109.005 and the average number of iterations is 7.99. The lowest error over the 100 runs is 97.2046. The run with the lowest error clustered 16 records incorrectly.

Now let use run the program 100 times with $k = 4$. That is, we let the k-means algorithm to produce four clusters every time. To do this, we execute the following command:

```
examples/kmean/kmean.exe --datafile=../Data/bezdekIris.data
--k=4 --numrun=100
```

Once the command is executed, we see the following output:

```
Number of records: 150
Number of attributes: 4
Number of numerical attributes: 4
Number of categorical attributes: 0

completed in 0.766 seconds
Clustering Summary:
Number of clusters: 4
Size of Cluster 0: 50
Size of Cluster 1: 32
Size of Cluster 2: 28
Size of Cluster 3: 40

Number of given clusters: 3
Cross Tabulation:
```

Cluster ID	Iris-setosa	Iris-versicolor	Iris-virginica
0	50	0	0
1	0	0	32
2	0	27	1
3	0	23	17

```
Number of runs: 100
Average number of iterations: 10.03
Average error: 89.7903
Best error: 83.6077
```

From the output we see that the average number of iterations increases to 10.03 compared to the previous multiple-run test with $k = 3$. However, the average error decreases to 83.6077. This result is expected since increasing the number of centers usually decreases the objective function. When we cluster the Iris dataset into 150 clusters, i.e., each cluster contains only one record, then the objective function is zero.

14.4 Summary

In this chapter, we introduced the standard k-means algorithm and its implementation. We also applied the k-means algorithm to a synthetic dataset and the Iris dataset with different parameters. Our experiments show that the k-means algorithm is sensitive to initial cluster centers and may terminate at local optimum solutions.

As the most popular and the simplest partitional clustering algorithm, the k-means algorithm has a long history. In fact, the algorithm was independently discovered by several people from different scientific fields (Jain, 2010). Since then many variations of the k-means algorithm have been proposed. For more information about the k-means algorithm and its variations, readers are referred to Gan et al. (2007) and references therein.

Chapter 15

The c-means Algorithm

The c-means algorithm is also referred to as the fuzzy c-means (FCM) algorithm, which was developed by Dunn (1974a) and improved by Bezdek (1981b). Since the c-means algorithm is a fuzzy clustering algorithm, it allows one record to belong to two or more clusters with some weights. The c-means algorithm is very similar to the k-means algorithm in other aspects. In this chapter, we implement the c-means algorithm and illustrate it with some examples.

15.1 Description of the Algorithm

Let $X = \{\mathbf{x}_0, \mathbf{x}_1, \cdots, \mathbf{x}_{n-1}\}$ be a dataset containing n records, each of which is described by d numeric attributes. Let U be a $n \times k$ fuzzy partition matrix that satisfies the following conditions:

$$0 \leq u_{ij} \leq 1, \quad 0 \leq j \leq k-1,\ 1 \leq i \leq n-1, \tag{15.1a}$$

$$\sum_{j=1}^{k} u_{ij} = 1, \quad 0 \leq i \leq n-1, \tag{15.1b}$$

$$\sum_{i=1}^{n} u_{ij} > 0, \quad 0 \leq j \leq k-1, \tag{15.1c}$$

where u_{ij} is the (i,j) entry of the matrix U.

Given the dataset X, the c-means algorithm finds a fuzzy partition of X by minimizing the following objective function:

$$J_\alpha = \sum_{j=0}^{k-1} \sum_{i=0}^{n-1} u_{ij}^\alpha D_{euc}(\mathbf{x}_i, \boldsymbol{\mu}_j)^2, \tag{15.2}$$

where $\alpha \in (1, \infty)$ is a weighting exponent, $\boldsymbol{\mu}_j$ is the center of cluster j, and $D_{euc}(\cdot, \cdot)$ is the Euclidean distance.

To minimize the objective function, the c-means algorithm employs an iterative process. That is, the c-means algorithm repeats updating the fuzzy cluster memberships given the cluster centers and updating the cluster centers given the fuzzy cluster memberships until some stop condition is met. At the beginning, the c-means algorithm selected k distinct records from the dataset as initial cluster centers. Suppose $\boldsymbol{\mu}_0^{(0)}, \boldsymbol{\mu}_1^{(0)}, \cdots, \boldsymbol{\mu}_{k-1}^{(0)}$ are the k initial cluster centers. Then the c-means algorithm updates the fuzzy cluster memberships according to the following formula:

$$u_{ij}^{(0)} = \left[\sum_{l=0}^{k-1} \left(\frac{D_{euc}(\mathbf{x}_i, \boldsymbol{\mu}_j^{(0)})}{D_{euc}(\mathbf{x}_i, \boldsymbol{\mu}_l^{(0)})} \right)^{\frac{2}{\alpha-1}} \right]^{-1} \quad (15.3)$$

for $j = 0, 1, \cdots, k-1$ and $i = 0, 1, \cdots, n-1$.

Once the fuzzy cluster memberships are updated, the c-means continues to update the cluster centers according to the following formula:

$$\boldsymbol{\mu}_j^{(1)} = \frac{\sum_{i=0}^{n-1} \left(u_{ij}^{(0)} \right)^\alpha \mathbf{x}_i}{\sum_{i=0}^{n-1} \left(u_{ij}^{(0)} \right)^\alpha} \quad (15.4)$$

for $j = 0, 1, \cdots, k-1$.

The c-means algorithm repeats the above steps until the change of the objective function values between two iterations is within the tolerance or the maximum number of iterations is reached.

Once a fuzzy partition U is obtained, the c-means algorithm produces a hard partition based on the fuzzy partition. Precisely, let $\gamma_0, \gamma_1, \cdots, \gamma_{n-1}$ be the hard partition. That is, γ_i is the index of the cluster to which record $\mathbf{x}_i$ belongs. Then the hard partition can be determined as follows:

$$\gamma_i = \underset{0 \le j \le k-1}{\operatorname{argmax}} u_{ij}, \quad i = 0, 1, \cdots, n-1. \quad (15.5)$$

15.2 Implementaion

It is straightforward to implement the c-means algorithm. Listing 15.1 shows the declaration of class **Cmean**, which implements the c-means algorithm. Class **Cmean** inherits from the base algorithm class **Algorithm**. In addition to the members inherited from the base class, class **Cmean** has several other data members and member functions.

Data member _clusters is a vector of shared pointers pointing to objects of class **CenterCluster**. Since the c-means algorithm produces clusters with only one center, we use class **CenterCluster** to represent these clusters. The fuzzy cluster memberships are not stored in objects of **CenterCluster**. Instead, the fuzzy cluster memberships are stored in data member _FCM, which is an object of uBLAS matrix.

Data member _CM is the hard partition obtained from the fuzzy partition according to Equation (15.5). Data member _numiter represent the actual number of iterations. Data member _dObj is the value of the objective function defined in Equation (15.2).

Data member _threshold is a very small positive real number used to terminate the repeating process. Data member _alpha is the parameter α in Equation (15.2). Data member _epsilon is a very small positive real number used to prevent divide-by-zero error. Data members _numclust and _maxiter represent the number of clusters and the maximum number of iterations, respectively. Data member _seed is a positive integer used to initialize the random number generator, which is used to select k distinct records from the dataset as initial cluster centers. Data member _distance is a shared pointer pointing to an object of a distance class.

Listing 15.1: Declaration of class **Cmean**.

```
class Cmean: public Algorithm {
private:
    void setupArguments();
    void performClustering() const;
    void fetchResults() const;
    void initialization() const;
    void iteration() const;
    void updateCenter() const;
    void updateFCM() const;
    void calculateObj() const;

    mutable std::vector<boost::shared_ptr<CenterCluster> >
        _clusters;
    mutable std::vector<Size> _CM;
    mutable ublas::matrix<Real> _FCM;
    mutable Size _numiter;
    mutable Real _dObj;

    Real _threshold;
    Real _alpha;
    Real _epsilon;
    Size _numclust;
    Size _maxiter;
    Size _seed;
    boost::shared_ptr<Distance> _distance;
};
```

Method **setupArguments** transfers parameters in _arguments, which is inherited from class **Algorithm**, into the algorithm. The method also validates the values of these parameters. For example, the method checks whether the dataset is numeric or not. In the method, member _distance is created as a shared pointer pointing to an object of **EuclideanDistance**. The c-means algorithm does not use the distance measure provided by users. The implementation of this method can be found in Listing B.17.

Method **performClustering** calls another two member functions: method **initialization** and method **iteration**. Method **initialization** does some initialization work. The implementation of method **initialization** is shown in Listing 15.2.

Method **initialization** first allocates spaces for members _FCM and _CM (lines 3–4). Then the method uses a Boost random number generator to select _numclust distinct records as initial cluster centers (lines 11–24). Once a record is selected, it is removed from the pool. In this way, no duplicate records are selected. Finally, the method calls member function `updateFCM` to update the fuzzy cluster memberships.

Listing 15.2: Implementation of function `initialization`.

```
void Cmean::initialization() const {
    Size numRecords = _ds->size();
    _FCM.resize(numRecords, _numclust);
    _CM.resize(numRecords, Null<Size>());

    std::vector<Size> index(numRecords,0);
    for(Size i=0;i<index.size();++i){
        index[i] = i;
    }

    boost::minstd_rand generator(_seed);
    for(Size i=0;i<_numclust;++i){
        boost::uniform_int<> uni_dist(0,numRecords-i-1);
        boost::variate_generator<boost::minstd_rand&,
            boost::uniform_int<> > uni(generator,uni_dist);
        Size r = uni();
        boost::shared_ptr<Record> cr =
            boost::shared_ptr<Record>(new Record(*(*_ds)[r]));
        boost::shared_ptr<CenterCluster> c = boost::shared_ptr<
            CenterCluster>(new CenterCluster(cr));
        c->set_id(i);
        _clusters.push_back(c);
        index.erase(index.begin()+r);
    }

    updateFCM();
}
```

After calling method `initialization`, method `performClustering` calls `iteration` to repeat updating the cluster centers and the fuzzy cluster memberships. The implementation of method `iteration` is shown in Listing 15.3.

Method `iteration` uses an infinite `while` loop to do the repeating process. In the `while` loop, method `updateCenter` and method `updateFCM` are called. Then the objective function value is calculated using method `calculateObj`. If the change of the objective function values between this iteration and the previous iteration is less than the tolerance, then the `while` loop is stopped. If the maximum number of iterations is reached, then the loop is also stopped.

Listing 15.3: Implementation of function `iteration`.

```
void Cmean::iteration() const {
    _numiter = 0;
    Real dPrevObj;
    while(true) {
        updateCenter();
        updateFCM();

        dPrevObj = _dObj;
        calculateObj();

        ++_numiter;

        if (std::fabs(_dObj - dPrevObj) < _threshold){
```

```
14            break;
15         }
16
17         if (_numiter >= _maxiter){
18            break;
19         }
20      }
21  }
```

Method `updateCenter` is implemented according to Equation (15.4). This method just loops through all clusters and all attributes to update the cluster centers. Note that we used the schema of the dataset to get and set individual components of a record (line 11, 14).

Listing 15.4: Implementation of function `updateCenter`.

```
1   void Cmean::updateCenter() const {
2       Real dSum1, dSum2, dTemp;
3       boost::shared_ptr<Schema> schema = _ds->schema();
4       for(Size k=0;k<_numclust;++k){
5           for(Size j=0;j<schema->size();++j){
6               dSum1 = 0.0;
7               dSum2 = 0.0;
8               for(Size i=0; i<_ds->size();++i){
9                   dTemp = std::pow(_FCM(i,k),_alpha);
10                  dSum1 += dTemp *
11                      (*schema)[j]->get_c_val((*_ds)(i,j));
12                  dSum2 += dTemp;
13              }
14              (*schema)[j]->set_c_val(
15                  (*_clusters[k]->center())[j],dSum1/dSum2);
16          }
17      }
18  }
```

Method `updateFCM` is implemented according to Equation (15.3). However, this method adds a small positive real number to some terms in Equation (15.3) in order to prevent divide-by-zero error. Specifically, the method updates the fuzzy cluster memberships according to the following equation:

$$\begin{aligned} u_{ij} &= \left[\sum_{l=0}^{k-1} \left(\frac{D_{euc}(\mathbf{x}_i, \boldsymbol{\mu}_j) + \epsilon}{D_{euc}(\mathbf{x}_i, \boldsymbol{\mu}_l) + \epsilon} \right)^{\frac{2}{\alpha-1}} \right]^{-1} \\ &= \left[\left(D_{euc}(\mathbf{x}_i, \boldsymbol{\mu}_j) + \epsilon \right)^{\frac{2}{\alpha-1}} \cdot \sum_{l=0}^{k-1} \frac{1}{(D_{euc}(\mathbf{x}_i, \boldsymbol{\mu}_l) + \epsilon)^{\frac{2}{\alpha-1}}} \right]^{-1} \end{aligned} \quad (15.6)$$

for $j = 0, 1, \cdots, k-1$ and $i = 0, 1, \cdots, n-1$.

Listing 15.5: Implementation of function `updateFCM`.

```
1   void Cmean::updateFCM() const {
2       Real dSum, dTemp;
3       std::vector<Real> dvTemp(_numclust);
4       boost::shared_ptr<Schema> schema = _ds->schema();
5       for(Size i=0;i<_ds->size();++i){
6           dSum = 0.0;
7           for(Size k=0;k<_numclust;++k){
8               dTemp = (*_distance)((*_ds)[i],
```

```
9                  _clusters[k]->center()) + _epsilon;
10          dvTemp[k] = std::pow(dTemp, 2/(_alpha-1));
11          dSum += 1 / dvTemp[k];
12       }
13       for(Size k=0;k<_numclust;++k){
14          _FCM(i,k) = 1.0 / (dvTemp[k] * dSum);
15       }
16    }
17 }
```

Method `fetchResults` is called to collect and transfer clustering results to data member `_results`, which is inherited from the base algorithm class `Algorithm`. Listing 15.6 shows the implementation of this member function. The function first populates the clusters with records and calculates the hard partition according to the fuzzy cluster memberships. Then the function puts the clusters, the hard partition, the fuzzy cluster memberships, the number of iterations, and the objective function value into `_results`.

Listing 15.6: Implementation of function `fetchResults`.

```
1  void Cmean::fetchResults() const {
2      Size s;
3      for(Size i=0;i<_ds->size();++i) {
4          Real dMax = MIN_REAL;
5          for(Size k=0;k<_numclust;++k) {
6              if (dMax < _FCM(i,k) ) {
7                  dMax = _FCM(i,k);
8                  s = k;
9              }
10         }
11         _CM[i] = s;
12         _clusters[s]->add((*_ds)[i]);
13     }
14
15     PClustering pc;
16     for(Size i=0;i<_clusters.size();++i){
17         pc.add(_clusters[i]);
18     }
19     _results.CM = _CM;
20     _results.insert("pc",pc);
21     _results.insert("fcm", _FCM);
22     _results.insert("numiter", _numiter);
23     _results.insert("dObj", _dObj);
24 }
```

The complete implementation of class `Cmean` can be found in Listing B.17.

15.3 Examples

In this section, we apply the c-means algorithm implemented in the previous section to cluster the synthetic dataset (see Figure 1.1) and the Iris dataset. Since the c-means algorithm requires several parameters, we use the Boost program options library to handle these parameters provided by users.

The program is compiled as part of the clustering library. The `Makefile.am` file for this program can be found in Listing B.113.

The c-means algorithm initializes the cluster centers by randomly selecting k distinct records from the dataset. In order to alleviate the impact of the initial cluster centers on the clustering results, we run the c-means algorithm multiple times. The code in Listing 15.7 shows how we runs the c-means algorithm multiple times.

The program calculates the average number of iterations and the average error (i.e., the objective function value) over the multiple runs. The run that has the lowest error is also saved. The seed provided by users is used only if the number of runs is one.

Listing 15.7: Code to run the c-means algorithm multiple times.

```
1   boost::timer t;
2   t.restart();
3
4   Results Res;
5   Real avgiter = 0.0;
6   Real avgerror = 0.0;
7   Real dMin = MAX_REAL;
8   Real error;
9   for(Size i=1; i<=numrun; ++i) {
10      Cmean ca;
11      Arguments &Arg = ca.getArguments();
12      Arg.ds = ds;
13      Arg.insert("alpha", alpha);
14      Arg.insert("epsilon", epsilon);
15      Arg.insert("threshold", threshold);
16      Arg.insert("numclust", numclust);
17      Arg.insert("maxiter", maxiter);
18      Arg.insert("seed", seed);
19      if (numrun == 1) {
20          Arg.additional["seed"] = seed;
21      } else {
22          Arg.additional["seed"] = i;
23      }
24
25      ca.clusterize();
26
27      const Results &tmp = ca.getResults();
28      avgiter += boost::any_cast<Size>(tmp.get("numiter"));
29      error = boost::any_cast<Real>(tmp.get("dObj"));
30      avgerror += error;
31      if (error < dMin) {
32          dMin = error;
33          Res = tmp;
34      }
35  }
36  avgiter /= numrun;
37  avgerror /= numrun;
38
39  double seconds = t.elapsed();
```

Note that a new instance of the c-means algorithm is created for each run (line 10). In fact, it is necessary to create a new instance for each run in order to prevent the previous run's affecting the current run. In Exercise A.10, readers are asked to override the `reset` in class `Cmean` so that a reseted object is the same as a new object.

248   *Data Clustering in C++: An Object-Oriented Approach*

Suppose we are in directory `ClusLib` and the *c*-mean program is in directory `ClusLib/examples/cmean`. To see a list of options the program can handle or the help message, we can execute the program with command line argument `--help` or without any arguments:

`examples/cmean/cmean.exe --help`

After the command is executed, we see the following help message:

```
Allowed options:
  --help                    produce help message
  --datafile arg            the data file
  --k arg (=3)              number of clusters
  --seed arg (=1)           seed used to choose random initial centers
  --maxiter arg (=100)      maximum number of iterations
  --numrun arg (=1)         number of runs
  --epsilon arg (=1e-6)     epsilon
  --alpha arg (=2.1)        alpha
  --threshold arg (=1e-12)  Objective function tolerance
```

From the help message we see that the program can take many arguments. The only required argument is the datafile. All other arguments have default values, which are shown in the round brackets after the option name.

Now let us apply the program to cluster the synthetic dataset. To do this, we execute the following command:

`examples/cmean/cmean.exe --datafile=../Data/600points.csv`

After the command is executed, we see the following output:

```
Number of records: 600
Number of attributes: 2
Number of numerical attributes: 2
Number of categorical attributes: 0

completed in 0.094 seconds
Clustering Summary:
Number of clusters: 3
Size of Cluster 0: 201
Size of Cluster 1: 199
Size of Cluster 2: 200

Number of given clusters: 3
Cross Tabulation:
Cluster ID    1    2    3
0             200  0    1
1             0    0    199
2             0    200  0
```

```
Number of run: 1
Average number of iterations: 17
Average error: 1021.29
Number of iterations for the best case: 17
Best error: 1021.29

Fuzzy cluster memberships of the first 5 records:
Record 0, 0.98715, 0.00813992, 0.00471031
Record 1, 0.921826, 0.0494792, 0.0286946
Record 2, 0.970043, 0.018652, 0.0113053
Record 3, 0.959935, 0.0262841, 0.0137809
Record 4, 0.915936, 0.059032, 0.0250324
...
```

From the output we see that one record was clustered incorrectly. From the fuzzy cluster memberships we see that the first five records belong to the first cluster since the first cluster has the highest weights.

Now let us apply the program to cluster the Iris data set. To do this, we execute the following command:

```
examples/cmean/cmean.exe --datafile=../Data/bezdekIris.data
```

After the command is executed, we see the following output:

```
Number of records: 150
Number of attributes: 4
Number of numerical attributes: 4
Number of categorical attributes: 0

completed in 0.125 seconds
Clustering Summary:
Number of clusters: 3
Size of Cluster 0: 60
Size of Cluster 1: 50
Size of Cluster 2: 40

Number of given clusters: 3
Cross Tabulation:
```

Cluster ID	Iris-setosa	Iris-versicolor	Iris-virginica
0	0	47	13
1	50	0	0
2	0	3	37

```
Number of run: 1
```

```
Average number of iterations: 72
Average error: 57.1124
Number of iterations for the best case: 72
Best error: 57.1124

Fuzzy cluster memberships of the first 5 records:
Record 0, 0.00407063, 0.99389, 0.00203941
Record 1, 0.0237269, 0.964727, 0.0115459
Record 2, 0.0200412, 0.969897, 0.0100623
Record 3, 0.0310576, 0.953841, 0.0151013
Record 4, 0.00632004, 0.990484, 0.0031957
...
```

From the output we see that sixteen records were clustered incorrectly. The error is 57.1124. To see if we can improve the cluster results by running the algorithm multiple times, we issue the following command:

```
examples/cmean/cmean.exe --datafile=../Data/bezdekIris.data
--numrun=100
```

After the command is executed, we see the following output:

```
Number of records: 150
Number of attributes: 4
Number of numerical attributes: 4
Number of categorical attributes: 0

completed in 7.14 seconds
Clustering Summary:
Number of clusters: 3
Size of Cluster 0: 50
Size of Cluster 1: 60
Size of Cluster 2: 40

Number of given clusters: 3
Cross Tabulation:
Cluster ID    Iris-setosa    Iris-versicolor    Iris-virginica
0             50             0                  0
1             0              47                 13
2             0              3                  37


Number of run: 100
Average number of iterations: 39.88
Average error: 57.1124
Number of iterations for the best case: 36
Best error: 57.1124
```

Fuzzy cluster memberships of the first 5 records:
Record 0, 0.99389, 0.00407063, 0.00203941
Record 1, 0.964727, 0.0237269, 0.0115459
Record 2, 0.969897, 0.0200412, 0.0100623
Record 3, 0.953841, 0.0310576, 0.0151013
Record 4, 0.990484, 0.00632004, 0.0031957
...

From the output we see that the best run still clustered sixteen records incorrectly. It seems that the best error is 57.1124 and the results cannot be improved further.

The above experiments used default values for most of the arguments. Let us test the impact of the parameter α on the cluster results. Parameter α can be any positive number in $(1, \infty)$. To do the test, we first run the program with $\alpha = 1.1$ and then run the program with $\alpha = 10$.

For the first experiment, we issue the following command:

examples/cmean/cmean.exe --datafile=../Data/bezdekIris.data
--numrun=100 --alpha=1.1

After the command is executed, we see the following output:

Number of records: 150
Number of attributes: 4
Number of numerical attributes: 4
Number of categorical attributes: 0

completed in 4.438 seconds
Clustering Summary:
Number of clusters: 3
Size of Cluster 0: 50
Size of Cluster 1: 38
Size of Cluster 2: 62

Number of given clusters: 3
Cross Tabulation:
Cluster ID   Iris-setosa   Iris-versicolor   Iris-virginica
0   50   0   0
1   0   2   36
2   0   48   14

Number of run: 100
Average number of iterations: 24.7
Average error: 104.921
Number of iterations for the best case: 16

Best error: 78.7051

Fuzzy cluster memberships of the first 5 records:
Record 0, 1, 9.48943e-32, 2.33471e-28
Record 1, 1, 7.28132e-22, 2.53628e-18
Record 2, 1, 9.51589e-23, 2.34265e-19
Record 3, 1, 1.538e-20, 5.44003e-17
Record 4, 1, 2.52524e-29, 5.61194e-26
...

From the output we see that the c-means algorithm with an α close to 1 behaves like the k-means algorithm. The weight for a cluster is close to 1 and the weights for other clusters are close to zero.

For the second experiment, we issue the following command:

examples/cmean/cmean.exe --datafile=../Data/bezdekIris.data
--numrun=100 --alpha=10

After the command is executed, we see the following output:

Number of records: 150
Number of attributes: 4
Number of numerical attributes: 4
Number of categorical attributes: 0

completed in 7.687 seconds
Clustering Summary:
Number of clusters: 3
Size of Cluster 0: 50
Size of Cluster 1: 52
Size of Cluster 2: 48

Number of given clusters: 3
Cross Tabulation:
Cluster ID	Iris-setosa	Iris-versicolor	Iris-virginica
0	50	0	0
1	0	45	7
2	0	5	43

Number of run: 100
Average number of iterations: 42.89
Average error: 0.0207559
Number of iterations for the best case: 24
Best error: 0.0207559

Fuzzy cluster memberships of the first 5 records:

```
Record 0, 0.501703, 0.256345, 0.241952
Record 1, 0.450013, 0.283181, 0.266806
Record 2, 0.452611, 0.281487, 0.265902
Record 3, 0.440619, 0.288004, 0.271377
Record 4, 0.485789, 0.264431, 0.24978
...
```

From the output we see that the weights are closer to each other when we increase the parameter α. The clustering results are also improved: the best run clustered 12 records incorrectly.

15.4 Summary

In this chapter, we implemented the c-means algorithm and illustrated the algorithm with several examples. The c-means algorithm implemented in this chapter is very similar to the fuzzy k-means algorithm (Bezdek, 1974; Gath and Geva, 1989). The c-means algorithm is one of the many fuzzy cluster algorithms. For more information about fuzzy clustering, readers are referred to Höppner et al. (1999), Valente de Oliveira and Pedrycz (2007), and Miyamoto et al. (2008).

Chapter 16

The k-prototypes Algorithm

The k-prototypes algorithm (Huang, 1998) is a clustering algorithm designed to cluster mixed-type datasets. The k-prototypes algorithm was based on the idea of the k-means algorithm and the k-modes algorithm (Huang, 1998; Chaturvedi et al., 2001). In this chapter, we implement the k-prototypes algorithm.

16.1 Description of the Algorithm

Let $X = \{\mathbf{x}_0, \mathbf{x}_1, \cdots, \mathbf{x}_{n-1}\}$ be a mixed-type dataset containing n records, each of which is described by d attributes. Without loss of generality, we assume that the first p attributes are numeric and the last $d - p$ attributes are categorical. Then the distance between two records $\mathbf{x}$ and $\mathbf{y}$ in X can be defined as (Huang, 1998):

$$D_{mix}(\mathbf{x}, \mathbf{y}, \beta) = \sum_{h=0}^{p-1} (x_h - y_h)^2 + \beta \sum_{h=p}^{d-1} \delta(x_h, y_h), \quad (16.1)$$

where x_h and y_h are the hth component of $\mathbf{x}$ and $\mathbf{y}$, respectively, β is a balance weight used to avoid favoring either type of attribute, and $\delta(\cdot, \cdot)$ is the simple matching distance defined as

$$\delta(x_h, y_h) = \begin{cases} 0, & \text{if } x_h = y_h, \\ 1, & \text{if } x_h \neq y_h. \end{cases}$$

The objective function that the k-prototypes algorithm tries to minimize is defined as

$$P_\beta = \sum_{j=0}^{k-1} \sum_{\mathbf{x} \in C_j} D_{mix}(\mathbf{x}, \boldsymbol{\mu}_j, \beta), \quad (16.2)$$

where $D_{mix}(\cdot, \cdot, \beta)$ is defined in Equation (16.1), k is the number of clusters, C_j is the jth cluster, and $\boldsymbol{\mu}_j$ is the center or prototype of cluster C_j.

To minimize the objective function defined in Equation (16.2), the algorithm proceeds iteratively. That is, the k-prototypes algorithm repeats updating the cluster memberships given the cluster centers and updating the cluster centers given the cluster memberships until some stop condition is met.

At the beginning, the k-prototypes algorithm initializes the k cluster centers by selecting k distinct records from the dataset randomly. Suppose $\boldsymbol{\mu}_0^{(0)}, \boldsymbol{\mu}_1^{(0)}, \cdots, \boldsymbol{\mu}_{k-1}^{(0)}$ are the k initial cluster centers. The k-prototypes algorithm updates the cluster memberships $\gamma_0, \gamma_1, \cdots, \gamma_{n-1}$ according to the following formula:

$$\gamma_i^{(0)} = \operatorname*{argmin}_{0 \le j \le k-1} D_{mix}(\mathbf{x}_i, \boldsymbol{\mu}_j^{(0)}, \beta), \qquad (16.3)$$

where $D_{mix}(\cdot, \cdot, \beta)$ is defined in Equation (16.1).

Once the cluster memberships are updated, the algorithm continues to update the cluster centers according to the following formula:

$$\mu_{jh}^{(1)} = \frac{1}{|C_j|} \sum_{\mathbf{x} \in C_j} x_h, \quad h = 0, 1, \cdots, p-1, \qquad (16.4a)$$

$$\mu_{jh}^{(1)} = \operatorname{mode}_h(C_j), \quad h = p, p+1, \cdots, d-1, \qquad (16.4b)$$

where $C_j = \left\{ \mathbf{x}_i \in X : \gamma_i^{(0)} = j \right\}$ for $j = 0, 1, \cdots, k-1$, and $\operatorname{mode}_h(C_j)$ is the most frequent categorical value of the hth attribute in cluster C_j. Let $A_{h0}, A_{h1}, \cdots, A_{h,m_h-1}$ be the distinct values the hth attribute can take, where m_h is the number of distinct values the hth attribute can take. Let $f_{ht}(C_j)$ be the number of records in cluster C_j, whose hth attribute takes value A_{ht} for $t = 0, 1, \cdots, m_h - 1$. That is,

$$f_{ht}(C_j) = |\{\mathbf{x} \in C_j : x_h = A_{ht}\}|, \quad , t = 0, 1, \cdots, m_h - 1.$$

Then

$$\operatorname{mode}_h(C_j) = \operatorname*{argmax}_{0 \le t \le m_h - 1} f_{ht}(C_j), \quad h = p, p+1, \cdots, d-1.$$

The k-prototypes algorithm repeats the above steps until the cluster memberships do not change or the maximum number of iterations is reached.

16.2 Implementation

The k-prototypes algorithm is very similar to the k-means algorithm except for updating cluster centers. To reuse the function in the k-means algorithm, we implement the k-prototypes algorithm as a class, which inherits from class **Kmean**. Listing 16.1 shows the declaration of class **Kprototype**.

Class `Kprototype` does not declare any data members and overrides only two methods of class `Kmean`. Since the k-prototypes algorithm updates cluster centers in a different way than the k-means algorithm does, the method `updateCenter` of class `Kmean` must be overridden. Although the k-prototypes and the k-means have the same parameters, the k-means algorithm requires a numeric dataset. As a result, the method `setupArguments` needs to be overridden.

Listing 16.1: Declaration of class `Kprototype`.

```
class Kprototype: public Kmean {
private:
    void setupArguments();
    void updateCenter() const;
};
```

The implementation of method `setupArguments` is shown in Listing 16.2. This function is identical to the `setupArguments` function of class `Kmean` except for the validation of the dataset. This function does not impose any requirements on the input dataset. The `setupArguments` function of class `Kmean`, however, checks for whether the input dataset is numeric or not.

Listing 16.2: Implementation of function `setupArguments`.

```
void Kprototype::setupArguments() {
    Algorithm::setupArguments();

    _distance = _arguments.distance;
    ASSERT(_distance, "distance is null");

    _numclust = boost::any_cast<Size>(
        _arguments.get("numclust"));
    ASSERT(_numclust>=2 && _numclust<=_ds->size(),
        "invalid numclust");

    _maxiter = boost::any_cast<Size>(
        _arguments.get("maxiter"));
    ASSERT(_maxiter>0, "invalid maxiter");

    _seed = boost::any_cast<Size>(
        _arguments.get("seed"));
    ASSERT(_seed>0, "invalid seed");
}
```

Listing 16.3 shows the implementation of method `updateCenter`. The function loops through all clusters and all attributes to update the cluster centers. The function does not assume the first p attributes are numeric and the last $d - p$ attributes are categorical. In fact, the numeric and categorical attributes can be mixed in order.

The function uses `can_cast_to_c` to determine whether an attribute is numeric or not (line 6). If an attribute is numeric, then the function updates the corresponding component of a cluster center using the average according to Equation (16.4a). If an attribute is categorical, then the function calculates the mode according to Equation (16.4b) and updates the corresponding component of a cluster center by the mode (line 17–39).

Listing 16.3: Implementation of function updateCenter.

```
void Kprototype::updateCenter() const {
    Real dTemp;
    boost::shared_ptr<Schema> schema = _ds->schema();
    for(Size k=0;k<_clusters.size();++k){
        for(Size j=0;j<schema->size();++j){
            if((*schema)[j]->can_cast_to_c()) {
                dTemp = 0.0;
                for(Size i=0; i<_clusters[k]->size();++i){
                    boost::shared_ptr<Record> rec =
                        (*_clusters[k])[i];
                    dTemp+= (*schema)[j]->get_c_val((*rec)[j]);
                }
                (*schema)[j]->set_c_val(
                    (*_clusters[k]->center())[j],
                    dTemp/_clusters[k]->size());
            } else {
                DAttrInfo da = (*schema)[j]->cast_to_d();
                std::map<Size, Size> freq;
                for(Size i=0;i<da.num_values(); ++i){
                    freq.insert(
                        std::pair<Size,Size>(i, 0));
                }

                for(Size i=0; i<_clusters[k]->size();++i){
                    boost::shared_ptr<Record> rec =
                        (*_clusters[k])[i];
                    freq[(*schema)[j]->get_d_val((*rec)[j])]
                        += 1;
                }

                Size nMax = 0;
                Size s;
                for(Size i=0;i<da.num_values(); ++i){
                    if(nMax < freq[i]) {
                        nMax = freq[i];
                        s = i;
                    }
                }
                da.set_d_val((*_clusters[k]->center())[j], s);
            }
        }
    }
}
```

The complete implementation of class Kprototype can be found in Listing B.31.

16.3 Examples

The k-prototypes algorithm is designed to cluster mixed-type datasets. Hence we can use the k-prototypes algorithm to cluster numeric datasets and categorical datasets as well. In this chapter, we apply the k-prototypes algorithm implemented in the previous section to the Iris dataset, the Soybean dataset, and the heart dataset (Frank and Asuncion, 2010). The Iris dataset is a numeric dataset, the Soybean dataset is a categorical dataset, and the heart dataset is a mixed-type dataset.

The program used to illustrate the k-prototypes algorithm is shown in Listing B.116. The program is compiled as part of the clustering library. The Makefile.am file is shown in Listing B.115. The k-prototypes algorithm requires several parameters. To see a list of parameters, we can execute the program with the argument --help or without any arguments:

examples/kprototype/kprototype.exe --help

After the command is executed, we see the following help message:

```
Allowed options:
  --help                 produce help message
  --datafile arg         the data file
  --normalize arg (=no)  normalize the data or not
  --k arg (=3)           number of clusters
  --beta arg (=1)        balance weight for distance
  --seed arg (=1)        seed used to choose random initial centers
  --maxiter arg (=100)   maximum number of iterations
  --numrun arg (=1)      number of runs
```

The parameter datafile is the only required parameter. All other parameters have default values. The parameter normalize is used to determine whether the input dataset should be normalized. The default case is to normalize the data so that all numeric attributes range from 0 to 1.

Now we cluster the Iris dataset using default options. Suppose we are in the directory ClusLib and the k-prototypes algorithm is in the directory ClusLib/examples/kprototype. Then we execute the following command:

example/kprototype/kprototype.exe
--datafile=../Data/bezdekIris.data

Once the command is executed, we see the following output:

```
Number of records: 150
Number of attributes: 4
Number of numerical attributes: 4
Number of categorical attributes: 0

completed in 0 seconds
Clustering Summary:
Number of clusters: 3
Size of Cluster 0: 32
Size of Cluster 1: 22
Size of Cluster 2: 96

Number of given clusters: 3
```

Cross Tabulation:
```
Cluster ID    Iris-setosa    Iris-versicolor    Iris-virginica
0             32             0                  0
1             18             4                  0
2             0              46                 50
```

Number of runs: 1
Average number of iterations: 6
Average error: 351.592
Best error: 351.592

Using the default options, the program clustered many records incorrectly due to the bad initial cluster centers.

To cluster the Soybean dataset, we issue the following command:

```
example/kprototype/kprototype.exe
--datafile=../Data/soybean-small.data --k=4
```

Once the command is executed, we see the following output:

```
Number of records: 47
Number of attributes: 35
Number of numerical attributes: 0
Number of categorical attributes: 35

completed in 0 seconds
Clustering Summary:
Number of clusters: 4
Size of Cluster 0: 10
Size of Cluster 1: 10
Size of Cluster 2: 14
Size of Cluster 3: 13
```

Number of given clusters: 4
Cross Tabulation:

Cluster ID	D1	D2	D3	D4
0	0	10	0	0
1	10	0	0	0
2	0	0	10	4
3	0	0	0	13

Number of runs: 1
Average number of iterations: 4
Average error: 201
Best error: 201

From the output we see that the k-prototypes algorithm clustered four records incorrectly. This might be due to the initial centers. To see if the algorithm can increase the accuracy by running multiple times, we execute the following command:

```
example/kprototype/kprototype.exe
--datafile=../Data/soybean-small.data --k=4 --numrun=50
```

Once the command is executed, we see the following output:

```
Number of records: 47
Number of attributes: 35
Number of numerical attributes: 0
Number of categorical attributes: 35

completed in 0.313 seconds
Clustering Summary:
Number of clusters: 4
Size of Cluster 0: 10
Size of Cluster 1: 10
Size of Cluster 2: 17
Size of Cluster 3: 10

Number of given clusters: 4
Cross Tabulation:
Cluster ID   D1    D2    D3    D4
0            10    0     0     0
1            0     10    0     0
2            0     0     0     17
3            0     0     10    0

Number of runs: 50
Average number of iterations: 3.94
Average error: 214.24
Best error: 199
```

This time all records were clustered correctly. The algorithm converges very fast as the average number of iterations is less than four.

Now we apply the k-prototypes algorithm to cluster the mixed-type dataset, i.e., the heart dataset. To do this, we execute the following command:

```
example/kprototype/kprototype.exe --datafile=../Data/heart.dat
--k=2 --numrun=50
```

Once the command is executed, we see the following output:

```
Number of records: 270
Number of attributes: 13
Number of numerical attributes: 6
```

Number of categorical attributes: 7

completed in 0.75 seconds
Clustering Summary:
Number of clusters: 2
Size of Cluster 0: 102
Size of Cluster 1: 168

Number of given clusters: 2
Cross Tabulation:
Cluster ID   2   1
0   56   46
1   64   104

Number of runs: 50
Average number of iterations: 9.36
Average error: 6.66663e+09
Best error: 6.65231e+09

The algorithm clustered about half of the records incorrectly. The error (i.e., the objective function) is very big. We can standardize the dataset before clustering the dataset. To do this, we execute the following command:

example/kprototype/kprototype.exe --datafile=../Data/heart.dat
--k=2 --numrun=50 --normalize=yes

Once the command is executed, we see the following output:

Number of records: 270
Number of attributes: 13
Number of numerical attributes: 6
Number of categorical attributes: 7

completed in 0.422 seconds
Clustering Summary:
Number of clusters: 2
Size of Cluster 0: 119
Size of Cluster 1: 151

Number of given clusters: 2
Cross Tabulation:
Cluster ID   2   1
0   93   26
1   27   124

Number of runs: 50
Average number of iterations: 5.08

Average error: 549.515
Best error: 541.157

From the output we see that the clustering results are better than that of the previous experiment. The error is smaller after the dataset is normalized.

16.4 Summary

In this chapter, we implemented the k-prototypes algorithm and illustrated it with different types of datasets. The k-prototypes algorithm integrates the k-means algorithm and the k-modes algorithm. The distance measure for the k-prototypes algorithm includes two components: the distance for numerical attributes and the distance for categorical attributes. The two components are balanced using a weight. In our implementation, we used the squared Euclidean distance for numerical attributes and the simple matching distance for the categorical attributes. In fact, the k-prototypes algorithm can use other mixed-type distance measures. For example, the general Minkowski distance (Ichino, 1988; Ichino and Yaguchi, 1994).

Chapter 17

The Genetic k-modes Algorithm

The k-modes algorithm is a center-based clustering algorithm designed to cluster categorical datasets (Huang, 1998; Chaturvedi et al., 2001). In the k-modes algorithm, centers of clusters are referred to as modes. One drawback of the k-modes algorithm is that the algorithm can guarantee only a locally optimal solution. The genetic k-modes algorithm(Gan et al., 2005) was developed to improve the k-modes algorithm by integrating the k-modes algorithm with the genetic algorithm (Holland, 1975). Although the genetic k-modes algorithm cannot guarantee a globally optimal solution, the genetic k-modes algorithm is more likely to find a globally optimal solution than the k-modes algorithm. In this chapter, we implement the genetic k-modes algorithm and illustrate the algorithm with examples.

17.1 Description of the Algorithm

Before we introduce the genetic k-modes algorithm, let us first give a brief introduction to the genetic algorithm, which was originally introduced by Holland (1975). In genetic algorithms, solutions (i.e., parameters) of a problem are encoded in chromosomes and a population is composed of many solutions. Each solution is associated with a fitness value.

A genetic algorithm evolves over generations. During each generation, three genetic operators (natural selection, crossover, and mutation) are applied to the current population to produce a new population. The natural selection operator selects a few chromosomes from the current population based on the principle of survival of the fittest. Then the selected chromosomes are modified by the crossover operator and the mutation operator before being put into the new population.

The genetic k-modes algorithm was developed based on the genetic k-means algorithm (Krishna and Narasimha, 1999). The genetic k-modes algorithm has five elements: coding, initialization, selection, mutation, and k-modes operator. In the genetic k-modes algorithm, the crossover operator is replaced by the k-modes operator.

In the k-modes algorithm, a solution (i.e., a partition) is coded as a vector of cluster memberships. Let $X = \{\mathbf{x}_0, \mathbf{x}_1, \cdots, \mathbf{x}_{n-1}\}$ be a dataset consisting of n records and $C_0, C_1, \cdots, C_{k-1}$ be a partition of X. Then the vector of cluster memberships corresponding to the partition is defined as

$$\gamma_i = j \quad \text{if} \quad \mathbf{x}_i \in C_j$$

for $i = 0, 1, \cdots, n-1$ and $j = 0, 1, \cdots, k-1$. A chromosome is said to be legal if each one of the corresponding k clusters is nonempty.

The loss function for a solution represented by $\Gamma = (\gamma_0, \gamma_1, \cdots, \gamma_{n-1})$ is defined as

$$L(\Gamma) = \sum_{i=0}^{n-1} D_{sim}(\mathbf{x}_i, \boldsymbol{\mu}_{\gamma_i}), \tag{17.1}$$

where $\boldsymbol{\mu}_j$ ($0 \leq j \leq k-1$) is the center of cluster C_j, and $D_{sim}(\cdot, \cdot)$ is the simple matching distance (see Section 9.4). The fitness value for the solution is defined as

$$F(\Gamma) = \begin{cases} cL_{max} - L(\Gamma) & \text{if } \Gamma \text{ is legal,} \\ e(\Gamma) F_{min} & \text{otherwise,} \end{cases} \tag{17.2}$$

where $L(\Gamma)$ is defined in Equation (17.1), $c \in (0, 3)$ is a constant, L_{max} is the maximum loss of chromosomes in the current population, F_{min} is the smallest fitness value of the legal chromosomes in the current population if legal chromosomes exist or 1 if otherwise, and $e(\Gamma)$ is the legality ratio defined as the ratio of the number of nonempty clusters over k.

The selection operator randomly selects a chromosome from the current population according to the following distribution:

$$P(\Gamma_i) = \frac{F(\Gamma_i)}{\sum_{r=0}^{N-1} F(\Gamma_i)},$$

where N is the number of chromosomes in the current population and $F(\cdot)$ is defined in Equation (17.2). Hence chromosomes with higher fitness values are more likely to be selected. The selection operator is applied N times in order to select N chromosomes for the new population.

The mutation operator replaces the ith component γ_i of a chromosome $\Gamma = (\gamma_0, \gamma_1, \cdots, \gamma_{n-1})$ with a cluster index randomly selected from $\{0, 1, \cdots, k-1\}$ according to the following distribution:

$$P(j) = \frac{c_m d_{max}(\mathbf{x}_i) - D_{sim}(\mathbf{x}_i, \boldsymbol{\mu}_j)}{\sum_{l=0}^{k-1} [c_m d_{max}(\mathbf{x}_i) - D_{sim}(\mathbf{x}_i, \boldsymbol{\mu}_l)]}, \tag{17.3}$$

where $c_m > 1$ is a constant and

$$d_{max}(\mathbf{x}_i) = \max_{0 \leq l \leq k-1} D_{sim}(\mathbf{x}_i, \boldsymbol{\mu}_l), \quad i = 0, 1, \cdots, n-1.$$

In the genetic algorithm, mutation occurs with some mutation probability

P_m. That is, for each component of each selected chromosome, the mutation operator is applied if a randomly generated standard uniform number is less than P_m.

Once a selected chromosome is mutated, the k-modes operator is applied to it. Let Γ be a chromosome. The k-modes operator changes Γ into another chromosome $\hat{\Gamma}$ based on the following two steps:

Updating cluster centers Given the cluster memberships Γ, the cluster center $\boldsymbol{\mu}_j$ of cluster C_j is updated to the modes of the cluster (see Section 16.1). Let $\hat{\boldsymbol{\mu}}_j$ ($j = 0, 1, \cdots, k-1$) be the updated cluster centers.

Updating cluster memberships Given the cluster centers $\hat{\boldsymbol{\mu}}_j$, the cluster memberships are updated as follows. The ith component γ_i of Γ is updated to

$$\hat{\gamma}_i = \operatorname*{argmin}_{0 \leq j \leq k-1} D_{sim}(\mathbf{x}_i, \hat{\boldsymbol{\mu}}_j), \quad i = 0, 1, \cdots, n-1.$$

If a cluster is empty, the distance of a record between the cluster is defined to be ∞. Hence illegal chromosomes remain illegal after the application of the k-modes operator.

17.2 Implementation

The genetic k-modes algorithm is implemented as a class called GKmode, whose declaration is shown in Listing 17.1. Class GKmode has four mutable data members, which can be changed by const member functions such as iteration. Data member _clusters is a vector of shared pointers pointing to objects of class CenterCluster. Data member _mP is a matrix, each row of which represents a chromosome (i.e., a vector of cluster memberships). Data member _dvFit is a vector of fitness values of the chromosomes in the population. Data member _bvLegal is a vector of booleans indicating whether a chromosome is legal or not.

Other six data members are not mutable. Data member _numclust is the number of clusters. Data member _numpop is the number chromosomes in the population. Data member _maxgen is the maximum number of generations the algorithm can evolve. The remaining three data members _c, _cm, and _Pm denote values of the parameters c in Equation (17.2), c_m in Equation (17.3), and the mutation probability P_m, respectively.

Listing 17.1: Declaration of class GKmode.

```
1  class GKmode : public Algorithm {
2  private:
3      void setupArguments();
4      void performClustering() const;
```

```
5        void fetchResults() const;
6        void initialization() const;
7        void iteration() const;
8        void selection(Size g) const;
9        void mutation(Size g) const;
10       void kmode(Size ind) const;
11       void calculateLoss(Size ind) const;
12       bool get_mode(Size &mode, Real &var,
13              Size ind, Size k, Size j) const;
14       Real calculateRatio(Size ind) const;
15       void generateRN(std::vector<Size>& nv, Size s) const;
16
17
18       mutable std::vector<boost::shared_ptr<CenterCluster> > 
19           _clusters;
20       mutable ublas::matrix<Size> _mP;
21       mutable std::vector<Real> _dvFit;
22       mutable std::vector<bool> _bvLegal;
23
24       Size _numclust;
25       Size _numpop;
26       Size _maxgen;
27       Real _c;
28       Real _cm;
29       Real _Pm;
30  };
```

Method **setupArgument** is used to transfer parameters in _arguments into the algorithm and validate these parameters. For example, this function ensures that the input dataset is categorical and that the constant c satisfies $0 < c < 3$.

Method **performClustering** calls another two methods: **initialization** and **iteration**. The implementation of method **initialization** is shown in Listing 17.2. The function first allocates space for the data member _mP and initializes the data members with null values (lines 2–7). Then the function randomly selects _numclust records from the dataset and puts them into different clusters (lines 15–18). Method **generateRN** is used to select _numclust records from the dataset. This step ensures that all clusters are nonempty. That is, all chromosomes are legal. Other entries of _mP are filled with random numbers (lines 19–23). Finally, the function allocates space for members _dvFit and _bvLegal.

Listing 17.2: Implementation of function **initialization**.

```
1   void GKmode::initialization() const {
2       _mP.resize(_numpop, _ds->size());
3       for(Size i=0; i<_numpop; ++i) {
4           for(Size j=0; j<_ds->size(); ++j) {
5               _mP(i,j) = Null<Size>();
6           }
7       }
8
9       std::vector<Size> nvTemp(_numclust, 0);
10      boost::minstd_rand generator(1u);
11      boost::uniform_int<> uni_dist(0, _numclust-1);
12      boost::variate_generator<boost::minstd_rand&,
13          boost::uniform_int<> > uni(generator, uni_dist);
14      for(Size i=0;i<_numpop;++i){
15          generateRN(nvTemp, i);
16          for(Size j=0; j<_numclust; ++j) {
```

```
17                    _mP(i, nvTemp[j]) = j;
18              }
19              for(Size j=0;j<_ds->size();++j) {
20                  if (_mP(i,j) == Null<Size>()) {
21                      _mP(i,j) = uni();
22                  }
23              }
24          }
25
26          _dvFit.resize(_numpop, 0.0);
27          _bvLegal.resize(_numpop, true);
28     }
```

The implementation of method `iteration` is shown in Listing 17.3. The function loops through _maxgen generations. During each generation, the function first calculates the losses and finds the maximum loss (lines 5–11) of the current population. Method `calculateLoss` is used to calculate the loss function defined in Equation (17.1). Then the function calculates the fitness values for all chromosomes (lines 13–25). Afterwards the three operators are applied to the selected chromosomes (lines 27–32). The loss functions are calculated again outside the loop (lines 35–37) since the population has been modified since the last calculation. Method `calculateRatio` (line 18) is used to calculate the legality ratio, which is defined to be the ratio of number of nonempty clusters over _numclust.

Note that we used member _dvFit to store fitness values as well as loss function values. Method `calculateLoss` calculates the loss function value for each chromosome in the current population and stores the value in _dvFit.

Listing 17.3: Implementation of function `iteration`.

```
1   void GKmode::iteration() const {
2       Size g = 0;
3       while (g < _maxgen) {
4           ++g;
5           Real Lmax = MIN_REAL;
6           for(Size i=0; i<_numpop; ++i) {
7               calculateLoss(i);
8               if( _bvLegal[i] && _dvFit[i]>Lmax) {
9                   Lmax = _dvFit[i];
10              }
11          }
12
13          Real dTemp = 0.0;
14          for(Size i=0; i<_numpop; ++i) {
15              if( _bvLegal[i]) {
16                  _dvFit[i] = _c*Lmax - _dvFit[i];
17              } else {
18                  _dvFit[i] = calculateRatio(i)*(_c-1)*Lmax;
19              }
20              dTemp += _dvFit[i];
21          }
22
23          for(Size i=0; i<_numpop; ++i) {
24              _dvFit[i] /= dTemp;
25          }
26
27          selection(g);
28          mutation(g);
29
30          for(Size i=0; i<_numpop; ++i) {
```

```
31              kmode( i );
32          }
33      }
34
35      for(Size i=0; i<_numpop; ++i) {
36          calculateLoss(i);
37      }
38  }
```

The implementation of method `calculateLoss` is shown in Listing 17.4. The function loops through all clusters and all attributes to calculate the loss function. The loss function defined in Equation (17.1) can be rewritten as

$$L(\Gamma) = \sum_{j=0}^{k-1} \sum_{h=0}^{d} \sum_{\substack{i=0 \\ \gamma_i = j}}^{n-1} \delta(x_{ih}, \mu_{jh}), \qquad (17.4)$$

where k is the number of clusters, d is the number of attributes, x_{ih} is the hth component of $\mathbf{x}_i$, μ_{jh} is the hth component of $\boldsymbol{\mu}_j$, and $\delta(\cdot, \cdot)$ is the simple matching distance defined in Equation (9.3).

Method `calculateLoss` is implemented according to Equation (17.4). The function stores the loss function value to member `_dvFit`. If a chromosome is illegal, then the loss function value is set to the maximum real number (line 20).

Listing 17.4: Implementation of function `calculateLoss`.
```
1   void GKmode::calculateLoss(Size ind) const {
2       Real dLoss = 0.0;
3       Size mode;
4       Real dVar;
5       bool isLegal;
6       for(Size k=0; k<_numclust; ++k){
7           for(Size j=0; j<_ds->num_attr(); ++j) {
8               isLegal = get_mode(mode, dVar, ind, k, j);
9               dLoss += dVar;
10              if(!isLegal) {
11                  break;
12              }
13          }
14      }
15
16      _bvLegal[ind] = isLegal;
17      if(isLegal) {
18          _dvFit[ind] = dLoss;
19      } else {
20          _dvFit[ind] = MAX_REAL;
21      }
22  }
```

Method `calculateLoss` calls another method `get_mode` to calculate the mode of an individual attribute for each cluster. This function takes five arguments and returns a boolean number. The first two arguments are of reference type and represent the mode and the loss, respectively. The third argument `ind` refers to the index of a chromosome. The last two arguments k and j represent the index of a cluster and the index of an attribute, respectively.

The function uses a vector to store the counts of each individual categorical value (lines 5–15). The mode is set to the categorical value that has the highest count (line 16–22). Variable nCount (line 8) is used to count the number of records in the cluster. If the variable has a final value of zero, then the chromosome is illegal (line 26–30).

Listing 17.5: Implementation of function get_mode.

```
1    bool GKmode::get_mode(Size &mode, Real &var,
2              Size ind, Size k, Size j) const {
3         boost::shared_ptr<Schema> schema = _ds->schema();
4         Size nValues = (*schema)[j]->cast_to_d().num_values();
5         std::vector<Size> nvFreq(nValues,0);
6
7         Size val;
8         Size nCount = 0;
9         for(Size i=0; i<_ds->size(); ++i) {
10             if (_mP(ind, i) == k) {
11                 val = (*schema)[j]->get_d_val( (*_ds)(i, j) );
12                 ++nvFreq[val];
13                 ++nCount;
14             }
15         }
16         val = 0;
17         for(Size i=0; i<nValues; ++i){
18             if(val < nvFreq[i]) {
19                 val = nvFreq[i];
20                 mode = i;
21             }
22         }
23
24         var = nCount - nvFreq[mode];
25
26         if (nCount == 0) {
27             return false;
28         } else {
29             return true;
30         }
31    }
```

The implementation of the selection operator selection is shown in Listing 17.6. The function takes one argument, which is used as a seed to initialize the random number generator (line 3). The function uses a local variable mP (line 9) to save a copy of the current population. Then the function loops through _numpop iterations. During each iteration, the function generates a uniform random number from $(0,1)$ and uses the random number to decide which chromosome should be selected based on the fitness values (lines 11-19).

Given a discrete distribution $P(j) = p_j$ $(j = 0, 1, \cdots, n-1)$ such that $p_j \geq 0$ and $p_0 + p_1 + \cdots + p_{n-1} = 1$, we use the inversion method to generate random numbers from this distribution (Devroye, 1986, Chapter III). First, we generate a uniform random number u from $(0,1)$ (line 12). Then we select the integer s from $\{0, 1, \cdots, n-1\}$ such that

$$F(s-1) < u \leq F(s),$$

where $F(-1) = 0$ and $F(i) = p_0 + p_1 + \cdots + p_i$ for $i = 0, 1, \cdots, n-1$.

Listing 17.6: Implementation of function **selection**.

```
void GKmode::selection(Size g) const {
    boost::minstd_rand generator(
        static_cast<unsigned int>(g+1));
    boost::uniform_real<> uni_dist(0, 1);
    boost::variate_generator<boost::minstd_rand&,
        boost::uniform_real<> > uni(generator, uni_dist);
    Real dRand, dTemp;
    Size s;
    ublas::matrix<Size> mP = _mP;
    for(Size i=0;i<_numpop;++i){
        dTemp = _dvFit[0];
        dRand = uni();
        s = 0;
        while(dTemp < dRand) {
            dTemp += _dvFit[++s];
        }
        for(Size j=0; j<_ds->size(); ++j) {
            _mP(i, j) = mP(s, j);
        }
    }
}
```

The implementation of the mutation operator **mutation** is shown in Listing 17.7. The function takes one argument, which is used to initialize the random number generator. The function first creates a uniform random number generator (lines 2–6). Then the function loops through all chromosomes in the current population and all components of chromosomes to do the mutation. For each component of a chromosome, the function first generates a uniform random number. If the random number is greater than or equal to the mutation probability, then the chromosome will not be mutated (lines 16–18). Otherwise, the chromosome is mutated (lines 20–57).

If mutation occurs for a component of the chromosome, then the function calculates the discrete distribution defined in Equation (17.3) (lines 20–43). After the discrete distribution dvProb is obtained, the inversion method is used again to select an integer from the distribution (lines 45–50). The selected integer is saved to a map mTemp (line 51). The mutation is done once all components of the chromosome are processed (lines 54–57).

Listing 17.7: Implementation of function **mutation**.

```
void GKmode::mutation(Size g) const {
    boost::minstd_rand generator(
        static_cast<unsigned int>(g+1));
    boost::uniform_real<> uni_dist(0, 1);
    boost::variate_generator<boost::minstd_rand&,
        boost::uniform_real<> > uni(generator, uni_dist);
    Real dRand, dMax, dVar, dTemp;
    Size mode;
    bool isLegal;
    std::map<Size, Size> mTemp;
    std::vector<Real> dvProb(_numclust, 0.0);
    for(Size i=0; i<_numpop; ++i) {
        mTemp.clear();
        for(Size j=0; j<_ds->size(); ++j) {
            dRand = uni();
            if (dRand >= _Pm ) {
                continue;
            }
```

```
19
20              dMax = MIN_REAL;
21              for(Size k=0; k<_numclust; ++k) {
22                  dvProb[k] = 0.0;
23                  for(Size d=0; d<_ds->num_attr(); ++d) {
24                      isLegal = get_mode(mode, dVar, i, k, d);
25                      if(!isLegal) {
26                          break;
27                      }
28                      dvProb[k] += dVar;
29                  }
30
31                  if(dvProb[k] >dMax) {
32                      dMax = dvProb[k];
33                  }
34              }
35
36              dTemp = 0.0;
37              for(Size k=0; k<_numclust; ++k) {
38                  dvProb[k] = _cm*dMax - dvProb[k];
39                  dTemp += dvProb[k];
40              }
41              for(Size k=0; k<_numclust; ++k) {
42                  dvProb[k] /= dTemp;
43              }
44
45              dRand = uni();
46              dTemp = dvProb[0];
47              Size k = 0;
48              while( dTemp < dRand) {
49                  dTemp += dvProb[++k];
50              }
51              mTemp.insert(std::pair<Size, Size>(j, k));
52          }
53
54          std::map<Size, Size>::iterator it;
55          for(it = mTemp.begin(); it!=mTemp.end(); ++it) {
56              _mP(i, it->first) = it->second;
57          }
58      }
59
60  }
```

The implementation of the k-modes operator kmode is shown in Listing 17.8. The function takes one argument, which is an index of a chromosome in the current population. The function first creates a local variable nmMode (lines 2–3) to hold the cluster centers. The the function calls get_mode to calculate the cluster centers (lines 7–14). Afterwards, the function checks for each record and assigns the record to the nearest cluster (lines 20–36).

Listing 17.8: Implementation of function kmode.

```
1   void GKmode::kmode(Size ind) const {
2       ublas::matrix<Size>
3           nmMode(_numclust, _ds->num_attr());
4
5       bool isLegal;
6       Real dVar;
7       for(Size k=0; k<_numclust; ++k) {
8           for(Size j=0; j<_ds->num_attr(); ++j) {
9               isLegal = get_mode(nmMode(k,j), dVar, ind, k, j);
10              if(!isLegal) {
11                  return;
12              }
```

```
                }
        }

        Real dDist;
        Real dMin;
        Size h;
        boost::shared_ptr<Schema> schema = _ds->schema();
        for(Size i=0; i<_ds->size(); ++i) {
            dMin = MAX_REAL;
            for(Size k=0; k<_numclust; ++k) {
                dDist = 0.0;
                for(Size j=0; j<_ds->num_attr(); ++j) {
                    if(nmMode(k, j) != (*schema)[j]->get_d_val(
                        (*(*_ds)[i])[j])) {
                        dDist += 1;
                    }
                }
                if(dMin > dDist) {
                    dMin = dDist;
                    h = k;
                }
            }
            _mP(ind, i) = h;
        }
}
```

The complete implementation of class `GKmode` can be found in Listing B.25.

17.3 Examples

In this section, we illustrate the genetic k-modes algorithm implemented in the previous section. The example program is shown in Listing B.118. The program is compiled as part of the clustering library. The `Makefile.am` file for this program is shown in Listing B.117. The program also uses the Boost program options library to handle command line arguments. Once we run the program with argument `--help` or without any arguments, we see the following help message:

```
Allowed options:
  --help                 produce help message
  --datafile arg         the data file
  --k arg (=3)           number of clusters
  --numpop arg (=50)     number of chromosomes in the population
  --maxgen arg (=100)    maximum number of generations
  --c arg (=1.5)         parameter c
  --cm arg (=1.5)        parameter c_m
  --pm arg (=0.2)        mutation probability
```

All the arguments have defaults values except for the argument `datafile`. For example, the default number of clusters is 3. The default number of chromosomes in the population is 50.

Now let us apply the genetic k-modes algorithm to cluster the Soybean dataset with default values. To do this, we execute the following command:

```
examples/gkmode/gkmode.exe --datafile=../Data/soybean-small.data
```

Once the command is executed, we see the following output:

```
Number of records: 47
Number of attributes: 35
Number of numerical attributes: 0
Number of categorical attributes: 35

completed in 10.844 seconds
Clustering Summary:
Number of clusters: 3
Size of Cluster 0: 27
Size of Cluster 1: 10
Size of Cluster 2: 10

Number of given clusters: 4
Cross Tabulation:
Cluster ID  D1   D2   D3   D4
0           0    0    10   17
1           0    10   0    0
2           10   0    0    0
```

From the result we see that the genetic k-modes algorithm clustered the dataset very accurately in that the algorithm did not split the given clusters. However, the algorithm is slow as it took the algorithm about 10 seconds to finish. The efficiency of the algorithm can be improved if we set some early stop criterion. In Exercise A.11, readers are asked to modify the function iteration so that the algorithm can stop early when it converges.

Let us run the program with 20 generations and 4 clusters by executing the following command:

```
examples/gkmode/gkmode.exe --datafile=../Data/soybean-small.data
--maxgen=20 --k=4
```

Once the command is executed, we see the following output:

```
Number of records: 47
Number of attributes: 35
Number of numerical attributes: 0
Number of categorical attributes: 35

completed in 2.781 seconds
Clustering Summary:
```

```
Number of clusters: 4
Size of Cluster 0: 10
Size of Cluster 1: 17
Size of Cluster 2: 10
Size of Cluster 3: 10

Number of given clusters: 4
Cross Tabulation:
Cluster ID    D1    D2    D3    D4
0             10    0     0     0
1             0     0     0     17
2             0     10    0     0
3             0     0     10    0
```

From the output we see that the run time was reduced to about one fourth of the run time of the previous experiment. The algorithm clustered all the records correctly.

Now let us do some experiments by changing the mutation probability. First, we run the program with a mutation probability of 0.1 by executing the following command:

```
examples/gkmode/gkmode.exe --datafile=../Data/soybean-small.data
--maxgen=20 --k=4 --pm=0.1
```

Once the command is executed, we see the following output:

```
Number of records: 47
Number of attributes: 35
Number of numerical attributes: 0
Number of categorical attributes: 35

completed in 1.766 seconds
Clustering Summary:
Number of clusters: 4
Size of Cluster 0: 10
Size of Cluster 1: 10
Size of Cluster 2: 12
Size of Cluster 3: 15

Number of given clusters: 4
Cross Tabulation:
Cluster ID    D1    D2    D3    D4
0             0     10    0     0
1             10    0     0     0
2             0     0     10    2
3             0     0     0     15
```

In this experiment, two records were clustered incorrectly. The run time of this experiment is less than that of the second experiment since the mutation probability is less than the previous one.

Second, we run the program with a mutation probability of 0.8 by executing the following command:

```
examples/gkmode/gkmode.exe --datafile=../Data/soybean-small.data
--maxgen=20 --k=4 --pm=0.8
```

Once the command is executed, we see the following output:

```
Number of records: 47
Number of attributes: 35
Number of numerical attributes: 0
Number of categorical attributes: 35

completed in 9.188 seconds
Clustering Summary:
Number of clusters: 4
Size of Cluster 0: 12
Size of Cluster 1: 9
Size of Cluster 2: 13
Size of Cluster 3: 13

Number of given clusters: 4
Cross Tabulation:
Cluster ID    D1    D2    D3    D4
0             10    1     1     0
1             0     9     0     0
2             0     0     1     12
3             0     0     8     5
```

In this experiment, we see that many records were clustered incorrectly and the run time increased by about four times. From the two experiments we see that the genetic k-modes algorithm is sensitive to the mutation probability. A mutation probability of 0.2 seems to be a good choice. Interested users can also test other parameters such as c and c_m.

17.4 Summary

In this chapter, we implemented the genetic k-modes algorithm. We also illustrated the algorithm with examples and tested the sensitivity of the mutation probability. The genetic k-modes algorithm increases the clustering accuracy of the k-modes algorithm. However, the run time of the genetic k-modes algorithm is significantly longer than that of the k-modes algorithm. One way to improve the genetic k-modes algorithm is to implement an early stop criterion.

The genetic k-modes algorithm implemented in this chapter is one of the search-based clustering algorithms. Other search-based clustering algorithms include the genetic k-means algorithm (GKA) (Krishna and Narasimha, 1999) and clustering algorithms based on the tabu search method (Gan et al., 2007).

Chapter 18

The FSC Algorithm

In fuzzy clustering algorithms such as the c-means algorithm, each record has a fuzzy membership associated with each cluster that indicates the degree of association of the record to the cluster. In the fuzzy subspace clustering (FSC) algorithm, each attribute has a fuzzy membership associated with each cluster that indicates the degree of importance of the attribute to the cluster. In this chapter, we implement the FSC algorithm (Gan, 2007; Gan et al., 2007).

18.1 Description of the Algorithm

The FSC algorithm is an extension of the k-means algorithm for subspace clustering. The FSC algorithm imposes weights on the distance measure of the k-means algorithm. Given a dataset $X = \{\mathbf{x}_0, \mathbf{x}_1, \cdots, \mathbf{x}_{n-1}\}$ consisting of n records, each is described by d numeric attribute. Recall that the objective function of the k-means algorithms is

$$E = \sum_{j=0}^{k-1} \sum_{\mathbf{x} \in C_j} D_{euc}(\mathbf{x}, \boldsymbol{\mu}_j)^2 = \sum_{j=0}^{k-1} \sum_{\mathbf{x} \in C_j} \sum_{r=0}^{d-1} (x_r - \mu_{jr})^2,$$

where $C_0, C_1, \cdots, C_{k-1}$ are k clusters, $\boldsymbol{\mu}_j$ ($0 \leq j \leq k-1$) is the center of cluster C_j, $D_{euc}(\cdot, \cdot)$ is the Euclidean distance, and x_r and μ_{jr} are the rth components of $\mathbf{x}$ and $\boldsymbol{\mu}_j$, respectively.

The objective function of the FSC algorithm is defined as

$$E_{\alpha,\epsilon} = \sum_{j=0}^{k-1} \sum_{\mathbf{x} \in C_j} \sum_{r=0}^{d-1} w_{jr}^\alpha (x_r - \mu_{jr})^2 + \epsilon \sum_{j=0}^{k-1} \sum_{r=0}^{d-1} w_{jr}^\alpha, \qquad (18.1)$$

where $\alpha \in (1, \infty)$ is a weight component or fuzzifier, ϵ is a very small positive real number used to prevent divide-by-zero error, and w_{jr} ($0 \leq j \leq k-1$, $0 \leq r \leq d-1$) is the (j,r) entry of the so called fuzzy dimension weight matrix W. A $k \times d$ weight matrix W satisfies the following conditions:

$$w_{jr} \in [0, 1], \quad 0 \leq j \leq k-1,\, 0 \leq r \leq d-1, \qquad (18.2a)$$

279

$$\sum_{r=0}^{d-1} w_{jr} = 1, \quad 0 \le j \le k-1. \tag{18.2b}$$

The FSC algorithm tries to minimize the objective function defined in Equation (18.1) using an iterative process. The iterative process is very similar to that of the k-means algorithm. That is, the FSC algorithm repeats updating the cluster centers given the fuzzy dimension weight matrix and updating the fuzzy dimension weight matrix given the cluster centers.

At the beginning, the FSC algorithm initializes the cluster centers by selecting k distinct records randomly and initializes the fuzzy dimension weight matrix equally, i.e., $w_{jr}^{(0)} = \frac{1}{d}$ for $j = 0, 1, \cdots, k-1$, $r = 0, 1, \cdots, d-1$. Suppose $\boldsymbol{\mu}_j^{(0)}$ ($j = 0, 1, \cdots, k-1$) are the initial cluster centers. Then the FSC algorithm updates the clusters $C_0, C_1, \cdots, C_{k-1}$ based on the initial cluster centers and the initial fuzzy dimension weight matrix as follows:

$$C_j^{(0)} = \left\{ \mathbf{x} \in X : D_{euc}\left(\mathbf{x}, \boldsymbol{\mu}_j^{(0)}, W^{(0)}\right) = \min_{0 \le l \le k-1} D_{euc}\left(\mathbf{x}, \boldsymbol{\mu}_l^{(0)}, W^{(0)}\right) \right\}, \tag{18.3}$$

for $j = 0, 1, \cdots, k-1$, where

$$D_{euc}\left(\mathbf{x}, \boldsymbol{\mu}_l^{(0)}, W^{(0)}\right) = \sum_{r=0}^{d-1} \left(w_{lr}^{(0)}\right)^\alpha \left(x_r - \mu_{lr}^{(0)}\right)^2, \quad 0 \le l \le k-1. \tag{18.4}$$

Then the FSC algorithm updates the fuzzy dimension weight matrix based on the cluster centers and the clusters according to the following formula:

$$w_{jr}^{(1)} = \left[\sum_{l=0}^{d-1} \left(\frac{\sum_{\mathbf{x} \in C_j^{(0)}} \left(x_r - \mu_{jr}^{(0)}\right)^2 + \epsilon}{\sum_{\mathbf{x} \in C_j^{(0)}} \left(x_l - \mu_{jl}^{(0)}\right)^2 + \epsilon} \right)^{\frac{1}{\alpha-1}} \right]^{-1}$$

$$= \frac{\left(\sum_{\mathbf{x} \in C_j^{(0)}} \left(x_r - \mu_{jr}^{(0)}\right)^2 + \epsilon\right)^{-\frac{1}{\alpha-1}}}{\sum_{l=0}^{d-1} \left(\sum_{\mathbf{x} \in C_j^{(0)}} \left(x_l - \mu_{jl}^{(0)}\right)^2 + \epsilon\right)^{-\frac{1}{\alpha-1}}} \tag{18.5}$$

for $j = 0, 1, \cdots, k-1$ and $r = 0, 1, \cdots, d-1$.

Once the fuzzy dimension weight matrix is updated. The FSC algorithm will update the cluster centers based on the fuzzy dimension weight matrix and the clusters according to the following formula:

$$\mu_{jr}^{(1)} = \frac{\sum_{\mathbf{x} \in C_j^{(0)}} x_r}{|C_j^{(0)}|} \tag{18.6}$$

for $j = 0, 1, \cdots, k-1$ and $r = 0, 1, \cdots, d-1$.

The FSC algorithm repeats the above three steps until the change of the objective function between two iterations is small or the maximum number of iterations is reached.

18.2 Implementation

The FSC algorithm is implemented as a class called **FSC**. The declaration of the class is shown in Listing 18.1. The class has several member functions and data members. All these members are protected. Four data members are declared **mutable**. Data member _clusters is a vector of shared pointers pointing to **SubspaceCluster**. Data member _CM is the vector of cluster memberships. Data members _dObj and _numiter represent the objective function value and the number of iterations, respectively.

Other data members are not mutable. Data member _seed is the seed to initialize the random number generator that is used to select the initial cluster centers. Data members _alpha and _epsilon represent the parameters α and ϵ in Equation (18.1), respectively. Data member _threshold is the tolerance of changes of objective functions. Data members _numclust and _maxiter represent the number of clusters and the maximum number of iterations, respectively.

Listing 18.1: Declaration of class **FSC**.

```
class FSC: public Algorithm {
protected:
    void setupArguments();
    void performClustering() const;
    void fetchResults() const;
    virtual void initialization() const;
    virtual void iteration() const;
    virtual void updateCenter() const;
    virtual void updateWeight() const;
    virtual void calculateObj() const;
    Real dist(const boost::shared_ptr<Record> &x,
             const boost::shared_ptr<SubspaceCluster> &c
             ) const;

    mutable std::vector<boost::shared_ptr<SubspaceCluster> >
        _clusters;
    mutable std::vector<Size> _CM;
    mutable Real _dObj;
    mutable Size _numiter;

    Size _seed;
    Real _alpha;
    Real _epsilon;
    Real _threshold;
    Size _numclust;
    Size _maxiter;
};
```

Class **FSC** declares nine member functions. Method _setupArguments just transfers the parameters in member _arguments into the algorithm and validates these parameters. The implementation of the function can be found in Listing B.23. Method _performClustering calls another two member functions: initialization and iteration.

Method initialization initializes cluster centers and cluster memberships. The implementation of this method is shown in Listing 18.2. The function first allocates space for _CM (line 4). Then the function randomly selects k distinct records from the dataset as cluster centers. Note that the fuzzy dimension weights are initialized equally in the constructor of the class **SubspaceCluster** (see Listing B.51). Afterwards, the function assigns records to clusters based on their distances to cluster centers. Member function _dist (line 30) is called to calculate the distances.

Listing 18.2: Implementation of function initialization.

```
void FSC::initialization() const {
    Size numRecords = _ds->size();
    std::vector<Integer> index(numRecords,0);
    _CM.resize(numRecords);
    for(Size i=0;i<index.size();++i){
        index[i] = i;
    }

    boost::minstd_rand generator(_seed);
    for(Size i=0;i<_numclust;++i){
        boost::uniform_int<> uni_dist(0,numRecords-i-1);
        boost::variate_generator<boost::minstd_rand&,
            boost::uniform_int<> > uni(generator,uni_dist);
        Integer r = uni();
        boost::shared_ptr<Record> cr =
            boost::shared_ptr<Record>(new Record(*(*_ds)[r]));
        boost::shared_ptr<SubspaceCluster> c =
            boost::shared_ptr<SubspaceCluster>(
            new SubspaceCluster(cr));
        c->set_id(i);
        _clusters.push_back(c);
        index.erase(index.begin()+r);
    }

    Integer s;
    Real dMin,dDist;
    for(Size i=0;i<numRecords;++i){
        dMin = MAX_REAL;
        for(Size j=0;j<_numclust;++j){
            dDist = dist((*_ds)[i], _clusters[j]);
            if (dDist<dMin){
                s = j;
                dMin = dDist;
            }
        }
        _clusters[s]->add((*_ds)[i]);
        _CM[i] = s;
    }
}
```

Member function dist is very simple. As we can see from its implementation in Listing 18.3, the function calculates the distance according to Equation (18.4). Method w() (line 7) is a member function of class **SubspaceCluster** and returns a component of the fuzzy dimension weight vector of a cluster.

Listing 18.3: Implementation of function dist.

```
Real FSC::dist(const boost::shared_ptr<Record> &x,
               const boost::shared_ptr<SubspaceCluster> &c
              ) const {
    Real dTemp = 0.0;
    boost::shared_ptr<Schema> schema = _ds->schema();
    for(Size j=0;j<schema->size();++j){
        dTemp += std::pow(c->w(j), _alpha) *
            std::pow((*schema)[j]->distance((*x)[j],
            (*c->center())[j]), 2.0);
    }

    return dTemp;
}
```

Once method initialization is called, method performClustering calls method iteration to do the clustering. Listing 18.4 shows the implementation of method iteration. The function first calls another member function updateWeight to update the fuzzy dimension weight matrix and then calls member function updateCenter to update the cluster centers. Member functions updateWeight and updateCenter update the fuzzy dimension weight matrix and the cluster centers according to Equation (18.5) and Equation (18.6), respectively.

After the first iteration, the function goes into an infinite while loop (lines 7–40) to repeat the process: updating the cluster memberships (lines 10–25), updating the fuzzy dimension weight matrix (line 27), and updating the cluster centers (line 28). Then the function calculates the value of the objective function (line 31) by calling member function calculateObj and checks whether the change of the objective function is less than the threshold. If the change of the objective function is less than the threshold, then the function terminates itself. The function also checks whether the maximum number of iterations is reached or not. If the maximum number of iterations is reached, then the function terminates itself.

Listing 18.4: Implementation of function iteration.

```
void FSC::iteration() const {
    Real dObjPre;

    updateWeight();
    updateCenter();
    _numiter = 1;
    while(true) {
        Integer s;
        Real dMin,dDist;
        for(Size i=0;i<_ds->size();++i) {
            dMin = MAX_REAL;
            for(Size k=0;k<_clusters.size();++k) {
                dDist = dist((*_ds)[i],_clusters[k]);
                if (dMin > dDist) {
                    dMin = dDist;
                    s = k;
                }
            }

            if (_CM[i] != s){
                _clusters[_CM[i]]->erase((*_ds)[i]);
                _clusters[s]->add((*_ds)[i]);
```

```
23                  _CM[i] = s;
24              }
25          }
26
27          updateWeight();
28          updateCenter();
29
30          dObjPre = _dObj;
31          calculateObj();
32          if(std::fabs(dObjPre - _dObj) < _threshold){
33              break;
34          }
35
36          ++_numiter;
37          if (_numiter > _maxiter){
38              break;
39          }
40      }
41  }
```

Once method `iteration` is finished, member function `fetchResults` is called to transfer clustering results to data member `_results`. The clustering results that are transferred to `_results` are cluster memberships, clusters, number of iterations, and the objective function value. The complete implementation of class `FSC` can be found in Listing B.23.

18.3 Examples

In this section, we apply the FSC algorithm implemented in the previous section to the synthetic dataset (see Figure 1.1) and the Iris dataset. Since the FSC algorithm initializes the cluster centers by randomly selecting k distinct records from a dataset, we need to run the algorithm multiple times by using different random seeds. The code in Listing 18.5 shows the part of the code to run the FSC algorithm multiple times. The complete code of the example can be found in Listing B.120.

Listing 18.5: Code to run the FSC algorithm multiple times.
```
1   boost::timer t;
2   t.restart();
3
4   Results Res;
5   Real avgiter = 0.0;
6   Real avgerror = 0.0;
7   Real dMin = MAX_REAL;
8   Real error;
9   for(Size i=1; i<=numrun; ++i) {
10      FSC ca;
11      Arguments &Arg = ca.getArguments();
12      Arg.ds = ds;
13      Arg.insert("alpha", alpha);
14      Arg.insert("epsilon", epsilon);
15      Arg.insert("threshold", threshold);
16      Arg.insert("numclust", numclust);
17      Arg.insert("maxiter", maxiter);
```

```
18        Arg.insert("seed", seed);
19        if (numrun == 1) {
20            Arg.additional["seed"] = seed;
21        } else {
22            Arg.additional["seed"] = i;
23        }
24
25        ca.clusterize();
26
27        const Results &tmp = ca.getResults();
28        avgiter += boost::any_cast<Size>(tmp.get("numiter"));
29        error = boost::any_cast<Real>(tmp.get("dObj"));
30        avgerror += error;
31        if (error < dMin) {
32            dMin = error;
33            Res = tmp;
34        }
35   }
36   avgiter /= numrun;
37   avgerror /= numrun;
38
39   double seconds = t.elapsed();
40   std::cout<<"completed in "<<seconds<<" seconds"<<std::endl;
```

The program runs the FSC algorithm **numrun** times and saves the run that has the lowest error or objective function value. The average number of iterations and the average error are also saved. For each run, the program declares a new local object of class **FSC** (line 10) in order to prevent previous runs from affecting the current run.

The Boost program options library is used in the example to handle command line options. When we execute the program with option `--help` or without any options, we see the following output:

```
Allowed options:
  --help                   produce help message
  --datafile arg           the data file
  --k arg (=3)             number of clusters
  --seed arg (=1)          seed used to choose random initial centers
  --maxiter arg (=100)     maximum number of iterations
  --numrun arg (=1)        number of runs
  --epsilon arg (=0)       epsilon
  --alpha arg (=2.1)       alpha
  --threshold arg (=1e-12) Objective function tolerance
```

From the help message we see that the program can take eight different arguments. The only required argument is `datafile`, which tells the location of the dataset. All other arguments have default values, which are shown in the round brackets after the argument name. For example, the default number of clusters is 3.

Now let us apply the FSC program to the synthetic dataset. Suppose that we are in the directory `ClusLib` and the program is in the directory `ClusLib/examples/fsc`. To cluster the synthetic dataset using the FSC algorithm, we can execute the following command:

```
examples/fsc/fsc.exe --datafile=../Data/600points.csv --seed=2
```

Once the command is executed, we see the following output:

```
Number of records: 600
Number of attributes: 2
Number of numerical attributes: 2
Number of categorical attributes: 0

completed in 0.015 seconds
Clustering Summary:
Number of clusters: 3
Size of Cluster 0: 201
Size of Cluster 1: 200
Size of Cluster 2: 199

Number of given clusters: 3
Cross Tabulation:
Cluster ID   1    2    3
0            200  0    1
1            0    200  0
2            0    0    199

Attribute Weights:
Cluster 0, 0.510489, 0.489511
Cluster 1, 0.542627, 0.457373
Cluster 2, 0.481934, 0.518066

Number of run: 1
Average number of iterations: 6
Average error: 282.729
Number of iterations for the best case: 6
Best error: 282.729
```

From the output we see that the FSC algorithm converged after 6 iterations and clustered one record incorrectly. Since the attribute weights for all three clusters are close to each other, the two attributes are equally important for these clusters. In addition, the FSC algorithm is very fast as it took the algorithm about 0.015 seconds to cluster the dataset.

Let us run the algorithm multiple times to see if we can improve the results. To do this, we issue the following command:

```
examples/fsc/fsc.exe --datafile=../Data/600points.csv
--numrun=100
```

Once the command is executed, we see the following output:

```
Number of records: 600
Number of attributes: 2
```

```
Number of numerical attributes: 2
Number of categorical attributes: 0

completed in 1.313 seconds
Clustering Summary:
Number of clusters: 3
Size of Cluster 0: 201
Size of Cluster 1: 199
Size of Cluster 2: 200

Number of given clusters: 3
Cross Tabulation:
Cluster ID     1    2    3
0              200  0    1
1              0    0    199
2              0    200  0

Attribute Weights:
Cluster 0, 0.510489, 0.489511
Cluster 1, 0.481934, 0.518066
Cluster 2, 0.542627, 0.457373

Number of run: 100
Average number of iterations: 5.89
Average error: 298.389
Number of iterations for the best case: 4
Best error: 282.729
```

From the output of this experiment and that of the previous one we see that 282.729 is the best error (or lowest objective function value) the algorithm can obtain. However, the average error over the 100 runs is 298.389, meaning that some runs produced bad clusterings due to the bad initial cluster centers.

Now let us apply the FSC algorithm to cluster the Iris dataset and run the algorithm multiple times. To do this, we execute the following command:

```
examples/fsc/fsc.exe --datafile=../Data/bezdekIris.data
--numrun=100
```

Once the command is executed, we see the following output:

```
Number of records: 150
Number of attributes: 4
Number of numerical attributes: 4
Number of categorical attributes: 0

completed in 0.766 seconds
Clustering Summary:
```

Number of clusters: 3
Size of Cluster 0: 50
Size of Cluster 1: 48
Size of Cluster 2: 52

Number of given clusters: 3
Cross Tabulation:

Cluster ID	Iris-setosa	Iris-versicolor	Iris-virginica
0	50	0	0
1	0	2	46
2	0	48	4

Attribute Weights:
Cluster 0, 0.0690569, 0.0605081, 0.25014, 0.620295
Cluster 1, 0.0918354, 0.330862, 0.106388, 0.470914
Cluster 2, 0.0907508, 0.228124, 0.0949321, 0.586193

Number of run: 100
Average number of iterations: 8.07
Average error: 3.36246
Number of iterations for the best case: 4
Best error: 2.5031

From the output we see that the best run clustered six records incorrectly. From the output attribute weights we see that the last two attributes are important to form the first cluster since the weights of the last two attributes are higher than the weights of the rest of the attributes. We also see that the second and the last attributes are important to form the second and third clusters.

It is very interesting to see the impact of the parameter α on the clustering results. The parameter α can be any real number in $(1, \infty)$. To test the impact, we first run the program with $\alpha = 1.2$ and then run the program with $\alpha = 10$.

To run the algorithm with $\alpha = 1.2$, we execute the following command:

```
examples/fsc/fsc.exe --datafile=../Data/bezdekIris.data
--numrun=100 --alpha=1.2
```

Once the command is executed, we see the following output:

Number of records: 150
Number of attributes: 4
Number of numerical attributes: 4
Number of categorical attributes: 0

completed in 0.781 seconds
Clustering Summary:
Number of clusters: 3

```
Size of Cluster 0: 43
Size of Cluster 1: 46
Size of Cluster 2: 61

Number of given clusters: 3
Cross Tabulation:
Cluster ID    Iris-setosa    Iris-versicolor    Iris-virginica
0             43             0                  0
1             0              27                 19
2             7              23                 31

Attribute Weights:
Cluster 0, 2.80013e-06, 1.3118e-06, 0.00346483, 0.996531
Cluster 1, 1.39861e-05, 0.998365, 7.52554e-07, 0.00162062
Cluster 2, 1.07266e-06, 0.999991, 4.16246e-09, 7.89051e-06

Number of run: 100
Average number of iterations: 8.18
Average error: 6.01705
Number of iterations for the best case: 6
Best error: 4.71667
```

From the output we see that when α is close to 1, the clusters are dominated by one attribute. As a result, the clustering results are not good.

To run the algorithm with $\alpha = 10$, we execute the following command:

```
examples/fsc/fsc.exe --datafile=../Data/bezdekIris.data
--numrun=100 --alpha=10
```

Once the command is executed, we see the following output:

```
Number of records: 150
Number of attributes: 4
Number of numerical attributes: 4
Number of categorical attributes: 0

completed in 0.828 seconds
Clustering Summary:
Number of clusters: 3
Size of Cluster 0: 50
Size of Cluster 1: 48
Size of Cluster 2: 52

Number of given clusters: 3
Cross Tabulation:
Cluster ID    Iris-setosa    Iris-versicolor    Iris-virginica
0             50             0                  0
```

1	0	2	46
2	0	48	4

```
Attribute Weights:
Cluster 0, 0.22411, 0.220519, 0.26229, 0.293081
Cluster 1, 0.226815, 0.26528, 0.23093, 0.276975
Cluster 2, 0.228265, 0.255486, 0.229525, 0.286723

Number of run: 100
Average number of iterations: 8.73
Average error: 7.0706e-05
Number of iterations for the best case: 10
Best error: 5.81597e-05
```

From the output we see that when α is large, the attributes have approximately the same weights. That is, all the attributes are equally important to form the clusters. Hence, when α is large, the FSC algorithm produces the same clustering results as the k-means algorithm does.

18.4 Summary

In this chapter, we implemented the FSC algorithm and illustrated the algorithm with several examples. The FSC algorithm is an extension of the k-means algorithm for subspace clustering. The FSC algorithm also applies the idea of fuzzy sets to attribute selection. Rather than treating an attribute as relevant or not relevant to a cluster, the FSC algorithm assigns a weight to the attribute to indicate the importance of the attribute. More information about the FSC algorithm and other relevant algorithms (e.g., mean shift for subspace clustering) can be found in Gan (2007) and Gan et al. (2007).

Chapter 19

The Gaussian Mixture Algorithm

Clustering based on Gaussian mixture models is a classical and powerful approach. Celeux and Govaert (1995) summarized sixteen Gaussian mixture models which result in sixteen clustering algorithms. These sixteen Gaussian mixture models are based on different assumptions on the component variance matrices. Four commonly used Gaussian mixture models are (Celeux and Govaert, 1995):

(a) No restriction is imposed on the component variance matrices $\Sigma_0, \Sigma_1, \cdots,$ and Σ_{k-1};

(b) $\Sigma_0 = \Sigma_1 = \cdots = \Sigma_{k-1} = \Sigma$;

(c) $\Sigma_0 = \Sigma_1 = \cdots = \Sigma_{k-1} = \text{Diag}(\sigma_0^2, \sigma_1^2, \cdots, \sigma_{d-1}^2)$, where $\sigma_0, \sigma_1, \cdots, \sigma_{d-1}$ are unknown;

(d) $\Sigma_0 = \Sigma_1 = \cdots = \Sigma_{k-1} = \text{Diag}(\sigma^2, \sigma^2, \cdots, \sigma^2)$, where σ is unknown.

In this chapter, we implement the clustering algorithm based on the first Gaussian mixture model, i.e., the most general one.

19.1 Description of the Algorithm

Let $X = \{\mathbf{x}_0, \mathbf{x}_1, \cdots, \mathbf{x}_{n-1}\}$ be a numeric dataset containing n records, each of which is described by d numeric attributes. In Gaussian mixture models, each record in the dataset X is assumed to be a sample drawn from a distribution characterized by the following probability density function (Celeux and Govaert, 1995):

$$f(\mathbf{x}) = \sum_{j=0}^{k-1} p_j \Phi(\mathbf{x}|\boldsymbol{\mu}_j, \Sigma_j), \tag{19.1}$$

where $\boldsymbol{\mu}_j$ is the mean of the jth component, Σ_j is the variance of the jth component, p_j is the mixing proportion of the jth component, and Φ is the probability density function of the multivariate Gaussian distribution.

The mixing proportions $p_0, p_1, \cdots, p_{k-1}$ in Equation (19.1) satisfy the following conditions:

(a) $p_j \in (0,1)$ for $j = 0, 1, \cdots, k-1$;

(b) The sum of the mixing proportions is equal to 1, i.e.,

$$\sum_{j=0}^{k-1} p_j = 1.$$

The probability density function of the multivariate Gaussian distribution is defined as

$$\Phi(\mathbf{x}|\boldsymbol{\mu}, \Sigma) = \frac{1}{\sqrt{(2\pi)^d |\Sigma|}} \exp\left(-\frac{1}{2}(\mathbf{x} - \boldsymbol{\mu})^T \Sigma^{-1} (\mathbf{x} - \boldsymbol{\mu})\right), \qquad (19.2)$$

where $\boldsymbol{\mu}$ is the mean, Σ is the variance matrix, and $|\Sigma|$ is the determinant of Σ. Here we assume that $\mathbf{x}$ and $\boldsymbol{\mu}$ are column vectors.

There are two approaches to cluster a dataset based on the Gaussian mixture model (Celeux and Govaert, 1995): the mixture approach and the classification approach. In the mixture approach, the likelihood is maximized over the mixture parameters (i.e., $\boldsymbol{\mu}_j$ and Σ_j). In the classification approach, the likelihood is maximized over the mixture parameters as well as over the identifying labels of the mixture component origin for each record.

In this chapter, we implement the Gaussian mixture model-based clustering algorithm based on the first approach, i.e., the mixture approach. In this approach, the parameters that need to be estimated are

$$\Theta = (p_0, p_1, \cdots, p_{k-1}, \boldsymbol{\mu}_0, \boldsymbol{\mu}_1, \cdots, \boldsymbol{\mu}_{k-1}, \Sigma_0, \Sigma_1, \cdots, \Sigma_{k-1}).$$

We use the EM algorithm (McLachlan and Krishnan, 1997) to estimate these parameters by maximizing the log-likelihood. Given the dataset X, the log-likelihood is defined as

$$\mathcal{L}(\Theta|X) = \sum_{i=0}^{n-1} \ln\left(\sum_{j=0}^{k-1} p_j \Phi(\mathbf{x}_i|\boldsymbol{\mu}_j, \Sigma_j)\right). \qquad (19.3)$$

To estimate the parameter Θ, the EM algorithm starts with an initial parameter $\Theta^{(0)}$ and repeats the E-step and the M-step until it converges or the maximum number of iterations is reached. In the E-step, the conditional probabilities $t_j(\mathbf{x}_i)$ ($0 \leq j \leq k-1$, $0 \leq i \leq n-1$) that $\mathbf{x}_i$ comes from the jth component are calculated according to the following equation (Celeux and Govaert, 1995):

$$t_j(\mathbf{x}_i) = \frac{p_j \Phi(\mathbf{x}_i|\boldsymbol{\mu}_j, \Sigma_j)}{\sum_{s=0}^{k-1} p_s \Phi(\mathbf{x}_i|\boldsymbol{\mu}_s, \Sigma_s)}, \qquad (19.4)$$

where $p_j, \boldsymbol{\mu}_j, \Sigma_j$ ($0 \leq j \leq k-1$) are current estimates of Θ.

In the M-step, the parameter Θ is estimated based on the conditional probabilities $t_j(\mathbf{x}_i)$ ($0 \le j \le k-1$, $0 \le i \le n-1$) according to the following equations (Celeux and Govaert, 1995):

$$p_j = \frac{\sum_{i=0}^{n-1} t_j(\mathbf{x}_i)}{\sum_{s=0}^{k-1} \sum_{i=0}^{n-1} t_s(\mathbf{x}_i)}, \qquad (19.5a)$$

$$\boldsymbol{\mu}_j = \frac{\sum_{i=0}^{n-1} t_j(\mathbf{x}_i)\mathbf{x}_i}{\sum_{i=0}^{n-1} t_j(\mathbf{x}_i)}, \qquad (19.5b)$$

$$\Sigma_j = \frac{\sum_{i=0}^{n-1} t_j(\mathbf{x}_i)(\mathbf{x}_i - \boldsymbol{\mu}_j) \cdot (\mathbf{x}_i - \boldsymbol{\mu}_j)^T}{\sum_{i=0}^{n-1} t_j(\mathbf{x}_i)}, \qquad (19.5c)$$

for $j = 0, 1, \cdots, k-1$.

In our implementation, the initial parameters $\Theta^{(0)}$ are chosen as follows:

$$p_j = \frac{1}{k}, \qquad (19.6a)$$

$$\boldsymbol{\mu}_j = \mathbf{x}_{i_j}, \qquad (19.6b)$$

$$\Sigma_j = \text{Diag}(\sigma_0^2, \sigma_1^2, \cdots, \sigma_{d-1}^2), \qquad (19.6c)$$

for $j = 0, 1, \cdots, k-1$, where i_j ($0 \le j \le k-1$) are random integers chosen from $\{0, 1, \cdots, n-1\}$ and

$$\sigma_s^2 = \frac{1}{n-1} \sum_{i=0}^{n-1} x_{is}^2 - \frac{1}{n(n-1)} \left(\sum_{i=0}^{n-1} x_{is} \right)^2, \quad s = 0, 1, \cdots, d-1.$$

Here x_{is} is the sth component of $\mathbf{x}_i$, and σ_s^2 is the sample variance of the sth attribute.

Once the parameter Θ is determined by the EM algorithm, the cluster memberships $\gamma_0, \gamma_1, \cdots, \gamma_{n-1}$ can be derived from Equation (19.4) as follows:

$$\gamma_i = \underset{0 \le j \le k-1}{\text{argmax}}\, t_j(\mathbf{x}_i), \quad i = 0, 1, \cdots, n-1. \qquad (19.7)$$

19.2 Implementation

The Gaussian mixture clustering algorithm is implemented as a class GMC, which is derived from the base class Algorithm. The declaration of class GMC is shown in Listing 19.1. The class declares several methods and many data members.

Data member _mu is used to hold the cluster centers. This member is a uBLAS matrix. Each row of this member represents a center. The matrix has k rows. Data member _sigma is a vector of uBLAS symmetric matrices, each of which represents a variance matrix for a Gaussian component.

Data member _p is the vector of mixing proportions of the k components. Data member _data is just the dataset in matrix form. Since the Gaussian mixture clustering algorithm is used to cluster numeric data and involves matrix operations, it is convenient to put the data in a matrix.

Data member _ll is a real number representing the log-likelihood defined in Equation (19.3). Data member _post is a matrix that stores the posterior probabilities defined in Equation (19.4). The entry in the ith row and the jth column of _post represents $t_j(\mathbf{x}_i)$.

Data member _clusters is a vector of shared pointers pointing to objects of CenterCluster. The center of a cluster is the mean of the corresponding component. Data member _CM is the vector of cluster memberships. Data member _converged is a boolean variable indicating whether the algorithm is converged or not. Data member _numiter is the actual number of iterations the algorithm went through.

Listing 19.1: Declaration of class GMC.

```
class GMC: public Algorithm {
protected:
    void setupArguments();
    void performClustering() const;
    void fetchResults() const;
    virtual void initialization() const;
    virtual void iteration() const;
    virtual void estep() const;
    virtual void mstep() const;

    mutable ublas::matrix<Real> _mu;
    mutable std::vector<ublas::symmetric_matrix<Real> > _sigma;
    mutable std::vector<Real> _p;
    mutable ublas::matrix<Real> _data;
    mutable Real _ll;
    mutable ublas::matrix<Real> _post;
    mutable std::vector<boost::shared_ptr<CenterCluster> >
        _clusters;
    mutable std::vector<Size> _CM;
    mutable bool _converged;
    mutable Size _numiter;

    Size _seed;
    Size _numclust;
    Size _maxiter;
    Real _threshold;
    Real _epsilon;
};
```

The data membership mentioned above are declared as mutable. Their members can be modified by const functions. The other five data members (lines 23–27) are not mutable. Their members are parameters and cannot be changed by the algorithm.

The Gaussian Mixture Algorithm

Data member _seed is the seed used to initialize the random number generator, which is used to generate random cluster centers from the dataset. Data member _numclust is the number of components or clusters. Data member _maxiter is the maximum number of iterations the algorithm is allowed to run. The actual number of iterations may be less than the maximum number of iterations since the algorithm may terminate earlier when it converges. Data member _threshold is a very small number used to stop the algorithm earlier if the increase of log-likelihood between two iterations is less than this number. Data member _epsilon is a regularization parameter that is added to the diagonal elements of the variance matrices in order to make these matrices positive definite.

Now let us look at the member functions. Method setupArguments overrides the same function in the base class. Listing 19.2 shows the implementation of this method. From the code we see that this method transfers parameters from member _arguments into the algorithm and validates these parameters. For example, the threshold must be positive and the epsilon must be nonnegative.

Listing 19.2: Implementation of method setupArguments.

```
 1  void GMC::setupArguments() {
 2      Algorithm::setupArguments();
 3      ASSERT(_ds->is_numeric(), "dataset is not numeric");
 4
 5      _seed = boost::any_cast<Size>(
 6          _arguments.get("seed"));
 7      ASSERT(_seed>0, "seed must be a positive number");
 8
 9      _numclust = boost::any_cast<Size>(
10          _arguments.get("numclust"));
11      ASSERT(_numclust>=2 && _numclust<=_ds->size(),
12          "invalid numclust");
13
14      _maxiter = boost::any_cast<Size>(
15          _arguments.get("maxiter"));
16      ASSERT(_maxiter>0, "invalide maxiter");
17
18      _threshold = boost::any_cast<Real>(
19          _arguments.get("threshold"));
20      ASSERT(_threshold>0, "threshold must be positive");
21
22      _epsilon = boost::any_cast<Real>(
23          _arguments.get("epsilon"));
24      ASSERT(_epsilon>=0, "epsilon must be nonnegative");
25  }
```

Method performClustering calls another two member functions: method initialization and method iteration. The implementation of member function initialization is shown in Listing 19.3. This method first allocates spaces for data members _mu, _sigma, _p, _data, and _post (lines 5–13). Member _p and member _data are also initialized. Component variances _sigma are initialized to be diagonal matrices, whose diagonal elements are attribute variances (lines 25–42). Cluster centers are initialized to be distinct records selected randomly (lines 50–68). Note that the selected records are copied and the copies are used as cluster centers.

Listing 19.3: Implementation of method `initialization`.

```cpp
void GMC::initialization() const {
    Size numRecords = _ds->size();
    Size numAttr = _ds->num_attr();

    _mu.resize(_numclust, numAttr);
    _sigma.resize(_numclust);
    _p.resize(_numclust);
    for(Size i=0; i<_numclust; ++i) {
        _sigma[i].resize(numAttr);
        _p[i] = 1.0 / _numclust;
    }
    _data.resize(numRecords, numAttr);
    _post.resize(numRecords, _numclust);

    std::vector<Real> mean(numAttr,0.0);
    boost::shared_ptr<Schema> schema = _ds->schema();
    for(Size i=0; i<numRecords; ++i) {
        for(Size j=0; j<numAttr; ++j) {
            Real val = (*schema)[j]->get_c_val((*_ds)(i,j));
            _data(i,j) = val;
            mean[j] += val;
        }
    }

    for(Size j=0; j<numAttr; ++j) {
        mean[j] /= numRecords;
    }

    for(Size i=0; i<numAttr; ++i) {
        for(Size j=0; j<i; ++j) {
            _sigma[0](i, j ) = 0.0;
        }
        _sigma[0](i, i) = ( ublas::inner_prod(
            ublas::column(_data, i),
            ublas::column(_data, i)) -
            numRecords*mean[i]*mean[i])
            / (numRecords - 1.0) + _epsilon;
    }

    for(Size i=1; i<_numclust; ++i) {
        _sigma[i] = _sigma[0];
    }

    _CM.resize(numRecords);
    std::vector<Integer> index(numRecords,0);
    for(Size i=0;i<index.size();++i){
        index[i] = i;
    }

    boost::minstd_rand generator(
        static_cast<unsigned int>(_seed));
    for(Size i=0;i<_numclust;++i){
        boost::uniform_int<> uni_dist(0,numRecords-i-1);
        boost::variate_generator<boost::minstd_rand&,
            boost::uniform_int<> > uni(generator, uni_dist);
        Integer r = uni();
        for(Size j=0; j<numAttr; ++j) {
            _mu(i, j) = _data(r,j);
        }

        boost::shared_ptr<Record> rec = (*_ds)[r];
        boost::shared_ptr<Record> cr = boost::shared_ptr
            <Record>(new Record(*rec));
        boost::shared_ptr<CenterCluster> c = boost::shared_ptr
            <CenterCluster>(new CenterCluster(cr));
        c->set_id(i);
```

```
67          _clusters.push_back(c);
68          index.erase(index.begin()+r);
69      }
70  }
```

Method `iteration` is very simple as we can see from its implementation shown in Listing 19.4. The method goes through a maximum number of iterations and calls methods `estep` and method `mstep` in each iteration. The method may terminate earlier if the algorithm is converged. The actual number of iterations is recorded.

Listing 19.4: Implementation of method `iteration`.

```
1   void GMC::iteration() const {
2       Real pre_ll = MIN_REAL;
3       _converged = false;
4       _numiter = 0;
5       for(Size iter=0; iter<_maxiter; ++iter) {
6           estep();
7
8           Real dTemp = _ll - pre_ll;
9           if( dTemp >=0 && dTemp < _threshold) {
10              _converged = true;
11              break;
12          }
13          pre_ll = _ll;
14
15          mstep();
16          ++_numiter;
17      }
18  }
```

Method `estep` and method `mstep` are the most important functions of class GMC. Method `estep` implements the calculation defined in Equation (19.4) and method `mstep` implements the calculation defined in Equation (19.5).

The implementation of method `estep` is shown in Listing 19.5. This method calculates the Cholesky decomposition of the variance matrix of each component (line 9) and calculates the logarithm of the determinant of the variance matrix using the diagonal elements of the Cholesky decomposition (lines 13–15). The function calculates the log-likelihood for each record given that the record comes from a component (lines 17–25) using the following formula:

$$\ln\left(p_j \Phi(\mathbf{x}_i | \boldsymbol{\mu}_j, \Sigma_j)\right)$$
$$= \ln(p_j) - \frac{1}{2} \cdot d \cdot \ln(2\pi) - \frac{1}{2} \cdot \ln(|\Sigma_j|) - \frac{1}{2}(\mathbf{x}_i - \boldsymbol{\mu}_j)^T \Sigma_j^{-1} (\mathbf{x}_i - \boldsymbol{\mu}_j).$$

Note that the last term is just the squared Mahalanobis distance (see Section 9.6).

Then the method calculates the posterior probabilities using Equation (19.4) and the total log-likelihood defined in Equation (19.3). During the calculation, the mean log-likelihood for each record over all components is subtracted from the individual log-likelihood (line 40). This is used to increase the stability of the algorithm since we are dealing with exponential numbers.

However, the subtracted number is canceled or added back in the posterior probabilities and the total log-likelihood.

Listing 19.5: Implementation of method `estep`.

```cpp
void GMC::estep() const {
    Size numRecords = _ds->size();
    Size numAttr = _ds->num_attr();

    ublas::matrix<Real> log_ll(numRecords, _numclust);
    ublas::triangular_matrix<Real> L(numAttr, numAttr);

    for(Size j=0; j<_numclust; ++j) {
        Size k = chol(_sigma[j], L);
        ASSERT(k==0, "invalid covariance matrix " << j);

        Real logDet = 0.0;
        for(Size i=0; i<numAttr; ++i) {
            logDet += 2.0 * std::log(L(i,i));
        }

        MahalanobisDistance md(_sigma[j]);
        for(Size i=0; i<numRecords; ++i) {
            Real dist = md((*_ds)[i], _clusters[j]->center());
            log_ll(i,j) =
                - 0.5 * dist * dist
                - 0.5 * logDet + std::log(_p[j])
                - 0.5 * numAttr * std::log(2*3.141592653589);
        }
    }

    ublas::vector<Real> mlog_ll(numRecords);
    ublas::vector<Real> ones(_numclust);
    for(Size i=0; i<_numclust; ++i) {
        ones(i) = 1.0 / _numclust;
    }

    ublas::axpy_prod(log_ll, ones, mlog_ll, true);
    Real temp = 0.0;
    Real density;
    _ll = 0.0;
    for(Size i=0; i<numRecords; ++i) {
        density = 0.0;
        for(Size j=0; j<_numclust; ++j) {
            _post(i,j) = std::exp(log_ll(i,j)-mlog_ll(i));
            density += _post(i,j);
        }
        for(Size j=0; j<_numclust; ++j) {
            _post(i,j) /= density;
        }
        _ll += std::log(density) + mlog_ll(i);
    }
}
```

The implementation of member function `mstep` is shown in Listing 19.6. This method first assigns the sum of posterior probabilities related to a component to the mixing proportion of the component (lines 6–9). The mixing proportions are normalized later (lines 38–40). The mean of a component is updated according to Equation (19.5b). Cluster centers are also updated since we need to use the cluster centers to calculate the Mahalanobis distance in method `estep`.

Then the method subtracts the mean of a component from the dataset (lines 21–26) and uses the centered data to estimate the variance for the component according to Equation (19.5c) (lines 27–35). Equation (19.5c) can be written as

$$\Sigma_j(r,s) = \frac{\sum_{i=0}^{n-1} t_j(\mathbf{x}_i)(x_{ir} - \mu_{jr}) \cdot (x_{is} - \mu_{js})}{\sum_{i=0}^{n-1} t_j(\mathbf{x}_i)}$$

$$= \frac{\sum_{i=0}^{n-1} \left[\sqrt{t_j(\mathbf{x}_i)}(x_{ir} - \mu_{jr})\right] \cdot \left[\sqrt{t_j(\mathbf{x}_i)}(x_{is} - \mu_{js})\right]}{\sum_{i=0}^{n-1} t_j(\mathbf{x}_i)}$$

for $r = 0, 1, \cdots, d-1$ and $s = 0, 1, \cdots, d-1$.

Note that the regularization parameter _epsilon is added to the diagonal elements of the variance matrix. This is used to make the variance matrix positive definite.

Listing 19.6: Implementation of method mstep.

```
void GMC::mstep() const {
    boost::shared_ptr<Schema> schema = _ds->schema();
    Size numRecords = _ds->size();
    Size numAttr = _ds->num_attr();
    Real psum = 0.0;
    for(Size j=0; j<_numclust; ++j) {
        _p[j] = ublas::sum( ublas::column(_post, j));
        psum += _p[j];
    }

    ublas::matrix<Real> centered(numRecords, numAttr);
    for(Size k=0; k<_numclust; ++k) {
        for(Size j=0; j<numAttr; ++j) {
            _mu(k, j) = ublas::inner_prod(
                ublas::column(_post, k),
                ublas::column(_data, j) ) / _p[k];
            (*schema)[j]->set_c_val(
                (*_clusters[k]->center())[j], _mu(k, j));
        }

        for(Size i=0; i<numRecords; ++i) {
            for(Size j=0; j<numAttr; ++j) {
                centered(i,j) = std::sqrt(_post(i, k)) *
                    (_data(i, j) - _mu(k, j));
            }
        }
        for(Size i=0; i<numAttr; ++i) {
            for(Size j=0; j<=i; ++j) {
                _sigma[k](i, j) =
                    ublas::inner_prod(
                        ublas::column(centered, i),
                        ublas::column(centered, j))/_p[k];
            }
            _sigma[k](i,i) += _epsilon;
        }
    }

    for(Size j=0; j<_numclust; ++j) {
        _p[j] /= psum;
```

```
40      }
41  }
```

The implementation of method `fetchResults` is shown in Listing 19.7. This method collects clustering results and transfers these results to data member _results to which users have access. The method first assigns records to clusters based on the posterior probabilities (lines 2–13). Then the function puts some selected clustering results into _results.

Listing 19.7: Implementation of method `fetchResults`.

```
1   void GMC::fetchResults() const {
2       for(Size i=0; i<_ds->size(); ++i) {
3           Real dMax = MIN_REAL;
4           Integer k;
5           for(Size j=0; j<_numclust; ++j) {
6               if(dMax < _post(i,j)){
7                   dMax = _post(i,j);
8                   k = j;
9               }
10          }
11          _CM[i] = k;
12          _clusters[k]->add((*_ds)[i]);
13      }
14
15      PClustering pc;
16      for(Size i=0;i<_clusters.size();++i){
17          pc.add(_clusters[i]);
18      }
19      _results.insert("pc", boost::any(pc));
20      _results.CM = _CM;
21      _results.insert("converged", boost::any(_converged));
22      _results.insert("p", boost::any(_p));
23      _results.insert("numiter", boost::any(_numiter));
24      _results.insert("ll", boost::any(_ll));
25      _results.insert("mu", boost::any(_mu));
26  }
```

The complete header file and source file of class `GMC` can be found in Listing B.26 and Listing B.27, respectively.

19.3 Examples

In this section, we apply the Gaussian mixture clustering algorithm implemented in the previous section to the synthetic dataset (see Figure 1.1) and the Iris dataset. The program is shown in Listing B.122. The program is compiled as part of the clustering library. One can see the `Makefile.am` file in Listing B.121.

We use the Boost program options library to parse command line options. Suppose we are in the directory `ClusLib` and the program is in the directory `ClusLib/examples/gmc`. When we execute the program using the following command:

```
examples/gmc/gmc.exe --help
```
we see the following output:
```
Allowed options:
  --help                        produce help message
  --datafile arg                the data file
  --k arg (=3)                  number of clusters
  --seed arg (=1)               seed used to choose random initial centers
  --maxiter arg (=100)          maximum number of iterations
  --numrun arg (=1)             number of runs
  --threshold arg (=1e-10)      Likelihood tolerance
  --epsilon arg (=0)            Regularization parameter
```

The program allows us to specify many command line arguments. The only required argument is the data file. All other arguments have default values. The code in Listing 19.8 shows how these arguments are used in the program. If the number of runs is equal to one, then the seed specified by users is used. Otherwise, the seed for each run ranges from 1 to the number of runs. If we run the algorithm multiple times with different random centers, we save the run with highest log-likelihood.

Listing 19.8: Code to run the Gaussian mixture clustering algorithm.

```
1   boost::timer t;
2   t.restart();
3
4   Results Res;
5   Real avgiter = 0.0;
6   Real avgll = 0.0;
7   Real maxll = MIN_REAL;
8   for(Size i=1; i<=numrun; ++i) {
9       GMC ca;
10      Arguments &Arg = ca.getArguments();
11      Arg.ds = ds;
12      Arg.insert("numclust", numclust);
13      Arg.insert("maxiter", maxiter);
14      if(numrun == 1) {
15          Arg.insert("seed", seed);
16      } else {
17          Arg.insert("seed", i);
18      }
19      Arg.insert("epsilon", epsilon);
20      Arg.insert("threshold", threshold);
21
22      ca.clusterize();
23      const Results &tmp = ca.getResults();
24      Real ll = boost::any_cast<Real>(tmp.get("ll"));
25      avgll += ll;
26      if (ll > maxll) {
27          maxll = ll;
28          Res = tmp;
29      }
30      avgiter += boost::any_cast<Size>(tmp.get("numiter"));
31  }
32  avgiter /= numrun;
33  avgll /= numrun;
34
35  double seconds = t.elapsed();
36  std::cout<<"completed_in_"<<seconds<<"_seconds"<<std::endl;
```

Now let use apply the Gaussian mixture clustering algorithm to the synthetic data. To do this, we issue the following command (in Cygwin):

```
examples/gmc/gmc.exe --datafile=../Data/600points.csv
```

That is, we use default values for all other arguments. When this command is executed, we see the following output:

```
Number of records: 600
Number of attributes: 2
Number of numerical attributes: 2
Number of categorical attributes: 0

completed in 0.109 seconds
Clustering Summary:
Number of clusters: 3
Size of Cluster 0: 200
Size of Cluster 1: 200
Size of Cluster 2: 200

Number of given clusters: 3
Cross Tabulation:
Cluster ID    1    2    3
0             200  0    0
1             0    0    200
2             0    200  0

Component size:
Cluster 0: 0.333499
Cluster 1: 0.333168
Cluster 2: 0.333333

Cluster Center:
Center 0 [2](0.0336113,0.0325094)
Center 1 [2](5.12493,5.05885)
Center 2 [2](9.97724,0.0133498)

Number of runs: 1
Average number of iterations: 18
Average likelihood: -2306.2
Best likelihood: -2306.2
Number of iterations for the best case: 18
```

From the output we see that the Gaussian mixture clustering algorithm is very efficient and accurate. The algorithm clustered all the records correctly.

Now let us run the algorithm multiple times with four clusters. To do this, we issue the following command:

The Gaussian Mixture Algorithm

```
examples/gmc/gmc.exe --datafile=../Data/600points.csv
--numrun=10 --k=4
```

Once this command is executed, we see the following output:

```
Number of records: 600
Number of attributes: 2
Number of numerical attributes: 2
Number of categorical attributes: 0

completed in 8.281 seconds
Clustering Summary:
Number of clusters: 4
Size of Cluster 0: 200
Size of Cluster 1: 200
Size of Cluster 2: 192
Size of Cluster 3: 8

Number of given clusters: 3
Cross Tabulation:
Cluster ID    1    2    3
0             0    200  0
1             200  0    0
2             0    0    192
3             0    0    8

Component size:
Cluster 0: 0.333334
Cluster 1: 0.333485
Cluster 2: 0.320318
Cluster 3: 0.0128642

Cluster Center:
Center 0 [2](9.97724,0.0133503)
Center 1 [2](0.0335426,0.0324172)
Center 2 [2](5.16863,5.02284)
Center 3 [2](4.03299,5.95243)

Number of runs: 10
Average number of iterations: 98.6
Average likelihood: -2302.18
Best likelihood: -2299.76
Number of iterations for the best case: 86
```

From the output we see that all the records were clustered correctly. The last cluster is small and contains only 8 points.

To apply the algorithm to the Iris dataset, we issue the following command:

```
examples/gmc/gmc.exe --datafile=../Data/bezdekIris.data
--epsilon=0.01
```

The value 0.01 for `--epsilon` was found by trial and error. If we use the default value or a value less than 0.01, the program will fail since some variance matrix is singular. Once the command is executed, we see the following output:

```
Number of records: 150
Number of attributes: 4
Number of numerical attributes: 4
Number of categorical attributes: 0

completed in 0.235 seconds
Clustering Summary:
Number of clusters: 3
Size of Cluster 0: 8
Size of Cluster 1: 42
Size of Cluster 2: 100

Number of given clusters: 3
Cross Tabulation:
Cluster ID    Iris-setosa    Iris-versicolor    Iris-virginica
0             8              0                  0
1             42             0                  0
2             0              50                 50

Component size:
Cluster 0: 0.0713094
Cluster 1: 0.262019
Cluster 2: 0.666671

Cluster Center:
Center 0 [4](5.28693,3.76632,1.52518,0.295851)
Center 1 [4](4.92955,3.33594,1.44481,0.232432)
Center 2 [4](6.26199,2.872,4.90597,1.67599)

Number of runs: 1
Average number of iterations: 100
Average likelihood: -219.822
Best likelihood: -219.822
Number of iterations for the best case: 100
```

From the output we see that many records were clustered incorrectly. This may be caused by the initial centers. Let us run the algorithm multiple times using the following command:

```
examples/gmc/gmc.exe --datafile=../Data/bezdekIris.data
--epsilon=0.01 --numrun=10
```

Once the above command is executed, we see the following output:

```
Number of records: 150
Number of attributes: 4
Number of numerical attributes: 4
Number of categorical attributes: 0

completed in 2.36 seconds
Clustering Summary:
Number of clusters: 3
Size of Cluster 0: 50
Size of Cluster 1: 51
Size of Cluster 2: 49

Number of given clusters: 3
Cross Tabulation:
Cluster ID    Iris-setosa    Iris-versicolor    Iris-virginica
0             50             0                  0
1             0              2                  49
2             0              48                 1

Component size:
Cluster 0: 0.333333
Cluster 1: 0.33843
Cluster 2: 0.328237

Cluster Center:
Center 0 [4](5.006,3.428,1.462,0.246)
Center 1 [4](6.57258,2.97027,5.53646,2.01542)
Center 2 [4](5.94178,2.77068,4.25596,1.32604)

Number of runs: 10
Average number of iterations: 97.5
Average likelihood: -208.631
Best likelihood: -197.201
Number of iterations for the best case: 100
```

From the output we see that the best run clustered only three records incorrectly. The experiments show that the Gaussian mixture clustering algorithm is also sensitive to initial centers.

19.4 Summary

In this chapter, we implemented one of the Gaussian mixture clustering algorithms summarized in (Celeux and Govaert, 1995). The Gaussian mixture clustering algorithms are model-based clustering algorithms. Model-based clustering is a major approach to cluster analysis and has a long history. Bock (1996) presented a survey of cluster analysis based on probabilistic models. Recent work on model-based clustering can be found in Zhong and Ghosh (2003).

Chapter 20

A Parallel k-means Algorithm

In this chapter, we implement a basic parallel k-means algorithm using message passing interface (MPI). First, we give a brief introduction to the MPI standard and the Boost MPI library. Then we introduce the parallel k-means algorithm. Finally, we present the implementation of the parallel k-means algorithm and examples of applying the algorithm.

20.1 Message Passing Interface

The Message Passing Interface (MPI) standard is a language-independent, flexible, and efficient API (application programming interface) specification used to program parallel computers (Gropp et al., 1999). There are many MPI libraries that implement the MPI standard. For example, OpenMPI, MPICH, and LAM/MPI are different implementations of the MPI standard. Almost all MPI libraries are implemented in C.

An MPI library (e.g., MPICH) provides functionalities to support MPI environment management, point-to-point communications, and collective communications. Point-to-point communication allows a process to send and receive messages from other processes. Collective communication allows processes to coordinate as a group.

The Boost MPI library is not a new MPI library but a C++ friendly interface to the MPI. That is, the Boost MPI library is not an implementation of the MPI standard. To use the Boost MPI library, we first need to install an MPI library such as MPICH2. Section C.5 introduces how to install the MPICH2 library and the Boost MPI library.

Suppose we have installed MPICH2 and the Boost MPI library following the instructions in Section C.5. Now let us consider a simple C++ MPI program to illustrate the aforementioned concepts. The code of the C++ MPI program is shown in Listing 20.1. Before we compile and execute the program, let us take a look at the code first.

The program includes the Boost header `mpi.hpp` (line 2), which contains all the headers of the Boost MPI library. Since data is serialized before transmitted from a process to another, the program includes two headers (i.e., **string.hpp** and **vector.hpp**) from the Boost serialization library (lines 3–4).

Listing 20.1: Program to illustrate MPI.

```
1   // mpia.cpp
2   #include<boost/mpi.hpp>
3   #include<boost/serialization/string.hpp>
4   #include<boost/serialization/vector.hpp>
5   #include<string>
6   #include<functional>
7   #include<iostream>
8
9   using namespace std;
10  using namespace boost::mpi;
11
12  int main(int argc, char* argv[]) {
13      environment env(argc, argv);
14      communicator world;
15
16      int rank = world.rank();
17      if(rank == 0) {
18          for(size_t p=1; p<world.size(); ++p) {
19              world.send(p, 0, string("a_msg_from_master"));
20          }
21      } else {
22          string msg;
23          world.recv(0, 0, msg);
24          cout<<"Process_"<<rank<<":_"<<msg<<endl;
25      }
26      cout<<endl;
27
28      vector<string> names;
29      if (rank == 0) {
30          names.push_back("zero_");
31          names.push_back("one_");
32          names.push_back("two_");
33          names.push_back("three_");
34      }
35      broadcast(world, names, 0);
36
37      string str, strsum, strsum2;
38      str = rank<4 ? names[rank] : "many";
39      reduce(world, str, strsum, plus<string>(), 0);
40      cout<<"Process_"<<rank<<":_"<<strsum<<endl;
41
42      all_reduce(world, str, strsum2, plus<string>());
43      cout<<"Process_"<<rank<<":_"<<strsum2<<endl;
44
45      return 0;
46  }
```

In the main function (lines 12–46), the program first creates an object of type **mpi::environment** using the arguments of the main function (line 13). The object initializes the MPI environment and enables communication among the processes. Since the construction of the object initialize MPI and the destruction of the object finalizes MPI, we do not need to worry about cleaning memory used by MPI. After the MPI environment is created, the program creates an **mpi::communicator** object using the default constructor (line 14). We can use the communicator object to determine the total number of processes and the unique number assigned to an individual process.

In line 16, the program uses the function `rank` to get the unique number (rank) associated with each process. The process with rank zero sends the message "a msg from master" to all other processes (lines 18–20). Processes with a non-zero rank receive the message and print the message to screen (lines 23–24). Functions `send` and `recv` are used to send and receive messages, respectively.

In line 28, the program declares a vector of strings. The process with rank zero populates the vector with four elements (29–34). Then the program uses the function `broadcast` to broadcast the vector of strings from the process with rank zero to all other processes so that all processes can use the vector. The `broadcast` method is one of the collective operations.

In line 39, the program summarizes the strings from each process into a single string at the process with rank zero. The `reduce` method is another collective operation. The value of `str` (line 38) is different for processes with different ranks. The value of `strsum` (line 39) for a process with rank zero is the sum of `str` from the other processes.

The `reduce` method does not broadcast the collective value to other processes. To broadcast the collective value to other Processes, we need to use the `all_reduce` method. In line 42, the program uses the method `all_reduce` to summarize the strings and to broadcast the sum of the strings to all processes.

Now let us compile the program. Compiling an MPI C++ program is a little different from compiling an ordinary C++ program. To compile an MPI C++ program, we need to use the command `mpic++` provided by the MPICH2 library. However, the compilation options are the same as those used with `g++`. To compile the MPI program shown in Listing 20.1, we use the following command:

```
mpic++ -o mpia -I/usr/local/include mpia.cpp -L/usr/local/lib
-lboost_mpi -lboost_serialization
```

The `-I` and `-L` options tell the compiler the locations of the Boost headers and libraries, respectively. Since the program uses the Boost MPI library and the Boost serialization library, we need to link those libraries using the `-l` option.

To execute the program, we use the command `mpirun` provided by the MPICH2 library. Before we execute the program, we need to make sure an MPD is running on each cluster node to be used by the program. An MPD is an MPICH2 daemon that controls, runs, and stops the processes on the cluster nodes. To start an MPD, we can use the command `mpd&`[1] provided by the MPICH2 library. The `&` symbol tells the computer to run the daemon in background.

Suppose a daemon is started. We can execute the program using the following command:

---

[1] For the first time using `mpd`, one needs to create a file called `.mpd.conf` in the home directory and change the file to read access. To do this, one can use the command `touch ~/.mpd.conf` and `chmod 600 ~/.mpd.conf`.

```
mpirun -n 3 ./mpia
```

The `-n` option specifies how many processes are used to run the program. We used a single computer to run the program. The computer has an AMD CPU with four cores. Once the program is executed, we see the following output:

```
Process 0: zero one two
Process 0: zero one two
Process 2: a msg from master

Process 2:
Process 2: zero one two
Process 1: a msg from master

Process 1:
Process 1: zero one two
```

The process with rank zero prints two messages: **strsum** (line 40) and **strsum2** (line 43). Processes with non-zero ranks print three messages: **msg** (line 24), **strsum** (line 40), and **strsum2** (line 43). The output shown above is what we expected.

In the above example, we introduced the usage of some functions provided by the Boost MPI library. For detailed explanation of those functions, readers are referred to the online documentation of the Boost MPI library.

20.2 Description of the Algorithm

Parallel data clustering is a process of using parallel processing to cluster data in order to reduce the execution time of the clustering algorithm. In this section, we describe a simple parallel k-means algorithm.

In Section 14.1, we introduced a sequential k-means algorithm. The sequential k-means algorithm calculates the distances between all records and all centers. A straightforward way to parallelize the k-means algorithm is to let each participating process handle n/p records, where p is the total number of processes and n is the number of records contained in the data set (Judd et al., 1998; Othman et al., 2005).

The major steps of the parallel k-means algorithm are given below:

(a) Master process: Read data from a data file and send blocks of data to client processes;

(b) Master process: Initialize cluster centers and broadcast those cluster centers to client processes;

(c) All process: Calculate the distances between each record of its data block and cluster centers, and assign the record to the nearest cluster;

(d) All process: Update the new cluster centers;

(e) All process: Repeat Step (c) and Step (d) until some stop condition is met;

(f) Master process: Collect cluster results and output clustering summary.

In the parallel k-means algorithm, each process is responsible for handling a data block. The parallel k-means algorithm spends some time on I/O operations, such as broadcasting data blocks from the master process to client processes. As a result, clustering a small data set with the parallel k-means algorithm on multiple CPUs or cores may take more time than clustering the small data set with the parallel k-means algorithm on a single CPU or core. This is the case when the time reduction due to parallel distance calculations does not offset the time overhead due to I/O operations.

20.3 Implementation

In this section, we introduce the implementation of the simple parallel k-means algorithm described in the previous section. To implement the parallel k-means algorithm, we just modify the sequential k-means algorithm implemented in Section 14.2. In what follows, we take a look at some functions of the parallel k-means program.

The parallel k-means algorithm is implemented as a class, whose declaration is shown in Listing 20.2. Class MPIKmean is very similar to class Kmean given in Listing 14.1. In class MPIKmean, we remove updateCenter and _distance from class Kmean and add _centers, _data, _numObj, _numAttr, and _world. A member function dist is also added to calculate the Euclidean distance.

Member _centers is a vector that contains all the cluster centers. The first _numAttr elements are the coordinates of the first cluster center. The second _numAttr elements are the coordinates of the second cluster centers, and so on. We encode all cluster centers into a vector so that we can transmit the cluster centers from one process to another easily. Similarly, member _data is a vector that contains the data block assigned to a process. Member _numObj is the number of records contained in _data. Member _numAttr is the number of attributes of the data set. Member _world is an object of boost::mpi::communicator.

Listing 20.2: Declaration of class `MPIKmean`.

```
class MPIKmean: public Algorithm {
protected:
    void setupArguments();
    void performClustering() const;
    void fetchResults() const;
    virtual void initialization() const;
    virtual void iteration() const;
    virtual Real dist(Size i, Size j) const;

    mutable std::vector<Real> _centers;
    mutable std::vector<Real> _data;
    mutable Size _numObj;
    mutable Size _numAttr;
    mutable std::vector<Size> _CM;

    mutable std::vector<boost::shared_ptr<CenterCluster> >
        _clusters;
    mutable Real _error;
    mutable Size _numiter;

    Size _numclust;
    Size _maxiter;
    Size _seed;
    boost::mpi::communicator _world;
};
```

The implementation of method `setupArguments` is shown in Listing 20.3. This method assigns values to members _numclust, _maxiter, and _seed. Those assignments are done in all processes. In the master process (i.e., the process with rank zero), the method transfers a dataset into the algorithm and checks whether the dataset is numeric or not. Note that the method does not set up a distance measure for the algorithm. In fact, we use the Euclidean distance for the parallel k-means algorithm.

Listing 20.3: Implementation of method `setupArguments`.

```
void MPIKmean::setupArguments() {
    _numclust = boost::any_cast<Size>(
        _arguments.get("numclust"));

    _maxiter = boost::any_cast<Size>(
        _arguments.get("maxiter"));
    ASSERT(_maxiter>0, "invalid_maxiter");

    _seed = boost::any_cast<Size>(
        _arguments.get("seed"));
    ASSERT(_seed>0, "invalid_seed");

    if(_world.rank() ==0) {
        Algorithm::setupArguments();
        ASSERT(_ds->is_numeric(), "dataset_is_not_numeric");

        ASSERT(_numclust>=2 && _numclust<=_ds->size(),
            "invalid_numclust");
    }
}
```

Method `performClustering` calls the two functions: `initialization` and `iteration`. Method `initialization` does some initialization work. The implementation of method `initialization` is shown in Listing 20.4.

As we can see in Listing 20.4, the master process is responsible for initializing the cluster centers (lines 5–37). Member _centers is populated in the master process (lines 30–33). Variable numRecords (line 2) is a local variable to the method. Variables numRecords and _numAttr are also assigned values in the master process (lines 6–7). Once the cluster centers are initialized, the method broadcasts the cluster centers and values of numRecords and _numAttr to client processes (lines 39–41).

Once the values are broadcasted, each process determines the data block size and the number of remaining records (lines 43–44). In the master process, the method first assigns the first data block to itself (lines 51–56). Then the master process assigns other data blocks to client processes (lines 59–71) using the send method. Each client process receives a data block sent from the master process (line 75).

Listing 20.4: Implementation of method initialization.

```
void MPIKmean::initialization() const {
    Size numRecords;
    Size rank = _world.rank();

    if (rank == 0) {
        numRecords = _ds->size();
        _numAttr = _ds->num_attr();
        _centers.resize(_numclust * _numAttr);

        std::vector<Integer> index(numRecords,0);
        for(Size i=0;i<index.size();++i){
            index[i] = i;
        }

        boost::shared_ptr<Schema> schema = _ds->schema();
        boost::minstd_rand generator(_seed);
        for(Size i=0;i<_numclust;++i){
            boost::uniform_int<> uni_dist(0,numRecords-i-1);
            boost::variate_generator<boost::minstd_rand&,
                boost::uniform_int<> >
                uni(generator, uni_dist);
            Integer r = uni();
            boost::shared_ptr<Record> cr = boost::shared_ptr
                <Record>(new Record(*(*_ds)[r]));
            boost::shared_ptr<CenterCluster> c =
                boost::shared_ptr<CenterCluster>(
                    new CenterCluster(cr));
            c->set_id(i);
            _clusters.push_back(c);
            for(Size j=0; j<_numAttr; ++j) {
                _centers[i*_numAttr + j] =
                    (*schema)[j]->get_c_val((*_ds)(r,j));
            }
            index.erase(index.begin()+r);
        }

    }

    boost::mpi::broadcast(_world, _centers, 0);
    boost::mpi::broadcast(_world, numRecords, 0);
    boost::mpi::broadcast(_world, _numAttr, 0);

    Size nDiv = numRecords / _world.size();
    Size nRem = numRecords % _world.size();
```

```
46      if(rank == 0) {
47          boost::shared_ptr<Schema> schema = _ds->schema();
48          _numObj = (nRem >0) ? nDiv+1: nDiv;
49          _data.resize(_numObj * _numAttr);
50          _CM.resize(_numObj);
51          for(Size i=0; i<_numObj; ++i) {
52              for(Size j=0; j<_numAttr; ++j) {
53                  _data[i*_numAttr +j] =
54                      (*schema)[j]->get_c_val((*_ds)(i, j));
55              }
56          }
57
58          Size nCount = _numObj;
59          for(Size p=1; p<_world.size(); ++p) {
60              Size s = (p< nRem) ? nDiv +1 : nDiv;
61              std::vector<Real> dv(s*_numAttr);
62              for(Size i=0; i<s; ++i) {
63                  for(Size j=0; j<_numAttr; ++j) {
64                      dv[i*_numAttr+j] =
65                          (*schema)[j]->get_c_val(
66                              (*_ds)(i+nCount,j));
67                  }
68              }
69              nCount += s;
70              _world.send(p, 0, dv);
71          }
72      } else {
73          _numObj = (rank < nRem) ? nDiv+1: nDiv;
74          _CM.resize(_numObj);
75          _world.recv(0,0,_data);
76      }
77  }
```

The implementation of method **iteration** is shown in Listing 20.5. The method first declares a single-element vector to hold the number of membership changes during an iteration (line 2)[2].

Inside the **while** loop, the method declares several local variables, including **nChangedLocal**, **newCenters**, and **newSize**. Each process will handle its own data block and calculate the distances between each record of its data block and each cluster center (lines 13–33). Variable **newCenters** contains the sum of all records of a data block in a cluster. Variable **newSize** contains the number of records of a data block in a cluster. Once all records in all data blocks are processed parallel, the method collects and broadcasts the total number of membership changes, new cluster centers, and new cluster sizes (lines 35–40). Then the method updates the cluster centers (lines 42–46) in all processes.

After the **while** loop, all client processes send the cluster membership vectors to the master process (line 55). The master process receives all the cluster membership vectors from client processes and updates its own cluster membership vector to contain the cluster memberships of all records in the input data set (lines 57–63).

Note that the method **vplus** (lines 36, 38, and 40) is a template **struct**

---

[2] If we declare nChanged as type Size and use std::plus<Size>() in line 36, then we will encounter the error "Attempting to use an MPI routine before initializing MPICH" in Cygwin. We use a single-element vector to work around this problem.

defined in the header **mpikmean.hpp** (see Listing B.123). The struct **vplus** implements operator "( )" to calculate the sum of two vectors on an element-wise basis.

Listing 20.5: Implementation of method `iteration`.

```
void MPIKmean::iteration() const {
    std::vector<Size> nChanged(1,1);

    _numiter = 1;
    while(nChanged[0] > 0) {
        nChanged[0] = 0;
        Size s;
        Real dMin,dDist;
        std::vector<Size> nChangedLocal(1,0);
        std::vector<Real> newCenters(_numclust*_numAttr,0.0);
        std::vector<Size> newSize(_numclust,0);

        for(Size i=0;i<_numObj;++i) {
            dMin = MAX_REAL;
            for(Size k=0;k<_numclust;++k) {
                dDist = dist(i, k);
                if (dMin > dDist) {
                    dMin = dDist;
                    s = k;
                }
            }

            for(Size j=0; j<_numAttr; ++j) {
                newCenters[s*_numAttr+j] +=
                    _data[i*_numAttr+j];
            }
            newSize[s] +=1;

            if (_CM[i] != s){
                _CM[i] = s;
                nChangedLocal[0]++;
            }
        }

        all_reduce(_world, nChangedLocal, nChanged,
                    vplus<Size>());
        all_reduce(_world, newCenters, _centers,
                    vplus<Real>());
        std::vector<Size> totalSize(_numclust,0);
        all_reduce(_world, newSize, totalSize, vplus<Size>());

        for(Size k=0; k<_numclust; ++k) {
            for(Size j=0; j<_numAttr; ++j) {
                _centers[k*_numAttr+j] /= totalSize[k];
            }
        }

        ++_numiter;
        if (_numiter > _maxiter){
            break;
        }
    }

    if(_world.rank() > 0) {
        _world.send(0,0,_CM);
    } else {
        for(Size p=1; p<_world.size(); ++p) {
            std::vector<Size> msg;
            _world.recv(p,0,msg);
            for(Size j=0; j<msg.size(); ++j) {
```

```
61            _CM.push_back(msg[j]);
62          }
63        }
64      }
65  }
```

Listing 20.6 shows the implementation of method `fetchResults`. The method calculates part of the objective function (see Equation (14.2)) in each process (lines 3–5) and collects those parts in the master process (line 7). In the master process, the method summarizes clustering results and transfers those results to member `_results` (lines 9–31).

Listing 20.6: Implementation of method `fetchResults`.

```
1   void MPIKmean::fetchResults() const {
2       std::vector<Real> error(1, 0.0), totalerror(1);
3       for(Size i=0;i<_numObj;++i) {
4           error[0] += dist(i,_CM[i]);
5       }
6
7       reduce(_world, error, totalerror, vplus<Real>(), 0);
8
9       if(_world.rank() == 0) {
10          boost::shared_ptr<Schema> schema = _ds->schema();
11          PClustering pc;
12          for(Size i=0;i<_numclust;++i){
13              for(Size j=0; j<_numAttr; ++j) {
14                  (*schema)[j]->set_c_val(
15                      (*_clusters[i]->center())[j],
16                      _centers[i*_numAttr+j]);
17              }
18              pc.add(_clusters[i]);
19          }
20
21          for(Size i=0; i<_CM.size(); ++i) {
22              _clusters[_CM[i]]->add((*_ds)[i]);
23          }
24
25          _results.CM = _CM;
26          _results.insert("pc", boost::any(pc));
27
28          _error = totalerror[0];
29          _results.insert("error", boost::any(_error));
30          _results.insert("numiter", boost::any(_numiter));
31      }
32  }
```

The complete implementation of the parallel k-means algorithm can be found in Listings B.123 and B.124. Method `dist` calculates the Euclidean distance between a record and a cluster center. One can see the implementation of this method in Listing B.124.

20.4 Examples

In this section, we illustrate the parallel k-means algorithm implemented in the previous section with some examples. The main function of our program is shown in Listing B.125. The MPI environment is initialized in the main function. The master process is responsible for reading data from a file (lines 60–64 of Listing B.125) and outputting a clustering summary (lines 104–132 of Listing B.125).

Suppose the clustering library is installed in /usr/local/include and /usr/local/lib (see Section 6.6) and the Boost libraries are also installed in these directories (see Section C.5). Then we can compile the parallel k-means program using the following command:

mpic++ -o mpikmean -I/usr/local/include -I. mpikmean.cpp mpimain.cpp -L/usr/local/lib -lboost_program_options -lboost_mpi -lboost_serialization -lClusLib

Once the compilation is finished, an executable is produced.

To execute the parallel k-means program, we use a relatively large dataset generated by the program described in Section 11.3.2. The dataset contains 15,000 records with three clusters, each of which contains 5,000 records. To execute the program with only one process, we issue the following command[3]:

mpirun -n 1 ./mpikmean --datafile=../../../Data/15000points.csv --k=10 --numrun=50

Once the command is finished, we see the following output:

```
Number of records: 15000
Number of attributes: 2
Number of numerical attributes: 2
Number of categorical attributes: 0

completed in 22.152 seconds
number of processes: 1
Clustering Summary:
Number of clusters: 10
Size of Cluster 0: 2323
Size of Cluster 1: 1275
Size of Cluster 2: 1308
Size of Cluster 3: 1729
Size of Cluster 4: 1166
Size of Cluster 5: 1370
Size of Cluster 6: 1540
Size of Cluster 7: 1315
Size of Cluster 8: 1731
Size of Cluster 9: 1243

Number of given clusters: 3
```

---

[3]If you encounter the error "error while loading shared libraries: ", then you need to copy Boost libraries from /usr/local/lib to /usr/lib.

Cross Tabulation:

Cluster ID	1	2	3
0	23220	1	
1	0	0	1275
2	13080	0	
3	0	17290	
4	0	0	1166
5	13700	0	
6	0	15400	
7	0	0	1315
8	0	17310	
9	0	0	1243

Number of runs: 50
Average number of iterations: 46.52
Average error: 12135
Best error: 11598.5

From the output we see that it took the program 22.152 seconds to cluster the data into 10 clusters 50 times.

If we execute the program with two processes using the following command:

mpirun -n 2 ./mpikmean --datafile=../../../Data/15000points.csv --k=10 --numrun=50

we see the following output:

Number of records: 15000
Number of attributes: 2
Number of numerical attributes: 2
Number of categorical attributes: 0

completed in 11.7 seconds
number of processes: 2
Clustering Summary:
Number of clusters: 10
Size of Cluster 0: 2323
Size of Cluster 1: 1275
Size of Cluster 2: 1308
Size of Cluster 3: 1729
Size of Cluster 4: 1166
Size of Cluster 5: 1370
Size of Cluster 6: 1540
Size of Cluster 7: 1315
Size of Cluster 8: 1731
Size of Cluster 9: 1243

```
Number of given clusters: 3
Cross Tabulation:
Cluster ID    1      2      3
0             23220  1
1             0      0      1275
2             13080  0
3             0      17290
4             0      0      1166
5             13700  0
6             0      15400
7             0      0      1315
8             0      17310
9             0      0      1243

Number of runs: 50
Average number of iterations: 46.52
Average error: 12135
Best error: 11598.5
```

From the output we see that it took the program 11.7 seconds to cluster the data into 10 clusters 50 times. We see that the run time is reduced when we increase the number of parallel processes.

Now let us execute the program with three processes using the following command:

```
mpirun -n 3 ./mpikmean --datafile=../../../Data/15000points.csv
--k=10 --numrun=50
```

Once the command is finished, we see the following output:

```
Number of records: 15000
Number of attributes: 2
Number of numerical attributes: 2
Number of categorical attributes: 0

completed in 8.268 seconds
number of processes: 3
Clustering Summary:
Number of clusters: 10
Size of Cluster 0: 2323
Size of Cluster 1: 1275
Size of Cluster 2: 1308
Size of Cluster 3: 1729
Size of Cluster 4: 1166
Size of Cluster 5: 1370
Size of Cluster 6: 1540
```

```
Size of Cluster 7: 1315
Size of Cluster 8: 1731
Size of Cluster 9: 1243

Number of given clusters: 3
Cross Tabulation:
Cluster ID    1       2       3
0             23220   1
1             0       0       1275
2             13080   0
3             0       17290
4             0       0       1166
5             13700   0
6             0       15400
7             0       0       1315
8             0       17310
9             0       0       1243

Number of runs: 50
Average number of iterations: 46.52
Average error: 12135
Best error: 11598.5
```

From the output we see that the run time is further reduced to 8.268 seconds. However, the amount of reduction is smaller than when we increased one process to two processes. The above experiments were done in a computer with a four-core AMD CPU.

20.5 Summary

In this chapter, we introduced and implemented a simple parallel k-means algorithm. We implemented the parallel k-means algorithm using the message passing interface and the Boost MPI library, which are also brief introduced in this chapter. Our illustrative examples show that the parallel version of the k-means algorithm does reduce the run time when we use multiple processes.

A number of researchers have studied parallel data clustering. For example, Kwok et al. (2002) proposed a parallel fuzzy c-means algorithm. Gürsoy and Cengiz (2001) and Gursoy (2004) developed a parallel version of the tree-based k-means algorithm (Alsabti et al., 1998) to cluster low-dimensional data on message passing computers. Jiménez and Vidal (2005) proposed two parallel clustering algorithms based on the k-means algorithm: α-Bisecting k-means and α-Bisecting spherical k-means. Hai and Susumu (2005) proposed a parallel clustering algorithm for categorical and mixed-type data. Song et al. (2008) proposed a parallel spectral clustering algorithm. Zhao et al. (2009) proposed a parallel k-means clustering algorithm based on MapReduce. Readers interested in parallel clustering can refer to those papers and the references therein.

Appendix A

Exercises and Projects

Exercise A.1. In class `Dataset`, we overloaded the operator "<<" (see Listing 7.7). Using class `Dataset` as an example, overload operator "<<" for class `AttrValue` and class `Record` so that the values stored in objects of classes `AttrValue` and `Record` can be outputted using streams. Use commas to separate the values in a record.

Exercise A.2. In the implementation of the overloaded operator "( )" of class `MinkowskiDistance` (cf. Section 9.2) and class `SimpleMatchingDistance` (cf. Section 9.4), we added code to check whether the schemata of two records are equal to each other. Implement a function in the base class `Distance` to do the check so that we can call the function rather than rewrite this same piece of code many times. This function should be a protected member function of class `Distance` and has the same arguments as the operator "( )".

Exercise A.3. The Manhattan distance and the maximum distance are defined in Equation (1.3) and Equation (1.4), respectively. Implement the two distance measures in class `ManhattanDistance` and class `MaximumDistance` that are derived from the base class `Distance`.

Exercise A.4. In Section 10.4, we used command line commands to compile and link the program shown in Listing 10.6. Using the `Makefile.am` file (see Listing B.107) for agglomerative hierarchical clustering as an example, write a `Makefile.am` file for the dummy clustering algorithm. Then modify the `Makefile.am` file (see Listing B.106) in folder `examples` and the `configure.ac` file (see Listing B.1) accordingly.

Exercise A.5. Modify the program shown in Listing 10.6 by removing the symbol "&" in line 67. Compile and link the modified program using the commands described in Section 10.4. Then execute the following command:

`./dummy.exe 5`

and see what the output is.

Exercise A.6. Class `DatasetReader` (see Listing B.99) recognizes symbols "RecordID", "Continuous", "Discrete", and "Class" in a schema file. Modify the class so that it can recognize an additional symbol "Ignore" and skip the column when reading a data file.

Exercise A.7. Modify the program in Listing 11.11 so that it outputs the original record identifiers and the original class labels. That is, output the dataset in the following format:

```
Data:
RecordID   Attribute(1)   Attribute(2)   Label
a          1.2            3              1
b          -0.8           3              2
c          3              4              1
d          0.01           2              2
```

Exercise A.8. The `DatasetReader` class described in Subsection 11.3.1 was designed to reader CSV files only. Modify the `DatasetReader` class so that it can read space-delimited files. Note that values might be separated by multiple spaces.

Exercise A.9. The agglomerative hierarchical clustering algorithms calculate the distance matrix internal using a distance measure provided by users (see Listing B.33). Modify the class `LW` so that it can take a distance matrix as input.

Exercise A.10. The `reset` method of class `Algorithm` (see Listing B.11) resets only the contents of member `_results`. Override the `reset` method in derived classes (e.g., class `Kmean`, class `Cmean`, etc.) so that the `reset` method will also reset all the clustering results in objects of these derived class. That is, calling the `reset` method will revert an object to its initial state.

Exercise A.11. The genetic k-modes algorithm implemented in Section 17.2 runs through all the generations no matter whether the algorithm converges or not. By converge, we mean that the populations between two generations do not change. Modify the function `iteration` so that the function will stop when the algorithm converges.

Exercise A.12. Vanjak and Mornar (2001) proposed a general object-oriented framework for iterative optimization algorithms. The framework includes program code used to output progress information and program code used to terminate the iterative process. In addition, the framework allows one to separate the implementation of the algorithm from program code used to output progress information and program code used to terminate the program execution. Many data clustering algorithms (e.g., the k-means algorithm) are formulated as iterative optimization algorithms. Implement the k-means algorithm using the framework.

Appendix B
Listings

B.1 Files in Folder ClusLib
B.1.1 Configuration File configure.ac

Listing B.1: configure.ac

```
1   AC_INIT([ClusLib], [3.141], [BUG-REPORT-ADDRESS])
2   AC_CONFIG_SRCDIR([cl/errors.hpp])
3   AC_CONFIG_AUX_DIR([config])
4   AC_CONFIG_HEADERS([cl/config.hpp])
5   AC_CONFIG_MACRO_DIR([m4])
6
7   AM_INIT_AUTOMAKE([-Wall -Werror foreign])
8
9   AC_ARG_WITH([boost-include],
10      AC_HELP_STRING([--with-boost-include=INCLUDE_PATH],
11                     [Supply the location of Boost header files]),
12      [boost_include_path="`cd ${withval} 2>/dev/null && pwd`"],
13      [boost_include_path=""])
14  if test [ -n "$boost_include_path" ] ; then
15      AC_SUBST([BOOST_INCLUDE],[" -I${boost_include_path}"])
16      AC_SUBST([CPPFLAGS],[" ${CPPFLAGS} -I${boost_include_path}"])
17  fi
18
19  AC_ARG_WITH([boost-lib],
20      AC_HELP_STRING([--with-boost-lib=LIB_PATH],
21                     [Supply the location of Boost libraries]),
22      [boost_lib_path="`cd ${withval} 2>/dev/null && pwd`"],
23      [boost_lib_path=""])
24  if test [ -n "$boost_lib_path" ] ; then
25      AC_SUBST([BOOST_LIB],[" -L${boost_lib_path}"])
26      AC_SUBST([LDFLAGS],[" ${LDFLAGS} -L${boost_lib_path}"])
27  fi
28
29  LT_INIT
30  AC_PROG_CC
31  AC_PROG_CXX
32  AC_PROG_CPP
33  AC_LANG(C++)
34
35  CHECK_BOOST
36
37  AC_CONFIG_FILES([Makefile
38                   cl/Makefile
39                   cl/algorithms/Makefile
40                   cl/clusters/Makefile
41                   cl/datasets/Makefile
42                   cl/distances/Makefile
43                   cl/patterns/Makefile
```

```
44              cl/utilities/Makefile
45              config/Makefile
46              examples/Makefile
47              examples/agglomerative/Makefile
48              examples/cmean/Makefile
49              examples/diana/Makefile
50              examples/fsc/Makefile
51              examples/gkmode/Makefile
52              examples/gmc/Makefile
53              examples/kmean/Makefile
54              examples/kprototype/Makefile
55              m4/Makefile
56              test-suite/Makefile])
57  AC_OUTPUT
```

B.1.2 m4 Macro File `acinclude.m4`

Listing B.2: `acinclude.m4`.

```
1   # CHECK_BOOST_DEVEL
2   # ----------------
3   # Check whether the Boost headers are available
4   AC_DEFUN([CHECK_BOOST_DEVEL],
5   [AC_MSG_CHECKING([for Boost development files])
6    AC_TRY_COMPILE(
7       [@%:@include <boost/version.hpp>
8        @%:@include <boost/shared_ptr.hpp>
9        @%:@include <boost/assert.hpp>
10       @%:@include <boost/current_function.hpp>],
11      [],
12      [AC_MSG_RESULT([yes])],
13      [AC_MSG_RESULT([no])
14       AC_MSG_ERROR([Boost development files not found])
15      ])
16  ])
17
18  # CHECK_BOOST_VERSION
19  # -------------------
20  # Check whether the Boost installation is up to date
21  AC_DEFUN([CHECK_BOOST_VERSION],
22  [AC_MSG_CHECKING([Boost version])
23   AC_REQUIRE([CHECK_BOOST_DEVEL])
24   AC_TRY_COMPILE(
25      [@%:@include <boost/version.hpp>],
26      [@%:@if BOOST_VERSION < 103100
27       @%:@error too old
28       @%:@endif],
29      [AC_MSG_RESULT([yes])],
30      [AC_MSG_RESULT([no])
31       AC_MSG_ERROR([outdated Boost installation])
32      ])
33  ])
34
35  # CHECK_BOOST_PROGRAM_OPTIONS
36  # ---------------------------
37  # Check whether the Boost program options is available
38  AC_DEFUN([CHECK_BOOST_PROGRAM_OPTIONS],
39  [AC_MSG_CHECKING([for Boost program options])
40   AC_REQUIRE([AC_PROG_CC])
41   original_LIBS=$LIBS
42   original_CXXFLAGS=$CXXFLAGS
43   CC_BASENAME=`basename $CC`
44   CC_VERSION=`echo "__GNUC__ __GNUC_MINOR__" | $CC -E -x c - | \
45       tail -n 1 | $SED -e "s/ //"`
```

```
46  for boost_lib in \
47          boost_program_options-$CC_BASENAME$CC_VERSION \
48          boost_program_options-$CC_BASENAME \
49          boost_program_options \
50          boost_program_options-mt-$CC_BASENAME$CC_VERSION \
51          boost_program_options-$CC_BASENAME$CC_VERSION-mt \
52          boost_program_options-x$CC_BASENAME$CC_VERSION-mt \
53          boost_program_options-mt-$CC_BASENAME \
54          boost_program_options-$CC_BASENAME-mt \
55          boost_program_options-mt ; do
56      LIBS="$original_LIBS -l$boost_lib"
57      CXXFLAGS="$original_CXXFLAGS"
58      boost_po_found=no
59      AC_LINK_IFELSE(
60          [@%:@include <boost/program_options.hpp>
61              using namespace boost::program_options;
62              int main(int argc, char** argv)
63              {
64                  options_description desc("Allowed options");
65                  desc.add_options()
66                      ("help", "help msg")("p","p");
67                  return 0;
68              }
69          ],
70          [boost_po_found=$boost_lib
71          break],
72          [])
73  done
74  LIBS="$original_LIBS"
75  CXXFLAGS="$original_CXXFLAGS"
76  if test "$boost_po_found" = no ; then
77      AC_MSG_RESULT([no])
78      AC_SUBST([BOOST_PROGRAM_OPTIONS_LIB],[""])
79      AC_MSG_WARN([Boost program options not found.])
80  else
81      AC_MSG_RESULT([yes])
82      AC_SUBST([BOOST_PROGRAM_OPTIONS_LIB],[$boost_lib])
83  fi
84  ])
85
86  # CHECK_BOOST
87  # --------------------
88  # Boost-related tests
89  AC_DEFUN([CHECK_BOOST],
90  [AC_REQUIRE([CHECK_BOOST_DEVEL])
91   AC_REQUIRE([CHECK_BOOST_VERSION])
92   AC_REQUIRE([CHECK_BOOST_PROGRAM_OPTIONS])
93  ])
```

B.1.3 Makefile

Listing B.3: Makefile.am in ClusLib.

```
1  SUBDIRS = cl config examples m4 test-suite
2
3  ACLOCAL_AMFLAGS = -I m4
```

B.2 Files in Folder cl
B.2.1 Makefile

Listing B.4: `Makefile.am` in cl.

```
 1  SUBDIRS = algorithms clusters datasets distances patterns \
 2      utilities
 3
 4  AM_CPPFLAGS = -I${top_srcdir} -I${top_builddir}
 5
 6  this_includedir = ${includedir}/${subdir}
 7  this_include_HEADERS = \
 8      cldefines.hpp \
 9      cluslib.hpp \
10      config.hpp \
11      errors.hpp \
12      types.hpp
13
14  libClusLib_la_SOURCES = \
15      errors.cpp
16
17
18  lib_LTLIBRARIES = libClusLib.la
19  libClusLib_la_LIBADD = \
20      algorithms/libAlgorithms.la \
21      clusters/libClusters.la \
22      datasets/libDatasets.la \
23      distances/libDistances.la \
24      patterns/libPatterns.la \
25      utilities/libUtilities.la
26
27  cluslib.hpp: Makefile.am
28      echo "// This file is generated. Please do not edit!" > $@
29      echo >> $@
30      echo "#include <cl/cldefines.hpp>" >> $@
31      echo >> $@
32      for i in $(filter-out config.hpp cluslib.hpp \
33      cldefines.hpp, $(this_include_HEADERS)); do \
34      echo "#include <${subdir}/$$i>" >> $@; \
35      done
36      echo >> $@
37      subdirs='$(SUBDIRS)'; for i in $$subdirs; do \
38      echo "#include <${subdir}/$$i/all.hpp>" >> $@; \
39      done
```

B.2.2 Macros and `typedef` Declarations

Listing B.5: Macros.

```
1  // cl/cldefines.hpp
2  #ifndef CLUSLIB_CLDEFINES_HPP
3  #define CLUSLIB_CLDEFINES_HPP
4
5  #include <boost/config.hpp>
6  #include <boost/version.hpp>
7  #include <limits>
8
9  #if BOOST_VERSION < 103100
```

```
10        #error using an old version of Boost, please update.
11  #endif
12
13  #define CL_VERSION "1.0.0"
14  #define CL_LIB_VERSION "1.0.0"
15
16  #if defined(HAVE_CONFIG_H)
17       #include<cl/config.hpp>
18  #endif
19
20  #define INTEGER int
21  #define BIGINTEGER long
22  #define REAL double
23
24  #define MIN_INTEGER std::numeric_limits<INTEGER>::min()
25  #define MAX_INTEGER std::numeric_limits<INTEGER>::max()
26  #define MIN_REAL -std::numeric_limits<REAL>::max()
27  #define MAX_REAL std::numeric_limits<REAL>::max()
28  #define MIN_POSITIVE_REAL std::numeric_limits<REAL>::min()
29  #define EPSILON std::numeric_limits<REAL>::epsilon()
30  #define NULL_INTEGER std::numeric_limits<INTEGER>::max()
31  #define NULL_SIZE std::numeric_limits<unsigned INTEGER>::max()
32  #define NULL_REAL std::numeric_limits<REAL>::max()
33
34  #endif
```

Listing B.6: Types.

```
1   // cl/types.hpp
2   #ifndef CLUSLIB_TYPES_HPP
3   #define CLUSLIB_TYPES_HPP
4
5   #include<cl/cldefines.hpp>
6   #include<cstddef>
7
8   namespace ClusLib {
9
10      typedef INTEGER Integer;
11      typedef BIGINTEGER BigInteger;
12      typedef unsigned INTEGER Natural;
13      typedef unsigned BIGINTEGER BigNatural;
14      typedef REAL Real;
15      typedef std::size_t Size;
16
17  }
18
19  #endif
```

B.2.3 Class Error

Listing B.7: The header file of class Error.

```
1   // cl/errors.hpp
2   #ifndef CLUSLIB_ERRORS_HPP
3   #define CLUSLIB_ERRORS_HPP
4
5   #include<boost/assert.hpp>
6   #include<boost/current_function.hpp>
7   #include<boost/shared_ptr.hpp>
8   #include<exception>
9   #include<sstream>
10
```

```cpp
namespace ClusLib {
    class Error: public std::exception {
    private:
        boost::shared_ptr<std::string> _msg;
    public:
        Error(const std::string& file,
              long line,
              const std::string& function,
              const std::string& msg = "");
        ~Error() throw() {}
        const char* what() const throw();
    };

}

#define FAIL(msg) \
    std::ostringstream ss; \
    ss << msg; \
    throw ClusLib::Error(__FILE__,__LINE__, \
            BOOST_CURRENT_FUNCTION,ss.str());

#define ASSERT(condition,msg) \
    if(!(condition)) { \
        std::ostringstream ss; \
        ss << msg; \
        throw ClusLib::Error(__FILE__,__LINE__, \
                BOOST_CURRENT_FUNCTION,ss.str()); \
    }

#endif
```

Listing B.8: The source file of class **Error**.

```cpp
// cl/errors.cpp
#include<cl/errors.hpp>
#include<stdexcept>

namespace {
    std::string format(const std::string& file,
            long line,
            const std::string& function,
            const std::string& msg) {
        std::ostringstream ss;
        ss<<function<<": ";
        ss<<"\n "<<file<<"("<<line<<"): \n"<<msg;
        return ss.str();
    }

}

namespace boost {
    void assertion_failed(char const * expr,
            char const * function,
            char const * file,
            long line){
        throw std::runtime_error(format(file,line,function,
            "Boost assertion failed: " + std::string(expr)));
    }
}


namespace ClusLib {
    Error::Error(const std::string& file,
                 long line,
                 const std::string& function,
                 const std::string& msg){
```

```
34        _msg = boost::shared_ptr<std::string>(new std::string(
35                  format(file,line,function,msg)));
36    }
37
38    const char* Error::what() const throw () {
39        return _msg->c_str();
40    }
41 }
```

B.3 Files in Folder cl/algorithms

B.3.1 Makefile

Listing B.9: Makefile.am in cl/algorithms.

```
1  noinst_LTLIBRARIES = libAlgorithms.la
2
3  AM_CPPFLAGS = -I${top_srcdir} -I${top_builddir}
4
5  this_includedir=${includedir}/${subdir}
6  this_include_HEADERS = \
7      all.hpp \
8      algorithm.hpp \
9      average.hpp \
10     centroid.hpp \
11     cmean.hpp \
12     complete.hpp \
13         diana.hpp \
14     fsc.hpp \
15     gkmode.hpp \
16         gmc.hpp \
17     kmean.hpp \
18     kprototype.hpp \
19     lw.hpp \
20         median.hpp \
21     single.hpp \
22     ward.hpp \
23     weighted.hpp
24
25 libAlgorithms_la_SOURCES = \
26     algorithm.cpp \
27     average.cpp \
28     centroid.cpp \
29     cmean.cpp \
30     complete.cpp \
31         diana.cpp \
32     fsc.cpp \
33     gkmode.cpp \
34         gmc.cpp \
35     kmean.cpp \
36     kprototype.cpp \
37     lw.cpp \
38         median.cpp \
39     single.cpp \
40     ward.cpp \
41     weighted.cpp
42
43
```

```
44  all.hpp: Makefile.am
45      echo "// This file is generated. Please do not edit!" > $@
46      echo >> $@
47      for i in $(filter-out all.hpp, $(this_include_HEADERS)); \
48      do \
49          echo "#include <${subdir}/$$i>" >> $@; \
50      done
51      echo >> $@
52      subdirs='$(SUBDIRS)'; for i in $$subdirs; do \
53          echo "#include <${subdir}/$$i/all.hpp>" >> $@; \
54      done
```

B.3.2 Class Algorithm

Listing B.10: The header file of class Algorithm.

```
1   // cl/algorithms/algorithm.hpp
2   #ifndef CLUSLIB_ALGORITHM_HPP
3   #define CLUSLIB_ALGORITHM_HPP
4
5   #include<cl/datasets/dataset.hpp>
6   #include<cl/distances/distance.hpp>
7   #include<map>
8   #include<boost/any.hpp>
9
10  namespace ClusLib {
11
12      class Additional {
13      public:
14          const boost::any& get(const std::string &name) const;
15          void insert(const std::string &name,
16                      const boost::any &val);
17
18          std::map<std::string, boost::any> additional;
19
20      protected:
21          Additional() {}
22      };
23
24      class Arguments : public Additional {
25      public:
26          boost::shared_ptr<Dataset> ds;
27          boost::shared_ptr<Distance> distance;
28      };
29
30      class Results : public Additional {
31      public:
32          void reset();
33
34          std::vector<Size> CM;
35      };
36
37      class Algorithm {
38      public:
39          virtual ~Algorithm() {}
40          virtual Arguments& getArguments();
41          virtual const Results& getResults() const;
42          virtual void reset() const;
43          virtual void clusterize();
44
45      protected:
46          virtual void setupArguments();
47          virtual void performClustering() const = 0;
48          virtual void fetchResults() const = 0;
```

```cpp
                boost::shared_ptr<Dataset> _ds;
                mutable Results _results;
                Arguments _arguments;
        };
}

#endif
```

Listing B.11: The source file of class **Algorithm**.

```cpp
// cl/algorithms/algorithm.cpp
#include<cl/algorithms/algorithm.hpp>

namespace ClusLib {

    const boost::any& Additional::get(const std::string &name)
        const {
            std::map<std::string, boost::any>::const_iterator it;
            it = additional.find(name);
            if (it == additional.end()) {
                FAIL(name << " not found");
            }

            return it->second;
        }

    void Additional::insert(const std::string &name,
        const boost::any &val) {
            additional.insert(std::pair<std::string,
                boost::any>(name, val));
        }

    void Results::reset() {
        CM.clear();
        additional.clear();
    }

    Arguments& Algorithm::getArguments() {
        return _arguments;
    }

    const Results& Algorithm::getResults() const {
        return _results;
    }

    void Algorithm::reset() const {
        _results.reset();
    }

    void Algorithm::clusterize() {
        setupArguments();
        performClustering();
        reset();
        fetchResults();
    }

    void Algorithm::setupArguments() {
        _ds = _arguments.ds;
        ASSERT(_ds, "dataset is null");
    }

}
```

B.3.3 Class Average

Listing B.12: The header file of class **Average**.

```cpp
// cl/algorithms/average.hpp
#ifndef CLUSLIB_AVERAGE_HPP
#define CLUSLIB_AVERAGE_HPP

#include<cl/algorithms/lw.hpp>

namespace ClusLib {

    class Average: public LW {
    private:
        void update_dm(Size p, Size q, Size r) const;
    };

}

#endif
```

Listing B.13: The source file of class **Average**.

```cpp
// cl/algorithms/average.cpp
#include<cl/algorithms/average.hpp>

namespace ClusLib {

    void Average::update_dm(Size p, Size q, Size r) 
        const {
        Real dist;
        std::set<Size>::iterator it;
        for(it = _unmergedClusters.begin();
            it != _unmergedClusters.end(); ++it) {
            if(*it == r) {
                continue;
            }

            dist = (_clusterSize[p]*_dm(p,*it) +
                    _clusterSize[q]*_dm(q,*it)) / _clusterSize[r];
            _dm.add_item(r,*it, dist);
        }
    }

}
```

B.3.4 Class Centroid

Listing B.14: The header file of class **Centroid**.

```cpp
// cl/algorithms/centroid.hpp
#ifndef CLUSLIB_CENTROID_HPP
#define CLUSLIB_CENTROID_HPP

#include<cl/algorithms/lw.hpp>

namespace ClusLib {

    class Centroid: public LW {
    private:
        void setupArguments();
```

```
            void update_dm(Size p, Size q, Size r) const;
    };

}

#endif
```

Listing B.15: The source file of class Centroid.

```
// cl/algorithms/centroid.cpp
#include<cl/algorithms/centroid.hpp>
#include<cl/distances/euclideandistance.hpp>
#include<cmath>

namespace ClusLib {

    void Centroid::setupArguments() {
        Algorithm::setupArguments();

        _distance = boost::shared_ptr<Distance>(new
            EuclideanDistance());
    }

    void Centroid::update_dm(Size p, Size q, Size r)
            const {
        Real dist;
        std::set<Size>::iterator it;
        Real sp = _clusterSize[p];
        Real sq = _clusterSize[q];
        for(it = _unmergedClusters.begin();
                it != _unmergedClusters.end(); ++it) {
            if(*it == r) {
                continue;
            }

            dist = std::pow(_dm(p,*it), 2.0)*sp/(sp+sq) +
                std::pow(_dm(q,*it), 2.0)*sq/(sp+sq) -
                std::pow(_dm(p,q), 2.0)*sp*sq/((sp+sq)*(sp+sq));
            _dm.add_item(r,*it, std::sqrt(dist));
        }
    }

}
```

B.3.5 Class Cmean

Listing B.16: The header file of class Cmean.

```
// cl/algorithms/cmean.hpp
#ifndef CLUSLIB_CMEAN_HPP
#define CLUSLIB_CMEAN_HPP

#include<cl/algorithms/algorithm.hpp>
#include<cl/algorithms/kmean.hpp>
#include<cl/types.hpp>
#include<cl/datasets/dataset.hpp>
#include<cl/clusters/centercluster.hpp>
#include<cl/clusters/pclustering.hpp>
#include<cl/distances/distance.hpp>
#include<cl/utilities/matrix.hpp>

namespace ClusLib {
```

```cpp
     class Cmean: public Algorithm {
     private:
         void setupArguments();
         void performClustering() const;
         void fetchResults() const;
         void initialization() const;
         void iteration() const;
         void updateCenter() const;
         void updateFCM() const;
         void calculateObj() const;

         mutable std::vector<boost::shared_ptr<CenterCluster> >
             _clusters;
         mutable std::vector<Size> _CM;
         mutable ublas::matrix<Real> _FCM;
         mutable Size _numiter;
         mutable Real _dObj;

         Real _threshold;
         Real _alpha;
         Real _epsilon;
         Size _numclust;
         Size _maxiter;
         Size _seed;
         boost::shared_ptr<Distance> _distance;
     };

}
#endif
```

Listing B.17: The source file of class **Cmean**.

```cpp
// cl/algorithms/cmean.cpp
#include<cl/algorithms/cmean.hpp>
#include<cl/errors.hpp>
#include<cl/distances/euclideandistance.hpp>
#include<iostream>
#include<cmath>
#include<boost/random.hpp>

namespace ClusLib {

    void Cmean::performClustering() const {
        initialization();
        iteration();
    }

    void Cmean::setupArguments() {
        Algorithm::setupArguments();
        ASSERT(_ds->is_numeric(), "not_a_numeric_dataset");

        _epsilon = boost::any_cast<Real>(
            _arguments.get("epsilon"));
        ASSERT(_epsilon>0, "epsilon_must_be_positive");

        _threshold = boost::any_cast<Real>(
            _arguments.get("threshold"));
        ASSERT(_threshold>EPSILON, "invalid_threshold");

        _alpha = boost::any_cast<Real>(_arguments.get("alpha"));
        ASSERT(_alpha>1, "invalid_alpha");

        _numclust = boost::any_cast<Size>(
            _arguments.get("numclust"));
        ASSERT(_numclust>=2 && _numclust<=_ds->size(),
```

```cpp
            "invalid_numclust");

    _maxiter = boost::any_cast<Size>(
        _arguments.get("maxiter"));
    ASSERT(_maxiter>0, "invalide_maxiter");

    _seed = boost::any_cast<Size>(
        _arguments.get("seed"));
    ASSERT(_seed>0, "invalide_seed");

    _distance = boost::shared_ptr<Distance>(new
        EuclideanDistance());
}

void Cmean::fetchResults() const {
    Size s;
    for(Size i=0;i<_ds->size();++i) {
        Real dMax = MIN_REAL;
        for(Size k=0;k<_numclust;++k) {
            if (dMax < _FCM(i,k) ) {
                dMax = _FCM(i,k);
                s = k;
            }
        }
        _CM[i] = s;
        _clusters[s]->add((*_ds)[i]);
    }

    PClustering pc;
    for(Size i=0;i<_clusters.size();++i){
        pc.add(_clusters[i]);
    }
    _results.CM = _CM;
    _results.insert("pc",pc);
    _results.insert("fcm", _FCM);
    _results.insert("numiter", _numiter);
    _results.insert("dObj", _dObj);
}

void Cmean::iteration() const {
    _numiter = 0;
    Real dPrevObj;
    while(true) {
        updateCenter();
        updateFCM();

        dPrevObj = _dObj;
        calculateObj();

        ++_numiter;

        if (std::fabs(_dObj - dPrevObj) < _threshold){
            break;
        }

        if (_numiter >= _maxiter){
            break;
        }
    }
}

void Cmean::updateCenter() const {
    Real dSum1, dSum2, dTemp;
    boost::shared_ptr<Schema> schema = _ds->schema();
    for(Size k=0;k<_numclust;++k){
        for(Size j=0;j<schema->size();++j){
            dSum1 = 0.0;
```

```
101                    dSum2 = 0.0;
102                    for(Size i=0; i<_ds->size();++i){
103                        dTemp = std::pow(_FCM(i,k),_alpha);
104                        dSum1 += dTemp *
105                            (*schema)[j]->get_c_val((*_ds)(i,j));
106                        dSum2 += dTemp;
107                    }
108                    (*schema)[j]->set_c_val(
109                        (*_clusters[k]->center())[j],dSum1/dSum2);
110                }
111            }
112        }
113
114        void Cmean::updateFCM() const {
115            Real dSum, dTemp;
116            std::vector<Real> dvTemp(_numclust);
117            boost::shared_ptr<Schema> schema = _ds->schema();
118            for(Size i=0;i<_ds->size();++i){
119                dSum = 0.0;
120                for(Size k=0;k<_numclust;++k){
121                    dTemp = (*_distance)((*_ds)[i],
122                        _clusters[k]->center()) + _epsilon;
123                    dvTemp[k] = std::pow(dTemp, 2/(_alpha-1));
124                    dSum += 1 / dvTemp[k];
125                }
126                for(Size k=0;k<_numclust;++k){
127                    _FCM(i,k) = 1.0 / (dvTemp[k] * dSum);
128                }
129            }
130        }
131
132        void Cmean::initialization() const {
133            Size numRecords = _ds->size();
134            _FCM.resize(numRecords, _numclust);
135            _CM.resize(numRecords, Null<Size>());
136
137            std::vector<Size> index(numRecords,0);
138            for(Size i=0;i<index.size();++i){
139                index[i] = i;
140            }
141
142            boost::minstd_rand generator(_seed);
143            for(Size i=0;i<_numclust;++i){
144                boost::uniform_int<> uni_dist(0,numRecords-i-1);
145                boost::variate_generator<boost::minstd_rand&,
146                    boost::uniform_int<> > uni(generator, uni_dist);
147                Size r = uni();
148                boost::shared_ptr<Record> cr =
149                    boost::shared_ptr<Record>(new Record(*(*_ds)[r]));
150                boost::shared_ptr<CenterCluster> c =
151                    boost::shared_ptr<CenterCluster>(
152                        new CenterCluster(cr));
153                c->set_id(i);
154                _clusters.push_back(c);
155                index.erase(index.begin()+r);
156            }
157
158            updateFCM();
159        }
160
161        void Cmean::calculateObj() const {
162            Real dSum = 0.0;
163            for(Size i=0; i<_ds->size(); ++i) {
164                for(Size j=0; j<_numclust; ++j) {
165                    Real dTemp = (*_distance)((*_ds)[i],
166                        _clusters[j]->center());
167                    dSum += std::pow(_FCM(i, j), _alpha) *
```

```
168                    std::pow(dTemp, 2.0);
169            }
170        }
171        _dObj = dSum;
172    }
173 }
```

B.3.6 Class Complete

Listing B.18: The header file of class Complete.

```
1  // cl/algorithms/complete.hpp
2  #ifndef CLUSLIB_COMPLETE_HPP
3  #define CLUSLIB_COMPLETE_HPP
4
5  #include<cl/algorithms/lw.hpp>
6
7  namespace ClusLib {
8
9      class Complete: public LW {
10     private:
11         void update_dm(Size p, Size q, Size r) const;
12     };
13
14 }
15
16 #endif
```

Listing B.19: The source file of class Complete.

```
1  // cl/algorithms/complete.cpp
2  #include<cl/algorithms/complete.hpp>
3
4  namespace ClusLib {
5
6      void Complete::update_dm(Size p, Size q, Size r)
7          const {
8          Real dist;
9          std::set<Size>::iterator it;
10         for(it = _unmergedClusters.begin();
11             it != _unmergedClusters.end(); ++it) {
12             if(*it == r) {
13                 continue;
14             }
15
16             dist = std::max(_dm(p,*it), _dm(q,*it));
17             _dm.add_item(r,*it,dist);
18         }
19     }
20
21 }
```

B.3.7 Class Diana

Listing B.20: The header file of class Diana.

```
1  // cl/algorithms/diana.hpp
2  #ifndef CLUSLIB_DIANA_HPP
3  #define CLUSLIB_DIANA_HPP
```

```
#include<cl/algorithms/algorithm.hpp>
#include<cl/clusters/hclustering.hpp>
#include<cl/utilities/nnmap.hpp>
#include<cl/datasets/dataset.hpp>
#include<cl/distances/distance.hpp>
#include<cl/patterns/nodevisitor.hpp>
#include<cl/types.hpp>
#include<set>

namespace ClusLib {

    class Diana: public Algorithm {
    protected:
        void setupArguments();
        void performClustering() const;
        void fetchResults() const;
        virtual void create_dm() const;
        virtual void initialization() const;
        virtual void division() const;
        virtual void do_split(Size ind) const;
        virtual void create_cluster(const std::set<Size> ele,
            Size ind) const;

        mutable iirMapA _dm;
        mutable std::set<Size> _unsplitClusters;
        mutable std::map<Size,Real> _clusterDiameter;
        mutable std::map<Size,boost::shared_ptr<LeafNode> > _leaf;
        mutable std::map<Size,boost::shared_ptr<InternalNode> >
            _internal;
        mutable std::set<Size> _clusterID;
        boost::shared_ptr<Distance> _distance;
    };
}

#endif
```

Listing B.21: The source file of class `Diana`.

```
// cl/algorithms/diana.cpp
#include<cl/algorithms/diana.hpp>
#include<cl/patterns/pcvisitor.hpp>
#include<cl/types.hpp>
#include<iostream>
#include<boost/pointer_cast.hpp>

namespace ClusLib {

    void Diana::setupArguments() {
        Algorithm::setupArguments();

        _distance = _arguments.distance;
        ASSERT(_distance, "distance_is_null");
    }

    void Diana::create_dm() const {
        Size n = _ds->size();
        for(Size i=0;i<n-1;++i){
            for(Size j=i+1;j<n;++j){
                _dm.add_item(i, j,
                    (*_distance)((*_ds)[i],(*_ds)[j]));
            }
        }
    }

    void Diana::performClustering() const {
```

```cpp
            create_dm();
            initialization();
            division();
        }

    void Diana::initialization() const {
        Size n = _ds->size();
        Size id = 2*n-2;
        boost::shared_ptr<InternalNode> pin(new InternalNode(id));
        for(Size s=0;s<n;++s){
            boost::shared_ptr<LeafNode> pln(new
                LeafNode((*_ds)[s], s));
            pln->set_level(0);
            pin->add(pln);
            _leaf.insert(std::pair<Size,
                boost::shared_ptr<LeafNode> >(s, pln));
        }
        _internal.insert(std::pair<Size,
            boost::shared_ptr<InternalNode> >(id, pin));
        _unsplitClusters.insert(id);

        Real dMax = MIN_REAL;
        for(Size i=0;i<n-1;++i){
            for(Size j=i+1;j<n;++j){
                if (dMax < _dm(i,j) ) {
                    dMax = _dm(i,j);
                }
            }
        }
        _clusterDiameter.insert(std::pair<Size, Real>(id,dMax));

        for(Size s=2*n-3; s>n-1; --s) {
            _clusterID.insert(s);
        }
    }

    void Diana::division() const {
        Size n = _ds->size();
        std::set<Size>::iterator it;
        Real dMax;
        Size ind;
        for(Size s=2*n-2; s>n-1; --s) {
            dMax= MIN_REAL;
            std::vector<Size> nvTemp(_unsplitClusters.begin(),
                            _unsplitClusters.end());
            for(Size i=0; i<nvTemp.size(); ++i) {
                if(dMax < _clusterDiameter[nvTemp[i]]) {
                    dMax = _clusterDiameter[nvTemp[i]];
                    ind = nvTemp[i];
                }
            }

            _internal[ind]->set_level(s-n+1);
            _internal[ind]->set_id(s);
            _internal[ind]->set_joinValue(dMax);
            do_split(ind);
        }
    }

    void Diana::do_split(Size ind) const {
        std::vector<boost::shared_ptr<Node> > data =
            _internal[ind]->data();
        Size n = data.size();

        Size ra;
        std::set<Size> splinter;
        std::set<Size> remaining;
```

```cpp
95      for(Size i=0; i<n; ++i) {
96          Size id = data[i]->get_id();
97          remaining.insert(id);
98      }
99
100     std::set<Size>::iterator it, it1;
101     Real dMax = MIN_REAL;
102     for(it = remaining.begin();
103         it != remaining.end(); ++it) {
104         Real dSum = 0.0;
105         for(it1 = remaining.begin();
106             it1 != remaining.end(); ++it1) {
107             if(*it == *it1) {
108                 continue;
109             }
110             dSum += _dm(*it, *it1);
111         }
112         if(dMax < dSum){
113             dMax = dSum;
114             ra = *it;
115         }
116     }
117     splinter.insert(ra);
118     remaining.erase(ra);
119
120     bool bChanged = true;
121     while(bChanged) {
122         bChanged = false;
123         for(it = remaining.begin();
124             it != remaining.end(); ++it) {
125             Real d1 = 0.0;
126             for(it1 = splinter.begin();
127                 it1 != splinter.end(); ++it1) {
128                 d1 += _dm(*it, *it1);
129             }
130             d1 /= splinter.size();
131
132             Real d2 = 0.0;
133             for(it1 = remaining.begin();
134                 it1 != remaining.end(); ++it1) {
135                 if(*it == *it1) {
136                     continue;
137                 }
138                 d2 += _dm(*it, *it1);
139             }
140             if(remaining.size() > 1) {
141                 d2 /= (remaining.size()-1.0);
142             }
143
144             if(d1 < d2) {
145                 bChanged = true;
146                 splinter.insert(*it);
147                 remaining.erase(it);
148                 break;
149             }
150         }
151     }
152
153     _unsplitClusters.erase(ind);
154     _internal[ind]->clear();
155     create_cluster(splinter, ind);
156     create_cluster(remaining, ind);
157 }
158
159 void Diana::create_cluster(const std::set<Size> ele,
160     Size ind) const {
161     std::set<Size>::iterator it;
```

```cpp
        Real dMax;
        if(ele.size() > 1) {
            boost::shared_ptr<InternalNode> pin(new
                InternalNode(0, _internal[ind]));
            _internal[ind]->add(pin);
            for(it = ele.begin(); it != ele.end(); ++it) {
                pin->add(_leaf[*it]);
            }

            it = _clusterID.end();
            --it;
            Size id = *it;
            _clusterID.erase(it);

            _internal.insert(std::pair<Size,
                boost::shared_ptr<InternalNode> >(id, pin));
            _unsplitClusters.insert(id);

            dMax = MIN_REAL;
            std::vector<Size> nvTemp(ele.begin(), ele.end());
            for(Size i=0; i<nvTemp.size(); ++i) {
                for(Size j=i+1; j<nvTemp.size(); ++j) {
                    if(dMax < _dm(nvTemp[i], nvTemp[j])) {
                        dMax = _dm(nvTemp[i], nvTemp[j]);
                    }
                }
            }
            _clusterDiameter.insert(
                std::pair<Size, Real>(id,dMax));
        } else {
            it = ele.begin();
            _internal[ind]->add(_leaf[*it]);
        }
    }
}

void Diana::fetchResults() const {
    Size n = _ds->size();
    HClustering hc(_internal[2*n-2]);
    _results.insert("hc", hc);
}
}
```

B.3.8 Class FSC

Listing B.22: The header file of class FSC.

```cpp
// cl/algorithms/fsc.hpp
#ifndef CLUSLIB_FSC_HPP
#define CLUSLIB_FSC_HPP

#include<cl/algorithms/algorithm.hpp>
#include<cl/datasets/dataset.hpp>
#include<cl/clusters/subspacecluster.hpp>

namespace ClusLib {

    class FSC: public Algorithm {
    protected:
        void setupArguments();
        void performClustering() const;
        void fetchResults() const;
        virtual void initialization() const;
        virtual void iteration() const;
        virtual void updateCenter() const;
```

```cpp
        virtual void updateWeight() const;
        virtual void calculateObj() const;
        Real dist(const boost::shared_ptr<Record> &x,
                  const boost::shared_ptr<SubspaceCluster> &c
                 ) const;

        mutable std::vector<boost::shared_ptr<SubspaceCluster> >
            _clusters;
        mutable std::vector<Size> _CM;
        mutable Real _dObj;
        mutable Size _numiter;

        Size _seed;
        Real _alpha;
        Real _epsilon;
        Real _threshold;
        Size _numclust;
        Size _maxiter;
    };
}

#endif
```

Listing B.23: The source file of class **FSC**.

```cpp
// cl/algorithms/fsc.cpp
#include<cl/algorithms/fsc.hpp>
#include<cl/clusters/pclustering.hpp>
#include<cl/errors.hpp>
#include<iostream>
#include<boost/random.hpp>

namespace ClusLib {

    void FSC::performClustering() const {
        initialization();
        iteration();
    }

    void FSC::setupArguments() {
        Algorithm::setupArguments();
        ASSERT(_ds->is_numeric(), "dataset_is_not_numeric");

        _epsilon = boost::any_cast<Real>(
            _arguments.get("epsilon"));
        ASSERT(_epsilon>=0, "invalid_epsilon");

        _threshold = boost::any_cast<Real>(
            _arguments.get("threshold"));
        ASSERT(_threshold>EPSILON, "invalid_threshold");

        _alpha = boost::any_cast<Real>(_arguments.get("alpha"));
        ASSERT(_alpha>1, "invalid_alpha");

        _numclust = boost::any_cast<Size>(
            _arguments.get("numclust"));
        ASSERT(_numclust>=2 && _numclust<=_ds->size(),
            "invalid_numclust");

        _maxiter = boost::any_cast<Size>(
            _arguments.get("maxiter"));
        ASSERT(_maxiter>0, "invalide_maxiter");

        _seed = boost::any_cast<Size>(
            _arguments.get("seed"));
        ASSERT(_seed>0, "invalide_seed");
```

```cpp
        }

        void FSC::fetchResults() const {
            PClustering pc;
            for(Size i=0;i<_clusters.size();++i){
                pc.add(_clusters[i]);
            }
            _results.CM = _CM;
            _results.insert("pc", boost::any(pc));
            _results.insert("numiter", boost::any(_numiter));
            _results.insert("dObj", boost::any(_dObj));
        }

        void FSC::iteration() const {
            Real dObjPre;

            updateWeight();
            updateCenter();
            _numiter = 1;
            while(true) {
                Integer s;
                Real dMin,dDist;
                for(Size i=0;i<_ds->size();++i) {
                    dMin = MAX_REAL;
                    for(Size k=0;k<_clusters.size();++k) {
                        dDist = dist((*_ds)[i],_clusters[k]);
                        if (dMin > dDist) {
                            dMin = dDist;
                            s = k;
                        }
                    }

                    if (_CM[i] != s){
                        _clusters[_CM[i]]->erase((*_ds)[i]);
                        _clusters[s]->add((*_ds)[i]);
                        _CM[i] = s;
                    }
                }

                updateWeight();
                updateCenter();

                dObjPre = _dObj;
                calculateObj();
                if(std::fabs(dObjPre - _dObj) < _threshold){
                    break;
                }

                ++_numiter;
                if (_numiter > _maxiter){
                    break;
                }
            }
        }

        void FSC::updateCenter() const {
            Real dTemp;
            boost::shared_ptr<Schema> schema = _ds->schema();
            for(Size k=0;k<_clusters.size();++k){
                for(Size j=0;j<schema->size();++j){
                    dTemp = 0.0;
                    for(Size i=0; i<_clusters[k]->size();++i){
                        boost::shared_ptr<Record> rec =
                            (*_clusters[k])[i];
                        dTemp += (*schema)[j]->get_c_val((*rec)[j]);
                    }
                    (*schema)[j]->set_c_val(
```

```cpp
                            (*_clusters[k]->center())[j],
                            dTemp/_clusters[k]->size());
                }
            }
        }

        void FSC::updateWeight() const {
            Real dVar, dSum;
            boost::shared_ptr<Schema> schema = _ds->schema();
            boost::shared_ptr<Record> c, r;
            std::vector<Real> w(schema->size(),0.0);
            for(Size k=0;k<_clusters.size();++k){
                c = _clusters[k]->center();
                dSum = 0.0;
                for(Size j=0;j<schema->size();++j){
                    dVar = 0.0;
                    for(Size i=0; i<_clusters[k]->size();++i){
                        r = (*_clusters[k])[i];
                        dVar += std::pow((*schema)[j]->distance(
                                    (*r)[j], (*c)[j]), 2.0);
                    }
                    w[j] = std::pow(1/(dVar + _epsilon),1/(_alpha-1));
                    dSum += w[j];
                }
                for(Size j=0;j<schema->size();++j) {
                    _clusters[k]->w(j) = w[j] / dSum;
                }
            }
        }

        void FSC::initialization() const {
            Size numRecords = _ds->size();
            std::vector<Integer> index(numRecords,0);
            _CM.resize(numRecords);
            for(Size i=0;i<index.size();++i){
                index[i] = i;
            }

            boost::minstd_rand generator(_seed);
            for(Size i=0;i<_numclust;++i){
                boost::uniform_int<> uni_dist(0,numRecords-i-1);
                boost::variate_generator<boost::minstd_rand&,
                    boost::uniform_int<> > uni(generator, uni_dist);
                Integer r = uni();
                boost::shared_ptr<Record> cr =
                    boost::shared_ptr<Record>(new Record(*(*_ds)[r]));
                boost::shared_ptr<SubspaceCluster> c =
                    boost::shared_ptr<SubspaceCluster>(
                        new SubspaceCluster(cr));
                c->set_id(i);
                _clusters.push_back(c);
                index.erase(index.begin()+r);
            }

            Integer s;
            Real dMin,dDist;
            for(Size i=0;i<numRecords;++i){
                dMin = MAX_REAL;
                for(Size j=0;j<_numclust;++j){
                    dDist = dist((*_ds)[i], _clusters[j]);
                    if (dDist<dMin){
                        s = j;
                        dMin = dDist;
                    }
                }
                _clusters[s]->add((*_ds)[i]);
                _CM[i] = s;
```

```cpp
            }
        }

        Real FSC::dist(const boost::shared_ptr<Record> &x,
                    const boost::shared_ptr<SubspaceCluster> &c
                    ) const {
            Real dTemp = 0.0;
            boost::shared_ptr<Schema> schema = _ds->schema();
            for(Size j=0;j<schema->size();++j){
                dTemp += std::pow(c->w(j), _alpha) *
                    std::pow((*schema)[j]->distance((*x)[j],
                    (*c->center())[j]), 2.0);
            }

            return dTemp;
        }

        void FSC::calculateObj() const {
            Real dTemp = 0.0;
            for(Size i=0; i<_ds->size(); ++i){
                dTemp += dist((*_ds)[i], _clusters[_CM[i]]);
            }

            _dObj = dTemp;
        }
    }
```

B.3.9 Class GKmode

Listing B.24: The header file of class GKmode.

```cpp
// cl/algorithms/gkmode.hpp
#ifndef CLUSLIB_GKMODE_HPP
#define CLUSLIB_GKMODE_HPP

#include<cl/algorithms/algorithm.hpp>
#include<cl/clusters/centercluster.hpp>
#include<cl/clusters/pclustering.hpp>
#include<cl/datasets/dataset.hpp>
#include<cl/distances/distance.hpp>
#include<cl/utilities/matrix.hpp>

namespace ClusLib {

    class GKmode : public Algorithm {
    private:
        void setupArguments();
        void performClustering() const;
        void fetchResults() const;
        void initialization() const;
        void iteration() const;
        void selection(Size g) const;
        void mutation(Size g) const;
        void kmode(Size ind) const;
        void calculateLoss(Size ind) const;
        bool get_mode(Size &mode, Real &var,
                    Size ind, Size k, Size j) const;
        Real calculateRatio(Size ind) const;
        void generateRN(std::vector<Size>& nv, Size s) const;


        mutable std::vector<boost::shared_ptr<CenterCluster> >
            _clusters;
        mutable ublas::matrix<Size> _mP;
```

```
34      mutable std::vector<Real> _dvFit;
35      mutable std::vector<bool> _bvLegal;
36
37      Size _numclust;
38      Size _numpop;
39      Size _maxgen;
40      Real _c;
41      Real _cm;
42      Real _Pm;
43  };
44
45  }
46
47  #endif
```

Listing B.25: The source file of class GKmode.

```
1   // cl/algorithms/gkmode.cpp
2   #include<cl/algorithms/gkmode.hpp>
3   #include<boost/random.hpp>
4   #include<cl/types.hpp>
5   #include<set>
6   #include<map>
7
8   namespace ClusLib {
9
10      void GKmode::setupArguments() {
11          Algorithm::setupArguments();
12          ASSERT(_ds->is_categorical(),
13              "dataset is not categorical");
14
15          _numclust = boost::any_cast<Size>(
16              _arguments.get("numclust"));
17          ASSERT(_numclust>=2 && _numclust<=_ds->size(),
18              "invalid numclust");
19
20          _numpop = boost::any_cast<Size>(
21              _arguments.get("numpop"));
22          ASSERT(_numpop>0, "invalide numpop");
23
24          _maxgen = boost::any_cast<Size>(
25              _arguments.get("maxgen"));
26          ASSERT(_maxgen>0, "invalide maxgen");
27
28          _c = boost::any_cast<Real>(_arguments.get("c"));
29          ASSERT(_c>0 && _c<3, "c must be in range (0, 3)");
30
31          _cm = boost::any_cast<Real>(_arguments.get("cm"));
32          ASSERT(_cm>0, "cm must be positive");
33
34          _Pm = boost::any_cast<Real>(_arguments.get("pm"));
35          ASSERT(_Pm>0, "pm must be positive");
36      }
37
38      void GKmode::performClustering() const {
39          initialization();
40          iteration();
41      }
42
43      void GKmode::fetchResults() const {
44          Real dMin = MAX_REAL;
45          Size s;
46          for(Size i=0; i<_numpop; ++i) {
47              if (dMin > _dvFit[i]) {
48                  dMin = _dvFit[i];
49                  s = i;
```

```cpp
            }
        }

        std::vector<Size> CM(_ds->size());
        for(Size i=0; i<_ds->size(); ++i){
            CM[i] = _mP(s, i);
        }
        _results.CM = CM;

        Real dVar;
        Size mode;
        PClustering pc;
        boost::shared_ptr<Schema> schema = _ds->schema();
        for(Size k=0; k<_numclust; ++k) {
            boost::shared_ptr<Record> r(new Record(schema));
            for(Size j=0; j<_ds->num_attr(); ++j) {
                if (get_mode(mode, s, k, j) ) {
                    (*schema)[j]->set_d_val((*r)[j], mode);
                }
            }
            boost::shared_ptr<CenterCluster> c(new
                CenterCluster(r));
            pc.add(c);
        }

        for(Size i=0; i<_ds->size(); ++i){
            pc[CM[i]]->add( (*_ds)[i]);
        }

        _results.insert("pc", boost::any(pc));
    }

    void GKmode::initialization() const {
        _mP.resize(_numpop, _ds->size());
        for(Size i=0; i<_numpop; ++i) {
            for(Size j=0; j<_ds->size(); ++j) {
                _mP(i,j) = Null<Size>();
            }
        }

        std::vector<Size> nvTemp(_numclust, 0);
        boost::minstd_rand generator(42u);
        boost::uniform_int<> uni_dist(0, _numclust-1);
        boost::variate_generator<boost::minstd_rand&,
            boost::uniform_int<> > uni(generator, uni_dist);
        for(Size i=0;i<_numpop;++i){
            generateRN(nvTemp, i);
            for(Size j=0; j<_numclust; ++j) {
                _mP(i, nvTemp[j]) = j;
            }
            for(Size j=0;j<_ds->size();++j) {
                if (_mP(i,j) == Null<Size>()) {
                    _mP(i,j) = uni();
                }
            }
        }

        _dvFit.resize(_numpop, 0.0);
        _bvLegal.resize(_numpop, true);
    }

    void GKmode::iteration() const {
        Size g = 0;
        while (g < _maxgen) {
            ++g;
            Real Lmax = MIN_REAL;
            for(Size i=0; i<_numpop; ++i) {
```

```cpp
                    calculateLoss(i);
                    if( _bvLegal[i] && _dvFit[i]>Lmax) {
                        Lmax = _dvFit[i];
                    }
                }

                Real dTemp = 0.0;
                for(Size i=0; i<_numpop; ++i) {
                    if( _bvLegal[i]) {
                        _dvFit[i] = _c*Lmax - _dvFit[i];
                    } else {
                        _dvFit[i] = calculateRatio(i)*(_c-1)*Lmax;
                    }
                    dTemp += _dvFit[i];
                }

                for(Size i=0; i<_numpop; ++i) {
                    _dvFit[i] /= dTemp;
                }

                selection(g);
                mutation(g);

                for(Size i=0; i<_numpop; ++i) {
                    kmode(i);
                }
            }

            for(Size i=0; i<_numpop; ++i) {
                calculateLoss(i);
            }
        }

        void GKmode::selection(Size g) const {
            boost::minstd_rand generator(
                static_cast<unsigned int>(g+1));
            boost::uniform_real<> uni_dist(0, 1);
            boost::variate_generator<boost::minstd_rand&,
                boost::uniform_real<> > uni(generator,uni_dist);
            Real dRand, dTemp;
            Size s;
            ublas::matrix<Size> mP = _mP;
            for(Size i=0;i<_numpop;++i){
                dTemp = _dvFit[0];
                dRand = uni();
                s = 0;
                while(dTemp < dRand) {
                    dTemp += _dvFit[++s];
                }
                for(Size j=0; j<_ds->size(); ++j) {
                    _mP(i, j) = mP(s, j);
                }
            }
        }

        void GKmode::mutation(Size g) const {
            boost::minstd_rand generator(
                static_cast<unsigned int>(g+1));
            boost::uniform_real<> uni_dist(0, 1);
            boost::variate_generator<boost::minstd_rand&,
                boost::uniform_real<> > uni(generator,uni_dist);
            Real dRand, dMax, dVar, dTemp;
            Size mode;
            bool isLegal;
            std::map<Size, Size> mTemp;
            std::vector<Real> dvProb(_numclust, 0.0);
            for(Size i=0; i<_numpop; ++i) {
```

```cpp
                mTemp.clear();
                for(Size j=0; j<_ds->size(); ++j) {
                    dRand = uni();
                    if (dRand >= _Pm ) {
                        continue;
                    }

                    dMax = MIN_REAL;
                    for(Size k=0; k<_numclust; ++k) {
                        dvProb[k] = 0.0;
                        for(Size d=0; d<_ds->num_attr(); ++d) {
                            isLegal = get_mode(mode, dVar, i, k, d);
                            if(!isLegal) {
                                break;
                            }
                            dvProb[k] += dVar;
                        }

                        if(dvProb[k] >dMax) {
                            dMax = dvProb[k];
                        }
                    }

                    dTemp = 0.0;
                    for(Size k=0; k<_numclust; ++k) {
                        dvProb[k] = _cm*dMax - dvProb[k];
                        dTemp += dvProb[k];
                    }
                    for(Size k=0; k<_numclust; ++k) {
                        dvProb[k] /= dTemp;
                    }

                    dRand = uni();
                    dTemp = dvProb[0];
                    Size k = 0;
                    while( dTemp < dRand) {
                        dTemp += dvProb[++k];
                    }
                    mTemp.insert(std::pair<Size, Size>(j, k));
                }

                std::map<Size, Size>::iterator it;
                for(it = mTemp.begin(); it!=mTemp.end(); ++it) {
                    _mP(i, it->first) = it->second;
                }
            }

        }

        void GKmode::kmode(Size ind) const {
            ublas::matrix<Size>
                nmMode(_numclust, _ds->num_attr());

            bool isLegal;
            Real dVar;
            for(Size k=0; k<_numclust; ++k) {
                for(Size j=0; j<_ds->num_attr(); ++j) {
                    isLegal = get_mode(nmMode(k,j), dVar, ind, k, j);
                    if(!isLegal) {
                        return;
                    }
                }
            }

            Real dDist;
            Real dMin;
            Size h;
```

```cpp
          boost::shared_ptr<Schema> schema = _ds->schema();
          for(Size i=0; i<_ds->size(); ++i) {
              dMin = MAX_REAL;
              for(Size k=0; k<_numclust; ++k) {
                  dDist = 0.0;
                  for(Size j=0; j<_ds->num_attr(); ++j) {
                      if(nmMode(k, j) != (*schema)[j]->get_d_val(
                              (*(*_ds)[i])[j])) {
                          dDist += 1;
                      }
                  }
                  if(dMin > dDist) {
                      dMin = dDist;
                      h = k;
                  }
              }
              _mP(ind, i) = h;
          }
      }

      void GKmode::calculateLoss(Size ind) const {
          Real dLoss = 0.0;
          Size mode;
          Real dVar;
          bool isLegal;
          for(Size k=0; k<_numclust; ++k){
              for(Size j=0; j<_ds->num_attr(); ++j) {
                  isLegal = get_mode(mode, dVar, ind, k, j);
                  dLoss += dVar;
                  if(!isLegal) {
                      break;
                  }
              }
          }

          _bvLegal[ind] = isLegal;
          if(isLegal) {
              _dvFit[ind] = dLoss;
          } else {
              _dvFit[ind] = MAX_REAL;
          }
      }

      bool GKmode::get_mode(Size &mode, Real &var,
              Size ind, Size k, Size j) const {
          boost::shared_ptr<Schema> schema = _ds->schema();
          Size nValues = (*schema)[j]->cast_to_d().num_values();
          std::vector<Size> nvFreq(nValues,0);

          Size val;
          Size nCount = 0;
          for(Size i=0; i<_ds->size(); ++i) {
              if (_mP(ind, i) == k) {
                  val = (*schema)[j]->get_d_val( (*_ds)(i, j) );
                  ++nvFreq[val];
                  ++nCount;
              }
          }
          val = 0;
          for(Size i=0; i<nValues; ++i){
              if(val < nvFreq[i]) {
                  val = nvFreq[i];
                  mode = i;
              }
          }

          var = nCount - nvFreq[mode];
```

```
            if (nCount == 0) {
                return false;
            } else {
                return true;
            }
        }

        Real GKmode::calculateRatio(Size ind) const {
            std::set<Size> vals;
            for(Size i=0; i<_ds->size(); ++i) {
                vals.insert(_mP(ind, i));
            }

            return (Real)vals.size() / _numclust;
        }

        void GKmode::generateRN(std::vector<Size>& nv,
                Size s) const {
            boost::minstd_rand generator(
                static_cast<unsigned int>(s+1));
            for(Size i=0;i<_numclust;++i){
                boost::uniform_int<> uni_dist(0, _ds->size()-i-1);
                boost::variate_generator<boost::minstd_rand&,
                    boost::uniform_int<> > uni(generator, uni_dist);
                nv[i] = uni();
            }
        }
    }
```

B.3.10 Class GMC

Listing B.26: The header file of class GMC.

```
// cl/algorithms/gmc.hpp
#ifndef CLUSLIB_GMC_HPP
#define CLUSLIB_GMC_HPP

#include<cl/algorithms/algorithm.hpp>
#include<cl/clusters/centercluster.hpp>
#include<cl/datasets/dataset.hpp>
#include<cl/distances/distance.hpp>
#include<cl/utilities/matrix.hpp>

namespace ClusLib {

    class GMC: public Algorithm {
    protected:
        void setupArguments();
        void performClustering() const;
        void fetchResults() const;
        virtual void initialization() const;
        virtual void iteration() const;
        virtual void estep() const;
        virtual void mstep() const;

        mutable ublas::matrix<Real> _mu;
        mutable std::vector<ublas::symmetric_matrix<Real> >
            _sigma;
        mutable std::vector<Real> _p;
        mutable ublas::matrix<Real> _data;
        mutable Real _ll;
        mutable ublas::matrix<Real> _post;
        mutable std::vector<boost::shared_ptr<CenterCluster> >
```

```
31              _clusters;
32       mutable std::vector<Size> _CM;
33       mutable bool _converged;
34       mutable Size _numiter;
35
36       Size _seed;
37       Size _numclust;
38       Size _maxiter;
39       Real _threshold;
40       Real _epsilon;
41    };
42 }
43
44 #endif
```

Listing B.27: The source file of class GMC.

```cpp
1  // cl/algorithms/gmc.cpp
2  #include<cl/algorithms/gmc.hpp>
3  #include<cl/clusters/pclustering.hpp>
4  #include<cl/distances/mahalanobisdistance.hpp>
5  #include<boost/random.hpp>
6  #include<cmath>
7
8  namespace ClusLib {
9
10     void GMC::setupArguments() {
11         Algorithm::setupArguments();
12         ASSERT(_ds->is_numeric(), "dataset is not numeric");
13
14         _seed = boost::any_cast<Size>(
15             _arguments.get("seed"));
16         ASSERT(_seed>0, "seed must be a positive number");
17
18         _numclust = boost::any_cast<Size>(
19             _arguments.get("numclust"));
20         ASSERT(_numclust>=2 && _numclust<=_ds->size(),
21             "invalid numclust");
22
23         _maxiter = boost::any_cast<Size>(
24             _arguments.get("maxiter"));
25         ASSERT(_maxiter>0, "invalid maxiter");
26
27         _threshold = boost::any_cast<Real>(
28             _arguments.get("threshold"));
29         ASSERT(_threshold>0, "threshold must be positive");
30
31         _epsilon = boost::any_cast<Real>(
32             _arguments.get("epsilon"));
33         ASSERT(_epsilon>=0, "epsilon must be nonnegative");
34     }
35
36     void GMC::performClustering() const {
37         initialization();
38         iteration();
39     }
40
41     void GMC::fetchResults() const {
42         for(Size i=0; i<_ds->size(); ++i) {
43             Real dMax = MIN_REAL;
44             Integer k;
45             for(Size j=0; j<_numclust; ++j) {
46                 if(dMax < _post(i,j)){
47                     dMax = _post(i,j);
48                     k = j;
49                 }
```

```
                }
                _CM[i] = k;
                _clusters[k]->add((*_ds)[i]);
            }

            PClustering pc;
            for(Size i=0;i<_clusters.size();++i){
                pc.add(_clusters[i]);
            }
            _results.insert("pc", boost::any(pc));
            _results.CM = _CM;
            _results.insert("converged", boost::any(_converged));
            _results.insert("p", boost::any(_p));
            _results.insert("numiter", boost::any(_numiter));
            _results.insert("ll", boost::any(_ll));
            _results.insert("mu", boost::any(_mu));
        }

        void GMC::iteration() const {
            Real pre_ll = MIN_REAL;
            _converged = false;
            _numiter = 0;
            for(Size iter=0; iter<_maxiter; ++iter) {
                estep();

                Real dTemp = _ll - pre_ll;
                if( dTemp >=0 && dTemp < _threshold) {
                    _converged = true;
                    break;
                }
                pre_ll = _ll;

                mstep();
                ++_numiter;
            }
        }

        void GMC::estep() const {
            Size numRecords = _ds->size();
            Size numAttr = _ds->num_attr();

            ublas::matrix<Real> log_ll(numRecords, _numclust);
            ublas::triangular_matrix<Real> L(numAttr, numAttr);

            for(Size j=0; j<_numclust; ++j) {
                Size k = chol(_sigma[j], L);
                ASSERT(k==0, "invalid_covariance_matrix_" << j);

                Real logDet = 0.0;
                for(Size i=0; i<numAttr; ++i) {
                    logDet += 2.0 * std::log(L(i,i));
                }

                MahalanobisDistance md(_sigma[j]);
                for(Size i=0; i<numRecords; ++i) {
                    Real dist = md((*_ds)[i], _clusters[j]->center());
                    log_ll(i,j) =
                        - 0.5 * dist * dist
                        - 0.5 * logDet + std::log(_p[j])
                        - 0.5 * numAttr * std::log(2*3.141592653589);
                }
            }

            ublas::vector<Real> mlog_ll(numRecords);
            ublas::vector<Real> ones(_numclust);
            for(Size i=0; i<_numclust; ++i) {
                ones(i) = 1.0 / _numclust;
```

```cpp
            }
            ublas::axpy_prod(log_ll, ones, mlog_ll, true);
            Real temp = 0.0;
            Real density;
            _ll = 0.0;
            for(Size i=0; i<numRecords; ++i) {
                density = 0.0;
                for(Size j=0; j<_numclust; ++j) {
                    _post(i,j) = std::exp(log_ll(i,j)-mlog_ll(i));
                    density += _post(i,j);
                }
                for(Size j=0; j<_numclust; ++j) {
                    _post(i,j) /= density;
                }
                _ll += std::log(density) + mlog_ll(i);
            }
        }

        void GMC::mstep() const {
            boost::shared_ptr<Schema> schema = _ds->schema();
            Size numRecords = _ds->size();
            Size numAttr = _ds->num_attr();
            Real psum = 0.0;
            for(Size j=0; j<_numclust; ++j) {
                _p[j] = ublas::sum( ublas::column(_post, j));
                psum += _p[j];
            }

            ublas::matrix<Real> centered(numRecords, numAttr);
            for(Size k=0; k<_numclust; ++k) {
                for(Size j=0; j<numAttr; ++j) {
                    _mu(k, j) = ublas::inner_prod(
                            ublas::column(_post, k),
                            ublas::column(_data, j) ) / _p[k];
                    (*schema)[j]->set_c_val(
                            (*_clusters[k]->center())[j], _mu(k, j));
                }

                for(Size i=0; i<numRecords; ++i) {
                    for(Size j=0; j<numAttr; ++j) {
                        centered(i,j) = std::sqrt(_post(i, k)) *
                                        (_data(i, j) - _mu(k, j));
                    }
                }
                for(Size i=0; i<numAttr; ++i) {
                    for(Size j=0; j<=i; ++j) {
                        _sigma[k](i, j) =
                            ublas::inner_prod(
                                ublas::column(centered, i),
                                ublas::column(centered,j))/_p[k];
                    }
                    _sigma[k](i,i) += _epsilon;
                }
            }

            for(Size j=0; j<_numclust; ++j) {
                _p[j] /= psum;
            }
        }

        void GMC::initialization() const {
            Size numRecords = _ds->size();
            Size numAttr = _ds->num_attr();

            _mu.resize(_numclust, numAttr);
            _sigma.resize(_numclust);
```

```cpp
            _p.resize(_numclust);
            for(Size i=0; i<_numclust; ++i) {
                _sigma[i].resize(numAttr);
                _p[i] = 1.0 / _numclust;
            }
            _data.resize(numRecords, numAttr);
            _post.resize(numRecords, _numclust);

            std::vector<Real> mean(numAttr,0.0);
            boost::shared_ptr<Schema> schema = _ds->schema();
            for(Size i=0; i<numRecords; ++i) {
                for(Size j=0; j<numAttr; ++j) {
                    Real val = (*schema)[j]->get_c_val((*_ds)(i,j));
                    _data(i,j) = val;
                    mean[j] += val;
                }
            }

            for(Size j=0; j<numAttr; ++j) {
                mean[j] /= numRecords;
            }

            for(Size i=0; i<numAttr; ++i) {
                for(Size j=0; j<i; ++j) {
                    _sigma[0](i, j) = 0.0;
                }
                _sigma[0](i, i) = ( ublas::inner_prod(
                    ublas::column(_data, i),
                    ublas::column(_data, i)) -
                    numRecords*mean[i]*mean[i])
                    / (numRecords - 1.0) + _epsilon;
            }

            for(Size i=1; i<_numclust; ++i) {
                _sigma[i] = _sigma[0];
            }

            _CM.resize(numRecords);
            std::vector<Integer> index(numRecords,0);
            for(Size i=0;i<index.size();++i){
                index[i] = i;
            }

            boost::minstd_rand generator(
                static_cast<unsigned int>(_seed));
            for(Size i=0;i<_numclust;++i){
                boost::uniform_int<> uni_dist(0,numRecords-i-1);
                boost::variate_generator<boost::minstd_rand&,
                    boost::uniform_int<> > uni(generator,uni_dist);
                Integer r = uni();
                for(Size j=0; j<numAttr; ++j) {
                    _mu(i, j) = _data(r,j);
                }

                boost::shared_ptr<Record> rec = (*_ds)[r];
                boost::shared_ptr<Record> cr = boost::shared_ptr
                    <Record>(new Record(*rec));
                boost::shared_ptr<CenterCluster> c = boost::shared_ptr
                    <CenterCluster>(new CenterCluster(cr));
                c->set_id(i);
                _clusters.push_back(c);
                index.erase(index.begin()+r);
            }
        }
}
```

B.3.11 Class Kmean

Listing B.28: The header file of class **Kmean**.

```cpp
// cl/algorithms/kmean.hpp
#ifndef CLUSLIB_KMEAN_HPP
#define CLUSLIB_KMEAN_HPP

#include<cl/algorithms/algorithm.hpp>
#include<cl/clusters/centercluster.hpp>
#include<cl/clusters/pclustering.hpp>
#include<cl/datasets/dataset.hpp>
#include<cl/distances/distance.hpp>
#include<cl/types.hpp>

namespace ClusLib {

    class Kmean: public Algorithm {
    protected:
        void setupArguments();
        void performClustering() const;
        void fetchResults() const;
        virtual void initialization() const;
        virtual void iteration() const;
        virtual void updateCenter() const;

        mutable std::vector<boost::shared_ptr<CenterCluster> >
            _clusters;
        mutable std::vector<Size> _CM;
        mutable Real _error;
        mutable Size _numiter;

        Size _numclust;
        Size _maxiter;
        Size _seed;
        boost::shared_ptr<Distance> _distance;
    };

}

#endif
```

Listing B.29: The source file of class **Kmean**.

```cpp
// cl/algorithms/kmean.cpp
#include<cl/algorithms/kmean.hpp>
#include<cl/errors.hpp>
#include<iostream>
#include<boost/random.hpp>

namespace ClusLib {

    void Kmean::performClustering() const {
        initialization();
        iteration();
    }

    void Kmean::setupArguments() {
        Algorithm::setupArguments();
        ASSERT(_ds->is_numeric(), "dataset_is_not_numeric");

        _distance = _arguments.distance;
        ASSERT(_distance, "distance_is_null");

```

```cpp
        _numclust = boost::any_cast<Size>(
            _arguments.get("numclust"));
        ASSERT(_numclust>=2 && _numclust<=_ds->size(),
            "invalid_numclust");

        _maxiter = boost::any_cast<Size>(
            _arguments.get("maxiter"));
        ASSERT(_maxiter>0, "invalide_maxiter");

        _seed = boost::any_cast<Size>(
            _arguments.get("seed"));
        ASSERT(_seed>0, "invalide_seed");
    }

    void Kmean::fetchResults() const {
        PClustering pc;
        for(Size i=0;i<_clusters.size();++i){
            pc.add(_clusters[i]);
        }
        _results.CM = _CM;
        _results.insert("pc", boost::any(pc));

        _error = 0.0;
        for(Size i=0;i<_ds->size();++i) {
            _error += (*_distance)((*_ds)[i],
                _clusters[_CM[i]]->center());
        }
        _results.insert("error", boost::any(_error));
        _results.insert("numiter", boost::any(_numiter));
    }

    void Kmean::iteration() const {
        bool bChanged = true;

        updateCenter();
        _numiter = 1;
        while(bChanged) {
            bChanged = false;
            Size s;
            Real dMin,dDist;
            for(Size i=0;i<_ds->size();++i) {
                dMin = MAX_REAL;
                for(Size k=0;k<_clusters.size();++k) {
                    dDist = (*_distance)((*_ds)[i],
                        _clusters[k]->center());
                    if (dMin > dDist) {
                        dMin = dDist;
                        s = k;
                    }
                }

                if (_CM[i] != s){
                    _clusters[_CM[i]]->erase((*_ds)[i]);
                    _clusters[s]->add((*_ds)[i]);
                    _CM[i] = s;
                    bChanged = true;
                }
            }

            updateCenter();
            ++_numiter;
            if (_numiter > _maxiter){
                break;
            }
        }
    }
```

```cpp
 88  void Kmean::updateCenter() const {
 89      Real dTemp;
 90      boost::shared_ptr<Schema> schema = _ds->schema();
 91      for(Size k=0;k<_clusters.size();++k){
 92          for(Size j=0;j<schema->size();++j){
 93              dTemp = 0.0;
 94              for(Size i=0; i<_clusters[k]->size();++i){
 95                  boost::shared_ptr<Record> rec =
 96                      (*_clusters[k])[i];
 97                  dTemp += (*schema)[j]->get_c_val((*rec)[j]);
 98              }
 99              (*schema)[j]->set_c_val(
100                  (*_clusters[k]->center())[j],
101                  dTemp/_clusters[k]->size());
102          }
103      }
104  }
105
106  void Kmean::initialization() const {
107      Size numRecords = _ds->size();
108      std::vector<Size> index(numRecords,0);
109      _CM.resize(numRecords);
110      for(Size i=0;i<index.size();++i){
111          index[i] = i;
112      }
113
114      boost::minstd_rand generator(_seed);
115      for(Size i=0;i<_numclust;++i){
116          boost::uniform_int<> uni_dist(0,numRecords-i-1);
117          boost::variate_generator<boost::minstd_rand&,
118              boost::uniform_int<> > uni(generator,uni_dist);
119          Size r = uni();
120          boost::shared_ptr<Record> cr = boost::shared_ptr
121              <Record>(new Record(*(*_ds)[r]));
122          boost::shared_ptr<CenterCluster> c = boost::shared_ptr
123              <CenterCluster>(new CenterCluster(cr));
124          c->set_id(i);
125          _clusters.push_back(c);
126          index.erase(index.begin()+r);
127      }
128
129      Size s;
130      Real dMin,dDist;
131      for(Size i=0;i<numRecords;++i){
132          dMin = MAX_REAL;
133          for(Size j=0;j<_numclust;++j){
134              dDist = (*_distance)((*_ds)[i],
135                  _clusters[j]->center());
136              if (dDist<dMin){
137                  s = j;
138                  dMin = dDist;
139              }
140          }
141          _clusters[s]->add((*_ds)[i]);
142          _CM[i] = s;
143      }
144  }
145  }
```

B.3.12 Class Kprototype

Listing B.30: The header file of class Kprototype.

```cpp
// cl/algorithms/kprototype.hpp
#ifndef CLUSLIB_KPROTOTYPE_HPP
#define CLUSLIB_KPROTOTYPE_HPP

#include<cl/algorithms/kmean.hpp>

namespace ClusLib {

    class Kprototype: public Kmean {
    private:
        void setupArguments();
        void updateCenter() const;
    };

}

#endif
```

Listing B.31: The source file of class Kprototype.

```cpp
// cl/algorithms/kprototype.cpp
#include<cl/algorithms/kprototype.hpp>
#include<cl/errors.hpp>
#include<iostream>
#include<map>

namespace ClusLib {

    void Kprototype::setupArguments() {
        Algorithm::setupArguments();

        _distance = _arguments.distance;
        ASSERT(_distance, "distance_is_null");

        _numclust = boost::any_cast<Size>(
            _arguments.get("numclust"));
        ASSERT(_numclust>=2 && _numclust<=_ds->size(),
            "invalid_numclust");

        _maxiter = boost::any_cast<Size>(
            _arguments.get("maxiter"));
        ASSERT(_maxiter>0, "invalide_maxiter");

        _seed = boost::any_cast<Size>(
            _arguments.get("seed"));
        ASSERT(_seed>0, "invalide_seed");
    }

    void Kprototype::updateCenter() const {
        Real dTemp;
        boost::shared_ptr<Schema> schema = _ds->schema();
        for(Size k=0;k<_clusters.size();++k){
            for(Size j=0;j<schema->size();++j){
                if((*schema)[j]->can_cast_to_c()) {
                    dTemp = 0.0;
                    for(Size i=0; i<_clusters[k]->size();++i){
                        boost::shared_ptr<Record> rec =
                            (*_clusters[k])[i];
                        dTemp+=(*schema)[j]->get_c_val((*rec)[j]);
                    }
```

```
                        (*schema)[j]->set_c_val(
                            (*_clusters[k]->center())[j],
                            dTemp/_clusters[k]->size());
                } else {
                    DAttrInfo da = (*schema)[j]->cast_to_d();
                    std::map<Size, Size> freq;
                    for(Size i=0;i<da.num_values(); ++i){
                        freq.insert(
                            std::pair<Size,Size>(i, 0));
                    }

                    for(Size i=0; i<_clusters[k]->size();++i){
                        boost::shared_ptr<Record> rec =
                            (*_clusters[k])[i];
                        freq[(*schema)[j]->get_d_val((*rec)[j])]
                            += 1;
                    }

                    Size nMax = 0;
                    Size s = 0;
                    for(Size i=0;i<da.num_values(); ++i){
                        if(nMax < freq[i]) {
                            nMax = freq[i];
                            s = i;
                        }
                    }
                    da.set_d_val((*_clusters[k]->center())[j], s);
                }
            }
        }
    }
}
```

B.3.13 Class LW

Listing B.32: The header file of class LW.

```
// cl/algorithms/lw.hpp
#ifndef CLUSLIB_LW_HPP
#define CLUSLIB_LW_HPP

#include<cl/algorithms/algorithm.hpp>
#include<cl/clusters/hclustering.hpp>
#include<cl/utilities/nnmap.hpp>
#include<cl/datasets/dataset.hpp>
#include<cl/distances/distance.hpp>
#include<cl/types.hpp>
#include<set>

namespace ClusLib {

    class LW: public Algorithm {
    protected:
        typedef std::map<Size, boost::shared_ptr<HClustering> >
            Forest;
        typedef std::map<Size, Size> SizeMap;

        void setupArguments();
        void performClustering() const;
        void fetchResults() const;
        virtual void create_dm() const;
        virtual void init_forest() const;
        virtual void linkage() const;
        virtual void update_dm(Size p, Size q, Size r)
```

```cpp
            const = 0;

        mutable iirMapA _dm;
        mutable std::set<Size> _unmergedClusters;
        mutable Forest _forest;
        mutable SizeMap _clusterSize;
        boost::shared_ptr<Distance> _distance;
    };
}

#endif
```

Listing B.33: The source file of class LW.

```cpp
// cl/algorithms/lw.cpp
#include<cl/algorithms/lw.hpp>
#include<cl/types.hpp>
#include<iostream>

namespace ClusLib {

    void LW::setupArguments() {
        Algorithm::setupArguments();

        _distance = _arguments.distance;
        ASSERT(_distance, "distance_is_null");
    }

    void LW::create_dm() const {
        Size n = _ds->size();
        for(Size i=0;i<n-1;++i){
            for(Size j=i+1;j<n;++j){
                _dm.add_item(i, j,
                    (*_distance)((*_ds)[i],(*_ds)[j]));
            }
        }
    }

    void LW::performClustering() const {
        create_dm();
        init_forest();
        linkage();
    }

    void LW::init_forest() const {
        Size n = _ds->size();
        for(Size s=0;s<n;++s){
            boost::shared_ptr<Node> pln(new
                LeafNode((*_ds)[s], s));
            pln->set_level(0);
            boost::shared_ptr<HClustering> phc(new
                HClustering(pln));
            _forest.insert(Forest::value_type(s, phc));
            _clusterSize.insert(SizeMap::value_type(s,1));
            _unmergedClusters.insert(s);
        }
    }

    void LW::linkage() const {
        Size n = _ds->size();
        std::set<Size>::iterator it;
        Real dMin, dTemp;
        Size m, s1, s2;
        for(Size s=0;s<n-1;++s){
            dMin = MAX_REAL;
            std::vector<Integer> nvTemp(_unmergedClusters.begin(),
```

```
53              _unmergedClusters.end());
54          m = nvTemp.size();
55          for(Size i=0;i<m;++i) {
56              for(Size j=i+1;j<m;++j){
57                  dTemp = _dm(nvTemp[i],nvTemp[j]);
58                  if(dTemp < dMin) {
59                      dMin = dTemp;
60                      s1 = nvTemp[i];
61                      s2 = nvTemp[j];
62                  }
63              }
64          }
65          boost::shared_ptr<Node> node =
66              _forest[s1]->joinWith(*_forest[s2],dMin);
67          node->set_id(n+s);
68          node->set_level(s+1);
69          boost::shared_ptr<HClustering> phc =
70              boost::shared_ptr<HClustering>(new
71                  HClustering(node));
72          _forest.insert(Forest::value_type(n+s, phc));
73          _clusterSize.insert(SizeMap::value_type(n+s,
74              _clusterSize[s1]+_clusterSize[s2]));
75          _unmergedClusters.erase(s1);
76          _unmergedClusters.erase(s2);
77          _unmergedClusters.insert(n+s);
78          update_dm(s1, s2, n+s);
79      }
80  }
81
82  void LW::fetchResults() const {
83      Size n = _ds->size();
84      _results.insert("hc",HClustering(_forest[2*n-2]->root()));
85  }
86 }
```

B.3.14 Class `Median`

Listing B.34: The header file of class `Median`.

```
1  // cl/algorithms/median.hpp
2  #ifndef CLUSLIB_MEDIAN_HPP
3  #define CLUSLIB_MEDIAN_HPP
4
5  #include<cl/algorithms/lw.hpp>
6
7  namespace ClusLib {
8
9      class Median: public LW {
10     private:
11         void setupArguments();
12         void update_dm(Size p, Size q, Size r) const;
13     };
14
15 }
16
17 #endif
```

Listing B.35: The source file of class `Median`.

```
1  // cl/algorithms/median.cpp
2  #include<cl/algorithms/median.hpp>
3  #include<cl/distances/euclideandistance.hpp>
```

```
4   #include<cmath>
5
6   namespace ClusLib {
7
8       void Median::setupArguments() {
9           Algorithm::setupArguments();
10
11          _distance = boost::shared_ptr<Distance>(new
12              EuclideanDistance());
13      }
14
15      void Median::update_dm(Size p, Size q, Size r) const {
16          Real dist;
17          std::set<Size>::iterator it;
18          for(it = _unmergedClusters.begin();
19              it != _unmergedClusters.end(); ++it) {
20              if(*it == r) {
21                  continue;
22              }
23
24              dist = 0.5*std::pow(_dm(p,*it), 2.0)+
25                  0.5*std::pow(_dm(q,*it), 2.0)-
26                  0.25*std::pow(_dm(p,q), 2.0);
27              _dm.add_item(r,*it, std::sqrt(dist));
28          }
29      }
30
31  }
```

B.3.15 Class Single

Listing B.36: The header file of class Single.
```
1   // cl/algorithms/single.hpp
2   #ifndef CLUSLIB_SINGLE_HPP
3   #define CLUSLIB_SINGLE_HPP
4
5   #include<cl/algorithms/lw.hpp>
6
7   namespace ClusLib {
8
9       class Single: public LW {
10      private:
11          void update_dm(Size p, Size q, Size r) const;
12      };
13
14  }
15
16  #endif
```

Listing B.37: The source file of class Single.
```
1   // cl/algorithms/single.cpp
2   #include<cl/algorithms/single.hpp>
3
4   namespace ClusLib {
5
6       void Single::update_dm(Size p, Size q, Size r) const {
7           Real dist;
8           std::set<Size>::iterator it;
9           for(it = _unmergedClusters.begin();
10              it != _unmergedClusters.end(); ++it) {
```

```
11              if(*it == r) {
12                  continue;
13              }
14
15              dist = std::min(_dm(p,*it), _dm(q,*it));
16              _dm.add_item(r,*it,dist);
17          }
18      }
19
20  }
```

B.3.16 Class Ward

Listing B.38: The header file of class **Ward**.

```
1   // cl/algorithms/ward.hpp
2   #ifndef CLUSLIB_WARD_HPP
3   #define CLUSLIB_WARD_HPP
4
5   #include<cl/algorithms/lw.hpp>
6
7   namespace ClusLib {
8
9       class Ward: public LW {
10      private:
11          void setupArguments();
12          void update_dm(Size p, Size q, Size r) const;
13      };
14
15  }
16
17  #endif
```

Listing B.39: The source file of class **Ward**.

```
1   // cl/algorithms/ward.cpp
2   #include<cl/algorithms/ward.hpp>
3   #include<cl/distances/euclideandistance.hpp>
4   #include<cmath>
5
6   namespace ClusLib {
7
8       void Ward::setupArguments() {
9           Algorithm::setupArguments();
10
11          _distance = boost::shared_ptr<Distance>(new
12              EuclideanDistance());
13      }
14
15      void Ward::update_dm(Size p, Size q, Size r) const {
16          Real dist;
17          std::set<Size>::iterator it;
18          Real sp = _clusterSize[p];
19          Real sq = _clusterSize[q];
20          for(it = _unmergedClusters.begin();
21              it != _unmergedClusters.end(); ++it) {
22              if(*it == r) {
23                  continue;
24              }
25
26              Real sk = _clusterSize[*it];
27              Real st = sp+sq+sk;
```

```
28          dist = std::pow(_dm(p,*it), 2.0)*(sp+sk)/st +
29                 std::pow(_dm(q,*it), 2.0)*(sk+sq)/st -
30                 std::pow(_dm(p,q), 2.0)*sk/st;
31          _dm.add_item(r,*it, std::sqrt(dist));
32      }
33   }
34
35 }
```

B.3.17 Class Weighted

Listing B.40: The header file of class **Weighted**.

```
1  // cl/algorithms/weighted.hpp
2  #ifndef CLUSLIB_WEIGHTED_HPP
3  #define CLUSLIB_WEIGHTED_HPP
4
5  #include<cl/algorithms/lw.hpp>
6
7  namespace ClusLib {
8
9      class Weighted: public LW {
10     private:
11         void update_dm(Size p, Size q, Size r) const;
12     };
13
14 }
15
16 #endif
```

Listing B.41: The source file of class **Weighted**.

```
1  // cl/algorithms/weighted.cpp
2  #include<cl/algorithms/weighted.hpp>
3
4  namespace ClusLib {
5
6      void Weighted::update_dm(Size p, Size q, Size r)
7          const {
8          Real dist;
9          std::set<Size>::iterator it;
10         for(it = _unmergedClusters.begin();
11             it != _unmergedClusters.end(); ++it) {
12             if(*it == r) {
13                 continue;
14             }
15
16             dist = (_dm(p,*it) + _dm(q,*it)) / 2;
17             _dm.add_item(r,*it, dist);
18         }
19     }
20
21 }
```

B.4 Files in Folder cl/clusters
B.4.1 Makefile

Listing B.42: `Makefile.am` in `cl/clusters`.

```
1   noinst_LTLIBRARIES = libClusters.la
2
3   AM_CPPFLAGS = -I${top_srcdir} -I${top_builddir}
4
5   this_includedir=${includedir}/${subdir}
6   this_include_HEADERS = \
7       all.hpp \
8       centercluster.hpp \
9       cluster.hpp \
10      hclustering.hpp \
11      pclustering.hpp \
12      subspacecluster.hpp
13
14  libClusters_la_SOURCES = \
15      centercluster.cpp \
16      hclustering.cpp \
17      pclustering.cpp \
18      subspacecluster.cpp
19
20
21  all.hpp: Makefile.am
22          echo "// This file is generated. Please do not edit!" > $@
23          echo >> $@
24          for i in $(filter-out all.hpp, $(this_include_HEADERS)); \
25          do \
26              echo "#include <${subdir}/$$i>" >> $@; \
27  done
28          echo >> $@
29          subdirs='$(SUBDIRS)'; for i in $$subdirs; do \
30              echo "#include <${subdir}/$$i/all.hpp>" >> $@; \
31  done
```

B.4.2 Class `CenterCluster`

Listing B.43: The header file of class `CenterCluster`.

```
1   // cl/clusters/centercluster.hpp
2   #ifndef CLUSLIB_CENTERCLUSTER_HPP
3   #define CLUSLIB_CENTERCLUSTER_HPP
4
5   #include<cl/clusters/cluster.hpp>
6   #include<cl/datasets/record.hpp>
7
8   namespace ClusLib {
9
10      class CenterCluster: public Cluster {
11      public:
12          CenterCluster() {}
13          CenterCluster(const boost::shared_ptr<Record>& center);
14          const boost::shared_ptr<Record>& center() const;
15          boost::shared_ptr<Record>& center();
16
17      protected:
```

```
18          boost::shared_ptr<Record> _center;
19      };
20
21  }
22
23
24  #endif
```

Listing B.44: The source file of class `CenterCluster`.

```
1   // cl/clusters/centercluster.cpp
2   #include<cl/clusters/centercluster.hpp>
3
4   namespace ClusLib {
5
6       CenterCluster::CenterCluster(
7           const boost::shared_ptr<Record>& center)
8           : _center(center) {
9       }
10
11      const boost::shared_ptr<Record>& CenterCluster::center()
12          const {
13          return _center;
14      }
15
16      boost::shared_ptr<Record>& CenterCluster::center() {
17          return _center;
18      }
19  }
```

B.4.3 Class Cluster

Listing B.45: The header file of class `Cluster`.

```
1   // cl/clusters/cluster.hpp
2   #ifndef CLUSLIB_CLUSTER_HPP
3   #define CLUSLIB_CLUSTER_HPP
4
5   #include<vector>
6   #include<cl/datasets/record.hpp>
7   #include<cl/utilities/container.hpp>
8
9   namespace ClusLib {
10
11      class Cluster: public Container<boost::shared_ptr<Record> > {
12      public:
13          virtual ~Cluster() {}
14
15          void set_id(Size id);
16          Size get_id() const;
17
18      protected:
19          Size _id;
20      };
21
22      inline void Cluster::set_id(Size id) {
23          _id = id;
24      }
25
26      inline Size Cluster::get_id() const {
27          return _id;
28      }
```

B.4.4 Class HClustering

Listing B.46: The header file of class HClustering.

```cpp
// cl/clusters/hclustering.hpp
#ifndef CLUSLIB_HCLUSTERING_HPP
#define CLUSLIB_HCLUSTERING_HPP

#include<cl/patterns/internalnode.hpp>
#include<cl/patterns/leafnode.hpp>
#include<cl/patterns/node.hpp>
#include<cl/utilities/dendrogram.hpp>
#include<cl/clusters/pclustering.hpp>

namespace ClusLib {

    class HClustering {
    public:
        HClustering() {}
        HClustering(const boost::shared_ptr<Node>& root);

        boost::shared_ptr<Node> joinWith(HClustering& hc,
            Real joinValue);
        const boost::shared_ptr<Node>& root() const;
        boost::shared_ptr<Node>& root();
        PClustering get_pc(Size maxclust) const;
        void save(const std::string &filename, Size p=100) const;

    private:
        boost::shared_ptr<Node> _root;
    };

    inline const boost::shared_ptr<Node>& HClustering::root()
        const {
        return _root;
    }

    inline boost::shared_ptr<Node>& HClustering::root() {
        return _root;
    }
}

#endif
```

Listing B.47: The source file of class HClustering.

```cpp
// cl/clusters/hclustering.cpp
#include<cl/clusters/hclustering.hpp>
#include<cl/errors.hpp>
#include<cl/patterns/pcvisitor.hpp>
#include<cl/patterns/joinvaluevisitor.hpp>
#include<cl/patterns/dendrogramvisitor.hpp>
#include<algorithm>

namespace ClusLib {

    HClustering::HClustering(const boost::shared_ptr<Node>& root)
        : _root(root) {
```

```cpp
    }

    boost::shared_ptr<Node> HClustering::joinWith(
            HClustering& hc, Real joinValue) {
        InternalNode* p = new InternalNode(joinValue);
        boost::shared_ptr<Node> node(p);

        boost::shared_ptr<Node>& cn1 = _root;
        const boost::shared_ptr<Node>& cn2 = hc.root();

        cn1->set_parent(node);
        cn2->set_parent(node);
        p->add(cn1);
        p->add(cn2);

        return node;

    }

    PClustering HClustering::get_pc(Size maxclust) const {
        ASSERT(maxclust>0, "invalide_maxclust");
        Size cutlevel = _root->get_level() - maxclust + 2;
        PClustering pc;
        PCVisitor pcv(pc, cutlevel);
        _root->accept(pcv);

        return pc;
    }

    void HClustering::save(const std::string &filename,
            Size p) const {
        JoinValueVisitor jvv;
        _root->accept(jvv);
        std::set<iirMapA::value_type, compare_iir> joinValues
            = jvv.get_joinValues();
        std::set<iirMapA::value_type, compare_iir >::iterator it;
        Real ljv, hjv;
        Size llevel, hlevel;
        it = joinValues.end();
        --it;
        hjv = it->second;
        hlevel = _root->get_level();
        if (p == 0) {
            it = joinValues.begin();
            ljv = it->second;
            llevel = 0;
        } else {
            it = joinValues.begin();
            for(Size i=0; i<joinValues.size() - p + 1; ++i) {
                ++it;
            }
            ljv = it->second;
            llevel = _root->get_level() - p + 1;
        }
        DendrogramVisitor dgv(hjv, llevel, hlevel);
        _root->accept(dgv);
        dgv.save(filename);
    }
}
```

B.4.5 Class `PClustering`

Listing B.48: The header file of class `PClustering`.

```cpp
// cl/clusters/pclustering.hpp
#ifndef CLUSLIB_PCLUSTERING_HPP
#define CLUSLIB_PCLUSTERING_HPP

#include<cl/clusters/cluster.hpp>
#include<cl/utilities/container.hpp>
#include<cl/utilities/nnmap.hpp>

namespace ClusLib {

    class PClustering:
        public Container<boost::shared_ptr<Cluster> > {
    public:
        friend std::ostream& operator<<(std::ostream& os,
            PClustering& pc);

        PClustering();
        void removeEmptyClusters();
        void createClusterID();
        void save(const std::string& filename);

    private:
        void print(std::ostream& os);
        void calculate();
        void crosstab();

        bool _bCalculated;
        Size _numclust;
        Size _numclustGiven;
        std::vector<Size> _clustsize;
        std::vector<std::string> _clustLabel;
        std::vector<Size> _CM;
        std::vector<Size> _CMGiven;
        iiiMapB _crosstab;
    };

}
#endif
```

Listing B.49: The source file of class `PClustering`.

```cpp
// cl/clusters/pclustering.cpp
#include<cl/clusters/pclustering.hpp>
#include<cl/errors.hpp>
#include<algorithm>
#include<fstream>
#include<iomanip>

namespace ClusLib {

    PClustering::PClustering(): _bCalculated(false),
        _numclustGiven(Null<Size>()) {
    }

    std::ostream& operator<<(std::ostream& os,
            PClustering& pc) {
        pc.print(os);
        return os;
    }

    void PClustering::removeEmptyClusters() {
```

```cpp
      std::vector<boost::shared_ptr<Cluster> >
          temp(_data.begin(), _data.end());
      _data.clear();
      for(iterator it=temp.begin(); it!=temp.end();++it){
          if((*it)->size() == 0) {
              continue;
          }
          _data.push_back(*it);
      }
  }

  void PClustering::createClusterID() {
      removeEmptyClusters();
      for(Size i=0;i<_data.size();++i){
          _data[i]->set_id(i);
      }
  }

  void PClustering::print(std::ostream& os) {
      calculate();

      os<<"Clustering Summary:\n";
      os<<"Number of clusters: "<<_numclust<<'\n';
      for(Size i=0;i<_numclust;++i){
          os<<"Size of Cluster "<<i<<": "<<_clustsize[i]<<'\n';
      }
      os<<'\n';
      if (_numclustGiven != Null<Size>()){
          os<<"Number of given clusters: "
              <<_numclustGiven<<'\n';
          os<<"Cross Tabulation:\n";
          std::vector<Size> w;
          w.push_back(13);
          os<<std::setw(w[0])<<std::left<<"Cluster ID";
          for(Size j=0;j<_numclustGiven;++j) {
              w.push_back(_clustLabel[j].size()+3);
              os<<std::setw(w[j+1])<<std::left<<_clustLabel[j];
          }
          os<<'\n';
          for(Size i=0;i<_numclust;++i){
              os<<std::setw(w[0])<<std::left<<i;
              for(Size j=0;j<_numclustGiven;++j) {
                  if(_crosstab.contain_key(i,j)){
                      os<<std::setw(w[j+1])<<std::left
                          <<_crosstab(i,j);
                  } else {
                      os<<std::setw(w[j+1])<<std::left<<0;
                  }
              }
              os<<'\n';
          }
      }
  }

  void PClustering::save(const std::string& filename) {
      std::ofstream of;
      of.open(filename.c_str());
      print(of);

      of<<"\nCluster Membership\n";
      of<<"Record_ID, Cluster_Index, Cluster_Index_Given\n";
      for(Size i=0; i<_CM.size();++i) {
          of<<i+1<<", "<<_CM[i];
          if(_numclustGiven == Null<Size>()){
              of<<", NA\n";
              continue;
          }
```

```cpp
                of<<","<<_CMGiven[i]<<'\n';
        }
        of.close();
    }

    void PClustering::crosstab() {
        Size c1, c2;
        for(Size i=0; i<_CM.size();++i) {
            c1 = _CM[i];
            c2 = _CMGiven[i];
            if (_crosstab.contain_key(c1,c2)) {
                _crosstab(c1,c2) += 1;
            } else {
                _crosstab.add_item(c1,c2,1);
            }
        }
    }

    void PClustering::calculate() {
        if(_bCalculated) {
            return;
        }

        createClusterID();
        _numclust = _data.size();
        boost::shared_ptr<Cluster> c;
        boost::shared_ptr<Record> r;

        _CM.resize(
            (*_data[0])[0]->schema()->idInfo()->num_values());
        for(Size i=0;i<_numclust;++i){
            c = _data[i];
            _clustsize.push_back(c->size());
            for(Size j=0;j<c->size();++j){
                r = (*c)[j];
                _CM[r->get_id()] = c->get_id();
            }
        }

        boost::shared_ptr<DAttrInfo> info =
            (*_data[0])[0]->schema()->labelInfo();
        if(!info) {
            _bCalculated = true;
            return;
        }

        _numclustGiven = info->num_values();
        for(Size i=0;i<_numclustGiven;++i){
            _clustLabel.push_back(info->int_to_str(i));
        }

        _CMGiven.resize(_CM.size());
        for(Size i=0;i<_numclust;++i){
            c = _data[i];
            for(Size j=0;j<c->size();++j){
                r = (*c)[j];
                _CMGiven[r->get_id()] = r->get_label();
            }
        }

        crosstab();
        _bCalculated = true;
    }
}
```

B.4.6 Class SubspaceCluster

Listing B.50: The header file of class SubspaceCluster.

```cpp
// cl/clusters/subspacecluster.hpp
#ifndef CLUSLIB_SUBSPACECLUSTER_HPP
#define CLUSLIB_SUBSPACECLUSTER_HPP

#include<cl/clusters/centercluster.hpp>
#include<vector>

namespace ClusLib {

    class SubspaceCluster : public CenterCluster {
    public:
        SubspaceCluster(const boost::shared_ptr<Record>& center);
        std::vector<Real>& w();
        const std::vector<Real>& w() const;
        Real& w(Size i);
        const Real& w(Size i) const;

    protected:
        std::vector<Real> _w;
    };
}

#endif
```

Listing B.51: The source file of class SubspaceCluster.

```cpp
// cl/clusters/subspacecluster.cpp
#include<cl/clusters/subspacecluster.hpp>

namespace ClusLib {

    SubspaceCluster::SubspaceCluster(const
        boost::shared_ptr<Record>& center)
        : CenterCluster(center) {
        _w.resize(_center->size(), 1.0 / _center->size());
    }

    std::vector<Real>& SubspaceCluster::w() {
        return _w;
    }

    const std::vector<Real>& SubspaceCluster::w() const {
        return _w;
    }

    Real& SubspaceCluster::w(Size i) {
        ASSERT(i>=0 && i< _w.size(), "index out of range");
        return _w[i];
    }

    const Real& SubspaceCluster::w(Size i) const {
        ASSERT(i>=0 && i< _w.size(), "index out of range");
        return _w[i];
    }
}
```

B.5 Files in Folder cl/datasets

B.5.1 Makefile

Listing B.52: `Makefile.am` in `cl/datasets`.

```
 1  noinst_LTLIBRARIES = libDatasets.la
 2
 3  AM_CPPFLAGS = -I${top_srcdir} -I${top_builddir}
 4
 5  this_includedir=${includedir}/${subdir}
 6  this_include_HEADERS = \
 7      all.hpp \
 8      attrinfo.hpp \
 9      attrvalue.hpp \
10      cattrinfo.hpp \
11      dataset.hpp \
12      dattrinfo.hpp \
13      record.hpp \
14      schema.hpp
15
16  libDatasets_la_SOURCES = \
17      attrinfo.cpp \
18      cattrinfo.cpp \
19      dataset.cpp \
20      dattrinfo.cpp \
21      record.cpp \
22      schema.cpp
23
24
25  all.hpp: Makefile.am
26      echo "// This file is generated. Please do not edit!" > $@
27      echo >> $@
28      for i in $(filter-out all.hpp, $(this_include_HEADERS)); \
29      do \
30          echo "#include <${subdir}/$$i>" >> $@; \
31      done
32      echo >> $@
33      subdirs='$(SUBDIRS)'; for i in $$subdirs; do \
34          echo "#include <${subdir}/$$i/all.hpp>" >> $@; \
35      done
```

B.5.2 Class `AttrValue`

Listing B.53: The header file of class `AttrValue`.

```
 1  // cl/datasets/attrvalue.hpp
 2  #ifndef CLUSLIB_ATTRVALUE_HPP
 3  #define CLUSLIB_ATTRVALUE_HPP
 4
 5  #include<boost/variant.hpp>
 6  #include<cl/types.hpp>
 7  #include<cl/utilities/null.hpp>
 8
 9  namespace ClusLib {
10
11      class AttrValue {
12      public:
13          friend class DAttrInfo;
```

```cpp
            friend class CAttrInfo;

            typedef boost::variant<Real, Size> value_type;
            AttrValue();

        private:
            value_type _value;
    };

    inline AttrValue::AttrValue(): _value(Null<Size>()) {
    }

}

#endif
```

B.5.3 Class AttrInfo

Listing B.54: The header file of class AttrInfo.

```cpp
// cl/datasets/attrinfo.hpp
#ifndef CLUSLIB_ATTRINFO_HPP
#define CLUSLIB_ATTRINFO_HPP

#include<cl/datasets/attrvalue.hpp>
#include<string>

namespace ClusLib {

    enum AttrType {
        Unknow,
        Continuous,
        Discrete
    };

    class DAttrInfo;
    class CAttrInfo;

    class AttrInfo {
    public:
        AttrInfo(const std::string &name, AttrType type);
        virtual ~AttrInfo() {}

        const std::string& name() const;
        std::string& name();
        AttrType type() const;

        virtual bool operator==(const AttrInfo& info) const;
        virtual bool operator!=(const AttrInfo& info) const;
        virtual AttrInfo* clone() const = 0;
        virtual Real distance(const AttrValue&,
                              const AttrValue&) const = 0;
        virtual void set_d_val(AttrValue&, Size) const;
        virtual Size get_d_val(const AttrValue&) const;
        virtual void set_c_val(AttrValue&, Real) const;
        virtual Real get_c_val(const AttrValue&) const;
        virtual void set_unknown(AttrValue&) const = 0;
        virtual bool is_unknown(const AttrValue&) const = 0;
        virtual DAttrInfo& cast_to_d();
        virtual const DAttrInfo& cast_to_d() const;
        virtual CAttrInfo& cast_to_c();
        virtual const CAttrInfo& cast_to_c() const;
        virtual bool can_cast_to_d() const;
        virtual bool can_cast_to_c() const;
```

```cpp
protected:
    bool equal_shallow(const AttrInfo&) const;

private:
    std::string _name;
    AttrType _type;
};

inline const std::string& AttrInfo::name() const {
    return _name;
}

inline std::string& AttrInfo::name() {
    return _name;
}

inline AttrType AttrInfo::type() const {
    return _type;
}

inline bool AttrInfo::operator==(const AttrInfo& info) const {
    return equal_shallow(info);
}

inline bool AttrInfo::operator!=(const AttrInfo& info) const {
    return !equal_shallow(info);
}

inline bool AttrInfo::can_cast_to_d() const {
    return false;
}

inline bool AttrInfo::can_cast_to_c() const {
    return false;
}

}

#endif
```

Listing B.55: The source file of class `AttrInfo`.

```cpp
// cl/datasets/attrinfo.cpp
#include<cl/datasets/attrinfo.hpp>
#include<cl/errors.hpp>

namespace ClusLib {

    AttrInfo::AttrInfo(const std::string &name, AttrType type)
        : _name(name), _type(type) {
    }

    bool AttrInfo::equal_shallow(const AttrInfo &info) const {
        if(_name != info.name()){
            return false;
        }

        if(_type != info.type()){
            return false;
        }

        return true;
    }

    void AttrInfo::set_d_val(AttrValue&, Size) const {
```

```
24            FAIL("can_not_be_called");
25        }
26
27        Size AttrInfo::get_d_val(const AttrValue&) const {
28            FAIL("can_not_be_called");
29            return 0;
30        }
31
32        void AttrInfo::set_c_val(AttrValue&, Real) const {
33            FAIL("can_not_be_called");
34        }
35
36        Real AttrInfo::get_c_val(const AttrValue&) const {
37            FAIL("can_not_be_called");
38            return 0.0;
39        }
40
41        DAttrInfo& AttrInfo::cast_to_d() {
42            FAIL("can_not_cast_an_AttrInfo_to_DAttrInfo");
43            return *(DAttrInfo*)NULL;
44        }
45
46        const DAttrInfo& AttrInfo::cast_to_d() const {
47            FAIL("can_not_cast_an_AttrInfo_to_DAttrInfo");
48            return *(DAttrInfo*)NULL;
49        }
50
51        CAttrInfo& AttrInfo::cast_to_c() {
52            FAIL("can_not_cast_an_AttrInfo_to_CAttrInfo");
53            return *(CAttrInfo*)NULL;
54        }
55
56        const CAttrInfo& AttrInfo::cast_to_c() const {
57            FAIL("can_not_cast_an_AttrInfo_to_CAttrInfo");
58            return *(CAttrInfo*)NULL;
59        }
60
61    }
```

B.5.4 Class CAttrInfo

Listing B.56: The header file of class CAttrInfo.

```
1   // cl/datasets/cattrinfo.hpp
2   #ifndef CLUSLIB_CATTRINFO_HPP
3   #define CLUSLIB_CATTRINFO_HPP
4
5   #include<cl/datasets/attrinfo.hpp>
6
7   namespace ClusLib {
8
9       class CAttrInfo : public AttrInfo {
10      public:
11          CAttrInfo(const std::string& name);
12
13          CAttrInfo& cast_to_c();
14          const CAttrInfo& cast_to_c() const;
15          bool can_cast_to_c() const;
16          CAttrInfo* clone() const;
17          Real distance(const AttrValue&, const AttrValue&) const;
18          void set_c_val(AttrValue&, Real) const;
19          Real get_c_val(const AttrValue&) const;
20          void set_unknown(AttrValue&) const;
21          bool is_unknown(const AttrValue&) const;
```

```cpp
        void set_min(Real);
        void set_max(Real);
        Real get_min() const;
        Real get_max() const;
        bool equal(const AttrInfo&) const;

    protected:
        Real _min;
        Real _max;
    };

    inline Real CAttrInfo::get_c_val(const AttrValue& av)
        const {
        return boost::get<Real>(av._value);
    }

    inline CAttrInfo& CAttrInfo::cast_to_c() {
        return *this;
    }

    inline const CAttrInfo& CAttrInfo::cast_to_c() const {
        return *this;
    }

    inline bool CAttrInfo::can_cast_to_c() const {
        return true;
    }

    inline void CAttrInfo::set_min(Real min) {
        _min = min;
    }

    inline void CAttrInfo::set_max(Real max) {
        _max = max;
    }

    inline Real CAttrInfo::get_min() const {
        return _min;
    }

    inline Real CAttrInfo::get_max() const {
        return _max;
    }
}
#endif
```

Listing B.57: The source file of class `CAttrInfo`.

```cpp
// cl/datasets/cattrinfo.cpp
#include<cl/datasets/cattrinfo.hpp>
#include<cl/errors.hpp>
#include<cl/utilities/null.hpp>
#include<boost/variant/get.hpp>

namespace ClusLib {

    CAttrInfo::CAttrInfo(const std::string& name)
        : AttrInfo(name, Continuous) {
        _min = Null<Real>();
        _max = Null<Real>();
    }

    bool CAttrInfo::equal(const AttrInfo& ai) const {
        if(!equal_shallow(ai)){
            return false;
```

```
            }

            if(!ai.can_cast_to_c()){
                return false;
            }

            return true;
        }

        CAttrInfo* CAttrInfo::clone() const {
            return new CAttrInfo(*this);
        }

        Real CAttrInfo::distance(const AttrValue& av1,
                                 const AttrValue& av2) const {
            if(is_unknown(av1) && is_unknown(av2)){
                return 0.0;
            }

            if(is_unknown(av1) ^ is_unknown(av2)){
                return 1.0;
            }

            return boost::get<Real>(av1._value) -
                   boost::get<Real>(av2._value);
        }

        void CAttrInfo::set_c_val(AttrValue& av, Real value) const {
            av._value = value;
        }

        void CAttrInfo::set_unknown(AttrValue& av) const {
            av._value = Null<Real>();
        }

        bool CAttrInfo::is_unknown(const AttrValue& av) const {
            return (boost::get<Real>(av._value) == Null<Real>());
        }

}
```

B.5.5 Class `DAttrInfo`

Listing B.58: The header file of class `DAttrInfo`.

```
// cl/datasets/dattrinfo.hpp
#ifndef CLUSLIB_DATTRINFO_HPP
#define CLUSLIB_DATTRINFO_HPP

#include<cl/datasets/attrinfo.hpp>
#include<vector>

namespace ClusLib {

    class DAttrInfo : public AttrInfo {
    public:
        DAttrInfo(const std::string& name);

        Size num_values() const;
        const std::string& int_to_str(Size i) const;
        Size str_to_int(const std::string&) const;
        Size add_value(const std::string&,
            bool bAllowDuplicate = true);
        void remove_value(const std::string&);
```

```cpp
        void remove_value(Size i);
        DAttrInfo* clone() const;
        Real distance(const AttrValue&, const AttrValue&) const;
        void set_d_val(AttrValue&, Size) const;
        Size get_d_val(const AttrValue&) const;
        void set_unknown(AttrValue&) const;
        bool is_unknown(const AttrValue&) const;
        DAttrInfo& cast_to_d();
        const DAttrInfo& cast_to_d() const;
        bool can_cast_to_d() const;
        bool operator==(const AttrInfo& info) const;
        bool operator!=(const AttrInfo& info) const;

    protected:
        typedef std::vector<std::string>::iterator iterator;
        typedef std::vector<std::string>::const_iterator
            const_iterator;

        bool equal(const AttrInfo&) const;

        std::vector<std::string> _values;
    };

    inline Size DAttrInfo::get_d_val(
        const AttrValue& av) const {
        return boost::get<Size>(av._value);
    }

    inline DAttrInfo& DAttrInfo::cast_to_d() {
        return *this;
    }

    inline const DAttrInfo&
        DAttrInfo::cast_to_d() const {
        return *this;
    }

    inline bool DAttrInfo::can_cast_to_d() const {
        return true;
    }

    inline bool DAttrInfo::operator==(const AttrInfo& info)
        const {
        return equal(info);
    }

    inline bool DAttrInfo::operator!=(const AttrInfo& info)
        const {
        return !equal(info);
    }

}
#endif
```

Listing B.59: The source file of class `DAttrInfo`.

```cpp
// cl/datasets/dattrinfo.cpp
#include<cl/datasets/dattrinfo.hpp>
#include<cl/errors.hpp>
#include<boost/variant/get.hpp>
#include<algorithm>

namespace ClusLib {

    DAttrInfo::DAttrInfo(const std::string& name)
```

```cpp
    : AttrInfo(name,Discrete) {
}

Size DAttrInfo::num_values() const {
    return _values.size();
}

const std::string& DAttrInfo::int_to_str(Size i) const {
    ASSERT(i>=0 && i<_values.size(), "index_out_of_range");
    return _values[i];
}

Size DAttrInfo::str_to_int(const std::string& s) const {
    for(Size i=0;i<_values.size();++i) {
        if(_values[i] == s)
            return i;
    }

    return Null<Size>();
}

Size DAttrInfo::add_value(const std::string& s,
    bool bAllowDuplicate) {
    Size ind = Null<Size>();
    for(Size i=0;i<_values.size();++i) {
        if(_values[i] == s) {
            ind = i;
            break;
        }
    }

    if(ind == Null<Size>()) {
        _values.push_back(s);
        return _values.size()-1;
    } else {
        if(bAllowDuplicate) {
            return ind;
        } else {
            FAIL("value_"<<s<<"_already_exists");
            return Null<Size>();
        }
    }
}

void DAttrInfo::remove_value(const std::string& val) {
    iterator it = std::find(_values.begin(), _values.end(),
        val);
    if(it != _values.end()){
        _values.erase(it);
    }
}

void DAttrInfo::remove_value(Size i) {
    if(i>=0 || i< _values.size()){
        _values.erase(_values.begin() + i);
    }
}

bool DAttrInfo::equal(const AttrInfo& info) const {
    if(!equal_shallow(info))
        return false;

    const DAttrInfo& nai = info.cast_to_d();
    if(nai.num_values() != _values.size())
        return false;

    for(Size i=0;i<_values.size();++i){
```

```cpp
           if(_values[i] != nai.int_to_str(i)) {
               return false;
           }
       }

       return true;
   }

   DAttrInfo* DAttrInfo::clone() const {
       return new DAttrInfo(*this);
   }

   Real DAttrInfo::distance(const AttrValue& av1,
                            const AttrValue& av2) const {
       if(is_unknown(av1) && is_unknown(av2)) {
           return 0.0;
       }

       if(is_unknown(av1) ^ is_unknown(av2)) {
           return 1.0;
       }

       if(boost::get<Size>(av1._value) ==
          boost::get<Size>(av2._value) ) {
           return 0.0;
       } else {
           return 1.0;
       }
   }

   void DAttrInfo::set_d_val(AttrValue& av, Size value) const {
       ASSERT(value>=0 && value<_values.size(),
           "invalid_value_"<<value);
       av._value = value;
   }

   void DAttrInfo::set_unknown(AttrValue& av) const {
       av._value = Null<Size>();
   }

   bool DAttrInfo::is_unknown(const AttrValue& av) const {
       return (boost::get<Size>(av._value) == Null<Size>());
   }
}
```

B.5.6 Class Record

Listing B.60: The header file of class Record.

```cpp
// cl/datasets/record.hpp
#ifndef CLUSLIB_RECORD_HPP
#define CLUSLIB_RECORD_HPP

#include<boost/shared_ptr.hpp>
#include<cl/types.hpp>
#include<cl/errors.hpp>
#include<cl/datasets/schema.hpp>
#include<cl/datasets/attrinfo.hpp>
#include<cl/utilities/container.hpp>
#include<vector>

namespace ClusLib {

    class Record: public Container<AttrValue> {
```

```cpp
    public:
        Record(const boost::shared_ptr<Schema>& schema);

        const boost::shared_ptr<Schema>& schema() const;
        AttrValue& labelValue();
        const AttrValue& labelValue() const;
        AttrValue& idValue();
        const AttrValue& idValue() const;
        Size get_id() const;
        Size get_label() const;

    private:
        boost::shared_ptr<Schema> _schema;
        AttrValue _label;
        AttrValue _id;
};

inline const boost::shared_ptr<Schema>& Record::schema()
            const {
    return _schema;
}

inline AttrValue& Record::labelValue() {
    return _label;
}

inline const AttrValue& Record::labelValue() const {
    return _label;
}

inline AttrValue& Record::idValue() {
    return _id;
}

inline const AttrValue& Record::idValue() const {
    return _id;
}

inline Size Record::get_id() const {
    return _schema->idInfo()->get_d_val(_id);
}

inline Size Record::get_label() const {
    return _schema->labelInfo()->get_d_val(_label);
}

}

#endif
```

Listing B.61: The source file of class **Record**.

```cpp
// cl/datasets/record.cpp
#include<cl/datasets/record.hpp>
#include<cl/utilities/null.hpp>
#include<boost/variant/get.hpp>

namespace ClusLib {

    Record::Record(const boost::shared_ptr<Schema>& schema)
        : _schema(schema) {
        _data.resize(schema->size());
        for(Size i=0;i<_schema->size();++i){
            (*_schema)[i]->set_unknown(_data[i]);
        }
    }
```

```
       void Schema::set_label(boost::shared_ptr<Record>& r,
                 const std::string& val) {
           Size label = _labelInfo->add_value(val);
           _labelInfo->set_d_val(r->labelValue(), label);
       }

       void Schema::set_id(boost::shared_ptr<Record>& r,
                 const std::string& val) {
           Size id = _idInfo->add_value(val, false);
           _idInfo->set_d_val(r->idValue(), id);
       }

}
```

B.5.7 Class Schema

Listing B.62: The header file of class Schema.

```
// cl/datasets/schema.hpp
#ifndef CLUSLIB_SCHEMA_HPP
#define CLUSLIB_SCHEMA_HPP

#include<cl/types.hpp>
#include<cl/datasets/attrinfo.hpp>
#include<cl/datasets/cattrinfo.hpp>
#include<cl/datasets/dattrinfo.hpp>
#include<cl/utilities/container.hpp>
#include<boost/shared_ptr.hpp>
#include<vector>

namespace ClusLib {

    class Record;

    class Schema: public Container<boost::shared_ptr<AttrInfo> > {
    public:
        virtual ~Schema() {}

        Schema* clone() const;
        boost::shared_ptr<DAttrInfo>& labelInfo();
        const boost::shared_ptr<DAttrInfo>& labelInfo() const;
        boost::shared_ptr<DAttrInfo>& idInfo();
        const boost::shared_ptr<DAttrInfo>& idInfo() const;
        void set_label(boost::shared_ptr<Record>& r,
                  const std::string& val);
        void set_id(boost::shared_ptr<Record>& r,
                  const std::string& val);
        bool is_labelled() const;

        virtual bool equal(const Schema& o) const;
        virtual bool equal_no_label(const Schema& o) const;
        virtual bool operator==(const Schema& o) const;
        virtual bool operator!=(const Schema& o) const;
        virtual bool is_member(const AttrInfo& info) const;

    protected:
        boost::shared_ptr<DAttrInfo> _labelInfo;
        boost::shared_ptr<DAttrInfo> _idInfo;
    };

    inline bool Schema::operator==(const Schema& o) const {
        return equal(o);
    }
```

```cpp
46
47      inline bool Schema::operator!=(const Schema& o) const {
48          return !equal(o);
49      }
50
51      inline boost::shared_ptr<DAttrInfo>& Schema::labelInfo() {
52          return _labelInfo;
53      }
54
55      inline const boost::shared_ptr<DAttrInfo>& Schema::labelInfo()
56          const {
57          return _labelInfo;
58      }
59
60      inline boost::shared_ptr<DAttrInfo>& Schema::idInfo() {
61          return _idInfo;
62      }
63
64      inline const boost::shared_ptr<DAttrInfo>& Schema::idInfo()
65          const {
66          return _idInfo;
67      }
68  }
69  #endif
```

Listing B.63: The source file of class **Schema**.

```cpp
1   // cl/datasets/schema.cpp
2   #include<cl/datasets/schema.hpp>
3   #include<cl/errors.hpp>
4
5   namespace ClusLib {
6
7       Schema* Schema::clone() const {
8           Schema* ret = new Schema();
9           for(Size i=0;i<_data.size();++i){
10              ret->add(
11                  boost::shared_ptr<AttrInfo>(_data[i]->clone()));
12          }
13          ret->labelInfo() = boost::shared_ptr<DAttrInfo>(
14              _labelInfo->clone());
15          ret->idInfo() = boost::shared_ptr<DAttrInfo>(
16              _idInfo->clone());
17          return ret;
18      }
19
20      bool Schema::is_labelled() const {
21          if(_labelInfo){
22              return true;
23          } else {
24              return false;
25          }
26      }
27
28      bool Schema::equal(const Schema& o) const {
29          if( is_labelled() ^ o.is_labelled()){
30              return false;
31          }
32
33          if( is_labelled() && *_labelInfo != *(o.labelInfo()) ){
34              return false;
35          }
36
37          return equal_no_label(o);
38      }
39
```

```
40  bool Schema::equal_no_label(const Schema& o) const {
41      if(_data.size() != o.size()){
42          return false;
43      }
44
45      for(Size i=0;i<_data.size();++i){
46          if(*(_data[i]) != *(o[i])) {
47              return false;
48          }
49      }
50
51      return true;
52  }
53
54  bool Schema::is_member(const AttrInfo& info) const {
55      for(Size i=0;i<_data.size();++i){
56          if(*(_data[i]) == info){
57              return true;
58          }
59      }
60
61      return false;
62  }
63  }
```

B.5.8 Class Dataset

Listing B.64: The header file of class Dataset.

```
1   // cl/datasets/dataset.hpp
2   #ifndef CLUSLIB_DATASET_HPP
3   #define CLUSLIB_DATASET_HPP
4
5   #include<cl/datasets/record.hpp>
6   #include<cl/datasets/schema.hpp>
7   #include<cl/utilities/container.hpp>
8   #include<vector>
9   #include<iostream>
10
11  namespace ClusLib {
12
13      class Dataset: public Container<boost::shared_ptr<Record> > {
14      public:
15          friend std::ostream& operator<<(std::ostream& os,
16                  const Dataset& ds);
17
18          Dataset(const boost::shared_ptr<Schema>&);
19          Dataset(const Dataset&);
20
21          Size num_attr() const;
22          const boost::shared_ptr<Schema>& schema() const;
23          AttrValue& operator()(Size i, Size j);
24          const AttrValue& operator()(Size i, Size j) const;
25          bool is_numeric() const;
26          bool is_categorical() const;
27          void save(const std::string& filename) const;
28          std::vector<Size> get_CM() const;
29          Dataset& operator=(const Dataset&);
30
31      protected:
32          void print(std::ostream& os) const;
33
34          boost::shared_ptr<Schema> _schema;
35      };
```

```cpp
        inline Size Dataset::num_attr() const {
            return _schema->size();
        }

        inline const boost::shared_ptr<Schema>& Dataset::schema()
            const {
            return _schema;
        }

        inline AttrValue& Dataset::operator()(Size i, Size j) {
            return (*_data[i])[j];
        }

        inline const AttrValue&
            Dataset::operator()(Size i, Size j) const {
            return (*_data[i])[j];
        }

}

#endif
```

Listing B.65: The source file of class **Dataset**.

```cpp
// cl/datasets/dataset.cpp
#include<cl/datasets/dataset.hpp>
#include<boost/lexical_cast.hpp>
#include<fstream>
#include<sstream>

namespace ClusLib {

    std::ostream& operator<<(std::ostream& os,
            const Dataset& ds) {
        ds.print(os);
        return os;
    }

    void Dataset::print(std::ostream& os) const {
        os<<"Number of records: "<<size()<<'\n';
        os<<"Number of attributes: "<<num_attr()<<'\n';
        Integer n = 0;
        for(Size i=0; i<num_attr(); ++i) {
            if((*_schema)[i]->can_cast_to_c()){
                ++n;
            }
        }
        os<<"Number of numerical attributes: "<<n<<'\n';
        os<<"Number of categorical attributes: "
            <<num_attr()-n<<'\n';
    }

    Dataset::Dataset(const boost::shared_ptr<Schema>& schema)
        : _schema(schema) {
    }

    Dataset::Dataset(const Dataset& other) {
        _schema =
            boost::shared_ptr<Schema>(other.schema()->clone());

        for(Size i=0;i<other.size();++i) {
            boost::shared_ptr<Record> tmp =
                boost::shared_ptr<Record>(new Record(_schema));
            for(Size j=0;j<_schema->size();++j){
                (*tmp)[j] = other(i,j);
```

```cpp
            }
            _data.push_back(tmp);
        }
    }

    Dataset& Dataset::operator=(const Dataset& other) {
        if(this != &other) {
            _schema = boost::shared_ptr<Schema>(
                other.schema()->clone());

            for(Size i=0;i<other.size();++i) {
                boost::shared_ptr<Record> tmp =
                    boost::shared_ptr<Record>(
                        new Record(_schema));
                for(Size j=0;j<_schema->size();++j){
                    (*tmp)[j] = other(i,j);
                }
                _data.push_back(tmp);
            }
        }

        return *this;
    }

    bool Dataset::is_numeric() const {
        bool ret = true;
        for(Size i=0;i<_schema->size();++i){
            if( !(*_schema)[i]->can_cast_to_c()) {
                ret = false;
            }
        }

        return ret;
    }

    bool Dataset::is_categorical() const {
        bool ret = true;
        for(Size i=0;i<_schema->size();++i){
            if( !(*_schema)[i]->can_cast_to_d()) {
                ret = false;
            }
        }

        return ret;
    }

    void Dataset::save(const std::string& filename) const {
        boost::shared_ptr<DAttrInfo> label = _schema->labelInfo();
        boost::shared_ptr<DAttrInfo> id = _schema->idInfo();
        std::stringstream ss;
        for(Size i=0; i<_data.size(); ++i) {
            ss<<id->int_to_str(_data[i]->get_id());
            for(Size j=0; j<_schema->size(); ++j) {
                ss<<",";
                if ((*_schema)[j]->can_cast_to_c()) {
                    ss<<(*_schema)[j]->get_c_val((*_data[i])[j]);
                } else {
                    Size val =
                        (*_schema)[j]->get_d_val((*_data[i])[j]);
                    ss<<(*_schema)[j]->cast_to_d().int_to_str(
                        val);
                }
            }
            if(label) {
                ss<<","<<label->int_to_str(_data[i]->get_label());
            }
            ss<<std::endl;
```

```
            }

            std::ofstream of(filename.c_str());
            ASSERT(of.good(), "can_not_open_file_" << filename);
            of<<ss.str();
            of.close();

            ss.str("");
            ss<<"This_is_the_schema_file_for_dataset_"
                <<filename<<std::endl;
            ss<<"///:_schema"<<std::endl;
            ss<<"1,_RecordID"<<std::endl;
            for(Size j=0; j<_schema->size(); ++j) {
                ss<<j+2<<",_";
                if ((*_schema)[j]->can_cast_to_c()) {
                    ss<<"Continuous";
                } else {
                    ss<<"Discrete";
                }
                ss<<std::endl;
            }
            if(label) {
                ss<<_schema->size()+2<<",_Class"<<std::endl;
            }

            Size ind = filename.find_last_of('.');
            ASSERT(ind != std::string::npos,
                filename << "_is_an_invalid_file_name");
            std::string schemafile =
                filename.substr(0,ind+1) + "names";

            of.open(schemafile.c_str());
            ASSERT(of.good(), "can_not_open_file_" << schemafile);
            of<<ss.str();
            of.close();
        }

        std::vector<Size> Dataset::get_CM() const {
            std::vector<Size> CM;
            boost::shared_ptr<DAttrInfo> label = _schema->labelInfo();
            if(!label) {
                return CM;
            }

            for(Size i=0; i<_data.size(); ++i) {
                CM.push_back(label->get_d_val(
                    _data[i]->labelValue()));
            }

            return CM;
        }

}
```

B.6 Files in Folder `cl/distances`
B.6.1 Makefile

Listing B.66: `Makefile.am` in `cl/distances`.
```
noinst_LTLIBRARIES = libDistances.la

AM_CPPFLAGS = -I${top_srcdir} -I${top_builddir}

this_includedir=${includedir}/${subdir}
this_include_HEADERS = \
    all.hpp \
        distance.hpp \
    euclideandistance.hpp \
    mahalanobisdistance.hpp \
    minkowskidistance.hpp \
    mixeddistance.hpp \
    simplematchingdistance.hpp

libDistances_la_SOURCES = \
    euclideandistance.cpp \
    mahalanobisdistance.cpp \
    minkowskidistance.cpp \
    mixeddistance.cpp \
    simplematchingdistance.cpp

all.hpp: Makefile.am
        echo "// This file is generated. Please do not edit!" > $@
        echo >> $@
        for i in $(filter-out all.hpp, $(this_include_HEADERS)); \
        do \
            echo "#include <${subdir}/$$i>" >> $@; \
        done
        echo >> $@
        subdirs='$(SUBDIRS)'; for i in $$subdirs; do \
            echo "#include <${subdir}/$$i/all.hpp>" >> $@; \
        done
```

B.6.2 Class `Distance`

Listing B.67: The header file of class `Distance`.
```
// cl/distances/distance.hpp
#ifndef CLUSLIB_DISTANCE_HPP
#define CLUSLIB_DISTANCE_HPP

#include<cl/types.hpp>
#include<cl/datasets/record.hpp>
#include<cl/errors.hpp>
#include<functional>
#include<boost/shared_ptr.hpp>
#include<vector>

namespace ClusLib {

    class Distance: std::binary_function<
        boost::shared_ptr<Record>,
        boost::shared_ptr<Record>, Real> {
```

```
17      public:
18          virtual ~Distance() {}
19          Distance(const std::string &name);
20
21          const std::string& name() const;
22          virtual Real operator()(const boost::shared_ptr<Record>&,
23              const boost::shared_ptr<Record>& ) const = 0;
24
25      protected:
26          std::string _name;
27      };
28
29      inline Distance::Distance(const std::string& name)
30          : _name(name) {
31      }
32
33      inline const std::string& Distance::name() const {
34          return _name;
35      }
36
37  }
38
39  #endif
```

B.6.3 Class EuclideanDistance

Listing B.68: The header file of class EuclideanDistance.

```
1   // cl/distances/euclideandistance.hpp
2   #ifndef CLUSLIB_EUCLIDEAN_DISTANCE
3   #define CLUSLIB_EUCLIDEAN_DISTANCE
4
5   #include<cl/distances/minkowskidistance.hpp>
6
7
8   namespace ClusLib {
9
10      class EuclideanDistance: public MinkowskiDistance {
11      public:
12          EuclideanDistance();
13          Real operator()(const boost::shared_ptr<Record>&,
14              const boost::shared_ptr<Record>& ) const;
15      };
16
17  }
18
19  #endif
```

Listing B.69: The source file of class EuclideanDistance.

```
1   // cl/distances/euclideandistance.cpp
2   #include<cl/distances/euclideandistance.hpp>
3
4   namespace ClusLib {
5
6       EuclideanDistance::EuclideanDistance()
7           : MinkowskiDistance(2.0) {
8           _name = "Euclidean_distance";
9       }
10
11      Real EuclideanDistance::operator()(
12          const boost::shared_ptr<Record> &x,
```

```cpp
13          const boost::shared_ptr<Record> &y) const {
14          return MinkowskiDistance::operator()(x,y);
15      }
16  }
```

B.6.4 Class `MahalanobisDistance`

Listing B.70: The header file of class `MahalanobisDistance`.

```cpp
1   // cl/distances/mahalanobisdistance.hpp
2   #ifndef CLUSLIB_MAHALANOBISDISTANCE_HPP
3   #define CLUSLIB_MAHALANOBISDISTANCE_HPP
4
5   #include<cl/distances/distance.hpp>
6   #include<cl/errors.hpp>
7   #include<cl/types.hpp>
8   #include<cl/utilities/matrix.hpp>
9
10  namespace ClusLib {
11
12      class MahalanobisDistance: public Distance {
13      public:
14          MahalanobisDistance(const
15              ublas::symmetric_matrix<Real> &sigma);
16          Real operator()(const boost::shared_ptr<Record>&,
17              const boost::shared_ptr<Record>&) const;
18
19      protected:
20          ublas::triangular_matrix<Real> _A;
21      };
22
23      inline MahalanobisDistance::MahalanobisDistance(const
24              ublas::symmetric_matrix<Real> &sigma)
25          : Distance("Mahalanobis_distance") {
26          ublas::triangular_matrix<Real> Lm;
27          Lm.resize(sigma.size1(), sigma.size1());
28          _A.resize(sigma.size1(), sigma.size1());
29          Size k = chol(sigma, Lm);
30          ASSERT(k==0, "invalide_covariance_matrix");
31          k = triangular_matrix_inverse(Lm, _A);
32          ASSERT(k==0, "inversing_triangular_matrix_failed");
33      }
34
35  }
36
37  #endif
```

Listing B.71: The source file of class `MahalanobisDistance`.

```cpp
1   // cl/distances/mahalanobisdistance.cpp
2   #include<cl/distances/mahalanobisdistance.hpp>
3   #include<cl/datasets/record.hpp>
4   #include<cmath>
5
6   namespace ClusLib {
7
8       Real MahalanobisDistance::operator()(
9           const boost::shared_ptr<Record> &x,
10          const boost::shared_ptr<Record> &y) const {
11          boost::shared_ptr<Schema> schema = x->schema();
12          ASSERT(*schema==*(y->schema()), "schema_does_not_match");
13          ASSERT(schema->size() == _A.size1(),
```

```
                    "record and matrix dimensions do not match");

      ublas::vector<Real> v(schema->size());
      for(Size i=0; i<schema->size(); ++i) {
          v(i) = (*schema)[i]->distance((*x)[i],(*y)[i]);
      }

      ublas::vector<Real> w;
      w.resize(v.size());
      ublas::axpy_prod(_A, v, w, true);

      return std::sqrt(ublas::inner_prod(w,w));
  }

}
```

B.6.5 Class MinkowskiDistance

Listing B.72: The header file of class MinkowskiDistance.

```
// cl/distances/minkowskidistance.hpp
#ifndef CLUSLIB_MINKOWSKIDISTANCE_HPP
#define CLUSLIB_MINKOWSKIDISTANCE_HPP

#include<cl/distances/distance.hpp>
#include<cl/errors.hpp>

namespace ClusLib {

    class MinkowskiDistance: public Distance {
    public:
        MinkowskiDistance();
        MinkowskiDistance(Real p);
        Real operator()(const boost::shared_ptr<Record>&,
            const boost::shared_ptr<Record>&) const;

    protected:
        Real _p;
    };

    inline MinkowskiDistance::MinkowskiDistance()
        : Distance("Minkowski distance"), _p(2.0) {
    }

    inline MinkowskiDistance::MinkowskiDistance(Real p)
        : Distance("Minkowski distance"), _p(p) {
        ASSERT(_p>=1, "invalid parameter");
    }

}

#endif
```

Listing B.73: The source file of class MinkowskiDistance.

```
// cl/distances/minkowskidistance.cpp
#include<cl/distances/minkowskidistance.hpp>
#include<cl/datasets/record.hpp>
#include<cmath>

namespace ClusLib {
```

```cpp
Real MinkowskiDistance::operator()(
    const boost::shared_ptr<Record> &x,
    const boost::shared_ptr<Record> &y) const {
    boost::shared_ptr<Schema> schema = x->schema();
    ASSERT(*schema==*(y->schema()), "schema_does_not_match");

    Real temp = 0.0;
    for(Size i=0;i<schema->size();++i){
        temp += std::pow(std::fabs(
            (*schema)[i]->distance((*x)[i],(*y)[i])),_p);
    }

    return std::pow(temp,1/_p);
}
```

B.6.6 Class MixedDistance

Listing B.74: The header file of class MixedDistance.

```cpp
// cl/distances/mixeddistance.hpp
#ifndef CLUSLIB_MIXEDDISTANCE_HPP
#define CLUSLIB_MIXEDDISTANCE_HPP

#include<cl/distances/euclideandistance.hpp>
#include<cl/distances/simplematchingdistance.hpp>
#include<cl/types.hpp>

namespace ClusLib {

    class MixedDistance: public Distance {
    public:
        MixedDistance();
        MixedDistance(Real beta);
        Real operator()(const boost::shared_ptr<Record>&,
            const boost::shared_ptr<Record>& ) const;

    protected:
        Real _beta;
    };

    inline MixedDistance::MixedDistance()
        : Distance("Mixed_Distance"), _beta(1.0) {
    }

    inline MixedDistance::MixedDistance(Real beta)
        : Distance("Mixed_Distance"), _beta(beta) {
    }
}
#endif
```

Listing B.75: The source file of class MixedDistance.

```cpp
// cl/distances/mixeddistance.cpp
#include<cl/distances/mixeddistance.hpp>
#include<cmath>
#include<iostream>

namespace ClusLib {

```

```cpp
Real MixedDistance::operator()(
    const boost::shared_ptr<Record> &x,
    const boost::shared_ptr<Record> &y) const {
    boost::shared_ptr<Schema> schema = x->schema();
    ASSERT(*schema==*(y->schema()), "schema does not match");

    Real d1 = 0.0;
    Real d2 = 0.0;
    for(Size i=0;i<schema->size();++i){
        if((*schema)[i]->can_cast_to_c()) {
            d1 += std::pow(std::fabs(
                (*schema)[i]->distance((*x)[i],(*y)[i])),2.0);
        } else {
            d2 += (*schema)[i]->distance((*x)[i],(*y)[i]);
        }
    }

    return d1 + _beta*d2;
}
```

B.6.7 Class SimpleMatchingDistance

Listing B.76: The header file of class SimpleMatchingDistance.

```cpp
// cl/distances/simplematchingdistance.hpp
#ifndef CLUSLIB_SIMPLEMATCHINGDISTANCE_HPP
#define CLUSLIB_SIMPLEMATCHINGDISTANCE_HPP

#include<cl/distances/distance.hpp>

namespace ClusLib {

    class SimpleMatchingDistance: public Distance {
    public:
        SimpleMatchingDistance();
        Real operator()(const boost::shared_ptr<Record>&,
            const boost::shared_ptr<Record>&) const;
    };

    inline SimpleMatchingDistance::SimpleMatchingDistance()
        : Distance("Simple Matching distance") {
    }

}

#endif
```

Listing B.77: The source file of class SimpleMatchingDistance.

```cpp
// cl/distances/simplematchingdistance.cpp
#include<cl/distances/simplematchingdistance.hpp>
#include<cl/datasets/record.hpp>
#include<cmath>

namespace ClusLib {

    Real SimpleMatchingDistance::operator()(
        const boost::shared_ptr<Record> &x,
        const boost::shared_ptr<Record> &y) const {
        boost::shared_ptr<Schema> schema = x->schema();
        ASSERT(*schema==*(y->schema()), "schema does not match");
```

```
13          Real temp = 0.0;
14          for(Size i=0;i<schema->size();++i){
15              temp += (*schema)[i]->distance((*x)[i],(*y)[i]);
16          }
17
18          return temp;
19      }
20  }
21
22  }
```

B.7 Files in Folder cl/patterns

B.7.1 Makefile

Listing B.78: `Makefile.am` in `cl/patterns`.

```
1   noinst_LTLIBRARIES = libPatterns.la
2
3   AM_CPPFLAGS = -I${top_srcdir} -I${top_builddir}
4
5   this_includedir=${includedir}/${subdir}
6   this_include_HEADERS = \
7       all.hpp \
8           dendrogramvisitor.hpp \
9       internalnode.hpp \
10          joinvaluevisitor.hpp \
11      leafnode.hpp \
12      node.hpp \
13          nodevisitor.hpp \
14          pcvisitor.hpp
15
16
17  libPatterns_la_SOURCES = \
18          dendrogramvisitor.cpp \
19      internalnode.cpp \
20          joinvaluevisitor.cpp \
21      leafnode.cpp \
22          pcvisitor.cpp
23
24  all.hpp: Makefile.am
25          echo "// This file is generated. Please do not edit!" > $@
26          echo >> $@
27          for i in $(filter-out all.hpp, $(this_include_HEADERS)); \
28          do \
29          echo "#include <${subdir}/$$i>" >> $@; \
30  done
31          echo >> $@
32          subdirs='$(SUBDIRS)'; for i in $$subdirs; do \
33          echo "#include <${subdir}/$$i/all.hpp>" >> $@; \
34  done
```

B.7.2 Class DendrogramVisitor

Listing B.79: The header file of class DendrogramVisitor.

```cpp
// cl/patterns/dendrogramvisitor.hpp
#ifndef CLUSLIB_DENDROGRAMVISITOR_HPP
#define CLUSLIB_DENDROGRAMVISITOR_HPP

#include<cl/patterns/leafnode.hpp>
#include<cl/patterns/internalnode.hpp>
#include<cl/patterns/nodevisitor.hpp>
#include<cl/utilities/dendrogram.hpp>
#include<iostream>
#include<map>

namespace ClusLib {

    class DendrogramVisitor : public NodeVisitor {
    public:
        DendrogramVisitor(Real hjv,
            Size llevel, Size hlevel);
        void visit(LeafNode& node);
        void visit(InternalNode& node);
        void save(const std::string &filename);

    private:
        Dendrogram _dg;
        Size _cutlevel;
        Size _count;
        Real _leftMargin;
        Real _bottomMargin;
        Real _boxx;
        Real _boxy;
        Real _height;
        Real _width;
        Real _hjv;
        Real _gap;
        bool _drawLabel;
        std::map<Size, std::pair<Size, Size> > _lines;
        std::map<Size, std::pair<Real, Real> > _points;

        Real get_x(Size id);
        void drawLink(Size id0, Size id1);
    };

}

#endif
```

Listing B.80: The source file of class DendrogramVisitor.

```cpp
// cl/patterns/dendrogramvisitor.cpp
#include<cl/patterns/dendrogramvisitor.hpp>
#include<cl/errors.hpp>
#include<sstream>

namespace ClusLib {

    DendrogramVisitor::DendrogramVisitor(Real hjv,
        Size llevel, Size hlevel) : _cutlevel(llevel), _count(0),
        _leftMargin(30), _bottomMargin(20),
        _hjv(hjv), _gap(15), _drawLabel(true) {
        ASSERT(hlevel>=llevel, "hlevel must >= llevel");

        Real x1, y1, x2, y2;
```

```
15          _boxx = 100;
16          _boxy = 100;
17          _width = 390;
18          _height = 540;
19          Size numLeaves = hlevel - llevel + 1;
20          if (numLeaves > 60) {
21              _drawLabel = false;
22          }
23          if (_gap*numLeaves > _height - _bottomMargin) {
24              _gap = (_height - _bottomMargin) / numLeaves;
25          } else {
26              _height = _gap*numLeaves + _bottomMargin;
27          }
28
29          _dg.setbox(_boxx, _boxy, _boxx+_width, _boxy+_height);
30      }
31
32      void DendrogramVisitor::visit(LeafNode& node) {
33          Real x = _boxx + _leftMargin;
34          Real y = _bottomMargin + _boxy + _gap*_count;
35          ++_count;
36          _dg.drawCircle(x, y, 1.5);
37          if(_drawLabel) {
38              std::stringstream ss;
39              ss<<node.get_id();
40              _dg.drawText(x,y,ss.str());
41          }
42          _points.insert(std::pair<Size, std::pair<Real, Real> >(
43              node.get_id(),
44              std::pair<Real, Real>(x, y)));
45      }
46
47      void DendrogramVisitor::visit(InternalNode& node) {
48          if(node.num_children() != 2){
49              FAIL("DendrogramVisitor only handles " <<
50                  "nodes with 2 children");
51          }
52
53          Real x,y;
54          if(node.get_level() > _cutlevel) {
55              _lines.insert(std::pair<Size, std::pair<Size, Size> >(
56                  node.get_id(),
57                  std::pair<Size, Size>(node[0]->get_id(),
58                      node[1]->get_id())));
59              x = (node.get_joinValue()) * (_width - _leftMargin)
60                  / _hjv + _leftMargin + _boxx;
61              _points.insert(std::pair<Size, std::pair<Real,Real> >(
62                  node.get_id(),
63                  std::pair<Real, Real>(x, Null<Real>())));
64              node[0]->accept(*this);
65              node[1]->accept(*this);
66          } else {
67              x = _boxx + _leftMargin;
68              y = _bottomMargin + _boxy + _gap*_count;
69              ++_count;
70              _dg.drawDot(x, y);
71              if(_drawLabel) {
72                  std::stringstream ss;
73                  ss<<node.get_id();
74                  _dg.drawText(x,y,ss.str());
75              }
76              _points.insert(std::pair<Size,std::pair<Real, Real> >(
77                  node.get_id(),
78                  std::pair<Real, Real>(x, y)));
79          }
80      }
81
```

```cpp
void DendrogramVisitor::save(const std::string &filename) {
    std::map<Size, std::pair<Size, Size> >::iterator it;
    Size topid = 0;
    for(it = _lines.begin(); it!=_lines.end(); ++it) {
        if (it->first > topid) {
            topid = it->first;
        }
    }
    _points[topid].second = get_x(topid);
    for(it = _lines.begin(); it!=_lines.end(); ++it) {
        drawLink(it->second.first, it->first);
        drawLink(it->second.second, it->first);
    }
    _dg.save(filename);
}

Real DendrogramVisitor::get_x(Size id) {
    Size id0 = _lines[id].first;
    Size id1 = _lines[id].second;

    Real x1, x2;
    if(_points[id0].second == Null<Real>()) {
        x1 = get_x(id0);
        _points[id0].second = x1;
        if(_points[id1].second == Null<Real>()) {
            x2 = get_x(id1);
            _points[id1].second = x2;
        } else {
            x2 = _points[id1].second;
        }
    } else {
        x1 = _points[id0].second;
        if(_points[id1].second == Null<Real>()) {
            x2 = get_x(id1);
            _points[id1].second = x2;
        } else {
            x2 = _points[id1].second;
        }
    }

    return 0.5*(x1 + x2);
}

void DendrogramVisitor::drawLink(Size id0, Size id1) {
    Real x1 = _points[id0].first;
    Real y1 = _points[id0].second;
    Real x2 = _points[id1].first;
    Real y2 = _points[id1].second;
    if (x1 == _boxx + _leftMargin) {
        x1 += 1.5;
    }
    _dg.drawLine(x1, y1, x2, y1);
    _dg.drawLine(x2, y1, x2, y2);
}
}
```

B.7.3 Class `InternalNode`

Listing B.81: The header file of class `InternalNode`.

```cpp
// cl/patterns/internalnode.hpp
#ifndef CLUSLIB_INTERNALNODE_HPP
#define CLUSLIB_INTERNALNODE_HPP

```

```cpp
#include<cl/patterns/node.hpp>
#include<cl/utilities/container.hpp>
#include<vector>

namespace ClusLib {

    class InternalNode: public Node,
                        public Container<boost::shared_ptr<Node> >{
    public:
        InternalNode(Size id = 0,
                const boost::shared_ptr<Node> p
                = boost::shared_ptr<Node>() );
        InternalNode(Real joinValue,
                Size id = 0,
                const boost::shared_ptr<Node> p
                = boost::shared_ptr<Node>() );

        void accept(NodeVisitor &v);
        Size num_children() const;
        Size num_records() const;
        Real get_joinValue ();
        void set_joinValue(Real joinValue);

    private:
        Real _joinValue;
    };

    inline Size InternalNode::num_children() const {
        return _data.size();
    }

    inline Real InternalNode::get_joinValue() {
        return _joinValue;
    }

    inline void InternalNode::set_joinValue(Real joinValue) {
        _joinValue = joinValue;
    }

}

#endif
```

Listing B.82: The source file of class `InternalNode`.

```cpp
// cl/patterns/internalnode.cpp
#include<cl/patterns/internalnode.hpp>

namespace ClusLib {

    InternalNode::InternalNode(Size id,
        const boost::shared_ptr<Node> p)
        :Node(p, id) {
    }

    InternalNode::InternalNode(Real joinValue,
            Size id,
            const boost::shared_ptr<Node> p)
        : Node(p,id), _joinValue(joinValue) {
    }

    void InternalNode::accept(NodeVisitor &v) {
        v.visit(*this);
    }

    Size InternalNode::num_records() const {
```

```
22        Size nSum = 0;
23        for(Size i=0;i<_data.size();++i){
24            nSum += _data[i]->num_children();
25        }
26
27        return nSum;
28    }
29
30 }
```

B.7.4 Class `LeafNode`

Listing B.83: The header file of class `LeafNode`.

```
1  // cl/patterns/leafnode.hpp
2  #ifndef CLUSLIB_LEAFNODE_HPP
3  #define CLUSLIB_LEAFNODE_HPP
4
5  #include<cl/patterns/node.hpp>
6  #include<cl/datasets/record.hpp>
7
8  namespace ClusLib {
9
10     class LeafNode: public Node {
11     public:
12         LeafNode(const boost::shared_ptr<Record>& r,
13                  Size id = 0,
14                  const boost::shared_ptr<Node>& p
15                      = boost::shared_ptr<Node>() );
16
17         void accept(NodeVisitor &v);
18         Size num_children() const;
19         Size num_records() const;
20         boost::shared_ptr<Record> get_data();
21
22     private:
23         boost::shared_ptr<Record> _data;
24
25     };
26
27     inline Size LeafNode::num_children() const {
28         return 0;
29     }
30
31     inline Size LeafNode::num_records() const {
32         return 1;
33     }
34
35     inline boost::shared_ptr<Record> LeafNode::get_data() {
36         return _data;
37     }
38 }
39
40 #endif
```

Listing B.84: The source file of class `LeafNode`.

```
1  // cl/patterns/leafnode.cpp
2  #include<cl/patterns/leafnode.hpp>
3
4  namespace ClusLib {
5
```

```
 6      LeafNode::LeafNode(const boost::shared_ptr<Record>& r,
 7          Size id,
 8          const boost::shared_ptr<Node>& p)
 9          : Node(p,id), _data(r) {
10      }
11
12      void LeafNode::accept(NodeVisitor &v) {
13          v.visit(*this);
14      }
15  }
```

B.7.5 Class Node

Listing B.85: The header file of class Node.

```
 1  // cl/patterns/node.hpp
 2  #ifndef CLUSLIB_NODE_HPP
 3  #define CLUSLIB_NODE_HPP
 4
 5  #include<cl/types.hpp>
 6  #include<cl/errors.hpp>
 7  #include<cl/patterns/nodevisitor.hpp>
 8  #include<boost/shared_ptr.hpp>
 9
10  namespace ClusLib {
11
12      class Node {
13      public:
14          virtual ~Node() {}
15
16          Size get_id() const;
17          void set_id(Size id);
18          Size get_level() const;
19          void set_level(Size level);
20          boost::shared_ptr<Node> get_parent();
21          void set_parent(const boost::shared_ptr<Node>& p);
22
23          virtual void accept(NodeVisitor &v) = 0;
24          virtual Size num_children() const = 0;
25          virtual Size num_records() const = 0;
26
27      protected:
28          Node(boost::shared_ptr<Node> p, Size id)
29              : _parent(p), _id(id) {}
30
31          boost::shared_ptr<Node> _parent;
32          Size _id;
33          Size _level;
34      };
35
36      inline Size Node::get_id() const {
37          return _id;
38      }
39
40      inline void Node::set_id(Size id) {
41          _id = id;
42      }
43
44      inline Size Node::get_level() const {
45          return _level;
46      }
47
48      inline void Node::set_level(Size level) {
49          _level = level;
```

```
50      }
51
52      inline boost::shared_ptr<Node> Node::get_parent() {
53          return _parent;
54      }
55
56      inline void Node::set_parent(
57          const boost::shared_ptr<Node>& p) {
58          _parent = p;
59      }
60  }
61
62  #endif
```

B.7.6   Class `NodeVisitor`

Listing B.86: The header file of class `NodeVisitor`.
```
1   // cl/patterns/nodevisitor.hpp
2   #ifndef CLUSLIB_NODEVISITOR_HPP
3   #define CLUSLIB_NODEVISITOR_HPP
4
5   #include<boost/shared_ptr.hpp>
6   #include<cl/types.hpp>
7
8   namespace ClusLib {
9
10      class LeafNode;
11      class InternalNode;
12
13      class NodeVisitor {
14      public:
15          virtual void visit(LeafNode& node) = 0;
16          virtual void visit(InternalNode& node) = 0;
17      };
18
19  }
20
21  #endif
```

B.7.7   Class `JoinValueVisitor`

Listing B.87: The header file of class `JoinValueVisitor`.
```
1   // cl/patterns/joinvaluevisitor.hpp
2   #ifndef CLUSLIB_JOINVALUEVISITOR_HPP
3   #define CLUSLIB_JOINVALUEVISITOR_HPP
4
5   #include<cl/utilities/nnmap.hpp>
6   #include<cl/patterns/leafnode.hpp>
7   #include<cl/patterns/internalnode.hpp>
8   #include<cl/patterns/nodevisitor.hpp>
9   #include<iostream>
10  #include<set>
11
12  namespace ClusLib {
13
14      class JoinValueVisitor : public NodeVisitor {
15      public:
16          friend std::ostream& operator<<(std::ostream& os,
```

```
17              const JoinValueVisitor& jv);
18          void visit(LeafNode& node);
19          void visit(InternalNode& node);
20          const std::set<iirMapA::value_type, compare_iir>&
21              get_joinValues() const;
22
23      private:
24          void print(std::ostream& os) const;
25
26          std::set<iirMapA::value_type, compare_iir> _joinValues;
27      };
28
29      inline const std::set<iirMapA::value_type, compare_iir>&
30          JoinValueVisitor::get_joinValues() const {
31          return _joinValues;
32      }
33  }
34
35  #endif
```

Listing B.88: The source file of class JoinValueVisitor.

```
1   // cl/patterns/joinvaluevisitor.cpp
2   #include<cl/patterns/joinvaluevisitor.hpp>
3   #include<cl/errors.hpp>
4   #include<algorithm>
5   #include<set>
6
7   namespace ClusLib {
8
9       void JoinValueVisitor::print(std::ostream& os) const {
10          std::set<iirMapA::value_type,compare_iir>::const_iterator
11              it;
12          for(it = _joinValues.begin(); it!=_joinValues.end();++it){
13              os<<(it->first).first+1<<","<<(it->first).second+1
14                  <<","<<it->second<<'\n';
15          }
16      }
17
18      std::ostream& operator<<(std::ostream& os,
19              const JoinValueVisitor& jv) {
20          jv.print(os);
21          return os;
22      }
23
24      void JoinValueVisitor::visit(LeafNode& node) {
25      }
26
27      void JoinValueVisitor::visit(InternalNode& node) {
28          if(node.num_children() != 2){
29              FAIL("JoinValueVisitor only handles " <<
30                  "nodes with 2 children");
31          }
32
33          _joinValues.insert(iirMapA::value_type(
34              nnPair(node[0]->get_id(), node[1]->get_id()),
35              node.get_joinValue()));
36
37          node[0]->accept(*this);
38          node[1]->accept(*this);
39      }
40  }
```

B.7.8 Class `PCVisitor`

Listing B.89: The header file of class `PCVisitor`.

```cpp
// cl/patterns/pcvisitor.hpp
#ifndef CLUSLIB_PCVISITOR_HPP
#define CLUSLIB_PCVISITOR_HPP

#include<cl/patterns/leafnode.hpp>
#include<cl/patterns/internalnode.hpp>
#include<cl/clusters/pclustering.hpp>
#include<cl/patterns/nodevisitor.hpp>
#include<cl/types.hpp>

namespace ClusLib {

    class CVisitor : public NodeVisitor {
    public:
        CVisitor();
        void visit(LeafNode& node);
        void visit(InternalNode& node);
        boost::shared_ptr<Cluster> get_cluster();

    private:
        boost::shared_ptr<Cluster> _cluster;
    };

    class PCVisitor : public NodeVisitor {
    public:
        PCVisitor(PClustering &pc, Size cutlevel);
        void visit(LeafNode& node);
        void visit(InternalNode& node);

    private:
        PClustering &_pc;
        Size _cutlevel;
    };

    inline boost::shared_ptr<Cluster> CVisitor::get_cluster() {
        return _cluster;
    }

}

#endif
```

Listing B.90: The source file of class `PCVisitor`.

```cpp
// cl/patterns/pcvisitor.cpp
#include<cl/patterns/pcvisitor.hpp>
#include<cl/patterns/joinvaluevisitor.hpp>
#include<cl/types.hpp>
#include<iostream>

namespace ClusLib {

    CVisitor::CVisitor() {
        _cluster = boost::shared_ptr<Cluster>(new Cluster());
    }

    void CVisitor::visit(LeafNode& node) {
        _cluster->add(node.get_data());
    }

    void CVisitor::visit(InternalNode& node) {
```

```
            Integer n = node.num_children();
            for(Integer i=0;i<n;++i){
                node[i]->accept(*this);
            }
        }

        PCVisitor::PCVisitor(PClustering &pc, Size cutlevel)
            : _pc(pc), _cutlevel(cutlevel) {
        }

        void PCVisitor::visit(LeafNode& node) {
            boost::shared_ptr<Cluster> c =
                boost::shared_ptr<Cluster>(new Cluster());
            c->add(node.get_data());
            _pc.add(c);
        }

        void PCVisitor::visit(InternalNode& node) {
            if(node.get_level() >= _cutlevel) {
                for(Size i=0;i<node.num_children();++i){
                    node[i]->accept(*this);
                }
            } else {
                CVisitor cv;
                node.accept(cv);
                _pc.add(cv.get_cluster());
            }
        }
    }
```

B.8 Files in Folder cl/utilities

B.8.1 Makefile

Listing B.91: Makefile.am in cl/utilities.

```
AM_CPPFLAGS = -I${top_srcdir} -I${top_builddir}

this_includedir=${includedir}/${subdir}
this_include_HEADERS = \
    all.hpp \
        matrix.hpp \
        container.hpp \
        dataadapter.hpp \
    datasetgenerator.hpp \
    datasetnormalizer.hpp \
    datasetreader.hpp \
        dendrogram.hpp \
        nnmap.hpp \
    null.hpp

noinst_LTLIBRARIES = libUtilities.la
libUtilities_la_SOURCES = \
        matrix.cpp \
    datasetgenerator.cpp \
    datasetnormalizer.cpp \
    datasetreader.cpp \
```

```
23              dendrogram.cpp
24
25  all.hpp: Makefile.am
26              echo "// This file is generated. Please do not edit!" > $@
27              echo >> $@
28              for i in $(filter-out all.hpp, $(this_include_HEADERS)); \
29              do \
30                  echo "#include <${subdir}/$$i>" >> $@; \
31  done
32              echo >> $@
33              subdirs='$(SUBDIRS)'; for i in $$subdirs; do \
34                  echo "#include <${subdir}/$$i/all.hpp>" >> $@; \
35  done
```

B.8.2 Class Container

Listing B.92: The header file of class **Container**.

```cpp
// cl/utilities/container.hpp
#ifndef CLUSLIB_CONTAINER_HPP
#define CLUSLIB_CONTAINER_HPP

#include<cl/types.hpp>
#include<cl/errors.hpp>
#include<vector>

namespace ClusLib {

    template<typename T>
    class Container {
    public:
        typedef typename std::vector<T>::iterator iterator;
        typedef typename std::vector<T>::const_iterator
            const_iterator;

        iterator begin();
        const_iterator begin() const;
        iterator end();
        const_iterator end() const;
        Size size() const;
        bool empty() const;
        void clear();

        const std::vector<T>& data() const;
        std::vector<T>& data();
        const T& operator[](Size i) const;
        T& operator[](Size i);
        void erase(const T& val);
        void add(const T&val);

    protected:
        ~Container() {}

        std::vector<T> _data;
    };

    template<typename T>
    inline typename Container<T>::iterator Container<T>::begin() {
        return _data.begin();
    }

    template<typename T>
    inline typename Container<T>::const_iterator
        Container<T>::begin() const{
```

```cpp
        return _data.begin();
}

template<typename T>
inline typename Container<T>::iterator Container<T>::end() {
        return _data.end();
}

template<typename T>
inline typename Container<T>::const_iterator
        Container<T>::end() const{
        return _data.end();
}

template<typename T>
inline Size Container<T>::size() const {
        return _data.size();
}

template<typename T>
inline bool Container<T>::empty() const {
        return _data.size() == 0;
}

template<typename T>
inline void Container<T>::clear() {
        _data.clear();
}

template<typename T>
inline const std::vector<T>& Container<T>::data() const {
        return _data;
}

template<typename T>
inline std::vector<T>& Container<T>::data() {
        return _data;
}

template<typename T>
inline const T& Container<T>::operator[](Size i) const {
        ASSERT(i>=0 && i< _data.size(), "index_out_of_range");
        return _data[i];
}

template<typename T>
inline T& Container<T>::operator[](Size i) {
        ASSERT(i>=0 && i< _data.size(), "index_out_of_range");
        return _data[i];
}

template<typename T>
inline void Container<T>::erase(const T& val) {
        for(iterator it=_data.begin(); it!=_data.end();++it){
            if(val == *it){
                _data.erase(it);
                break;
            }
        }
}

template<typename T>
inline void Container<T>::add(const T&val) {
        _data.push_back(val);
}
}
```

```
114  #endif
```

B.8.3 Class DataAdapter

Listing B.93: The header file of class DataAdapter.
```
1   // cl/utilities/dataadapter.hpp
2   #ifndef CLUSLIB_DATAADAPTER_HPP
3   #define CLUSLIB_DATAADAPTER_HPP
4
5   #include<cl/datasets/dataset.hpp>
6
7   namespace ClusLib {
8
9       class DataAdapter {
10      public:
11          virtual ~DataAdapter() {}
12          virtual void fill(boost::shared_ptr<Dataset> &ds) = 0;
13      };
14  }
15
16  #endif
```

B.8.4 Class DatasetGenerator

Listing B.94: The header file of class DatasetGenerator.
```
1   // cl/utilities/datasetgenerator.hpp
2   #ifndef CLUSLIB_DATASETGENERATOR_HPP
3   #define CLUSLIB_DATASETGENERATOR_HPP
4
5   #include<cl/utilities/dataadapter.hpp>
6   #include<cl/utilities/matrix.hpp>
7   #include<vector>
8   #include<boost/random.hpp>
9
10  namespace ClusLib {
11
12      class DatasetGenerator: public DataAdapter {
13      public:
14          DatasetGenerator(ublas::matrix<Real> mu,
15              std::vector<ublas::symmetric_matrix<Real> > sigma,
16              std::vector<Size> records,
17              Size seed = 1);
18          void fill(boost::shared_ptr<Dataset> &ds);
19
20      protected:
21          typedef boost::variate_generator<boost::minstd_rand,
22              boost::normal_distribution<> > gen_type;
23
24          void generate(Size ind);
25          ublas::matrix<Real> _data;
26
27          Size _seed;
28          std::vector<Size> _records;
29          ublas::matrix<Real> _mu;
30          std::vector<ublas::symmetric_matrix<Real> > _sigma;
31
32          gen_type _generator;
33      };
```

```
34    }
35
36    #endif
```

Listing B.95: The source file of class `DatasetGenerator`.

```cpp
// cl/utilities/datasetgenerator.cpp
#include<cl/utilities/datasetgenerator.hpp>
#include<sstream>

namespace ClusLib {

    DatasetGenerator::DatasetGenerator(ublas::matrix<Real> mu,
            std::vector<ublas::symmetric_matrix<Real> > sigma,
            std::vector<Size> records,
            Size seed): _mu(mu), _sigma(sigma), _records(records),
            _seed(seed), _generator(boost::minstd_rand(seed),
            boost::normal_distribution<>() )
    {
        ASSERT(_mu.size1() > 0 && _mu.size2()>0,
            "empty_input");
        ASSERT(_mu.size1()==records.size() &&
            _mu.size1()==sigma.size(),
            "number_of_clusters_is_not_consistent");
        ASSERT(_mu.size2()==sigma[0].size1(),
            "number_of_attributes_is_not_consistent");
        ASSERT(_seed>0, "seed_must_be_postive");

        Size N = 0;
        for(Size i=0; i<_records.size(); ++i) {
            N += _records[i];
        }
        _data.resize(N, _mu.size2());
    }

    void DatasetGenerator::generate(Size ind) {
        ublas::triangular_matrix<Real>
            T(_mu.size2(), _mu.size2());
        Size k = chol(_sigma[ind], T);
        ASSERT(k==0, "can_not_decompose_sigma_"<<ind);

        Size nStart = 0;
        for(Size i=0; i<ind; ++i) {
            nStart += _records[i];
        }

        ublas::vector<Real> v(_mu.size2());
        ublas::vector<Real> w(_mu.size2());
        for(Size i=0; i<_records[ind]; ++i) {
            for(Size j=0; j<v.size(); ++j) {
                v(j) = _generator();
            }
            ublas::axpy_prod(T, v, w);
            for(Size j=0; j<v.size(); ++j) {
                _data(nStart+i, j) = w(j) + _mu(ind, j);
            }
        }
    }

    void DatasetGenerator::fill(boost::shared_ptr<Dataset> &ds) {
        for(Size i=0; i<_records.size(); ++i) {
            generate(i);
        }

        boost::shared_ptr<Schema> schema(new Schema());
        boost::shared_ptr<DAttrInfo> labelInfo(
```

```
61              new DAttrInfo("Label"));
62          boost::shared_ptr<DAttrInfo> idInfo(
63              new DAttrInfo("Identifier"));
64          schema->labelInfo() = labelInfo;
65          schema->idInfo() = idInfo;
66          std::stringstream ss;
67          for(Size s =0; s<_data.size2(); ++s) {
68              ss.str("");
69              ss<<"Attribute_"<<s+1;
70              boost::shared_ptr<CAttrInfo> cai(
71                  new CAttrInfo(ss.str()));
72              schema->add(cai);
73          }
74
75          ds = boost::shared_ptr<Dataset>(new Dataset(schema));
76          Size nCount = 0;
77          for(Size i=0; i<_records.size(); ++i) {
78              ss.str("");
79              ss<<i+1;
80              std::string label = ss.str();
81              for(Size j=0; j<_records[i]; ++j) {
82                  ss.str("");
83                  ss<<nCount;
84                  std::string id = ss.str();
85                  boost::shared_ptr<Record> r(new Record(schema));
86                  schema->set_id(r, id);
87                  schema->set_label(r, label);
88
89                  for(Size s=0; s<_data.size2(); ++s) {
90                      (*schema)[s]->set_c_val((*r)[s],
91                          _data(nCount, s));
92                  }
93                  ds->add(r);
94                  ++nCount;
95              }
96          }
97      }
98  }
```

B.8.5 Class DatasetNormalizer

Listing B.96: The header file of class DatasetNormalizer.

```
1   // cl/utilities/datasetnormalizer.hpp
2   #ifndef CLUSLIB_DATASETNORMALIZER_HPP
3   #define CLUSLIB_DATASETNORMALIZER_HPP
4
5   #include<cl/utilities/dataadapter.hpp>
6   #include<cl/utilities/matrix.hpp>
7   #include<vector>
8   #include<boost/random.hpp>
9
10  namespace ClusLib {
11
12      class DatasetNormalizer: public DataAdapter {
13      public:
14          DatasetNormalizer(const boost::shared_ptr<Dataset> &ds);
15          void fill(boost::shared_ptr<Dataset> &ds);
16
17      protected:
18          void get_minmax();
19
20          std::vector<Real> _dvMin;
21          std::vector<Real> _dvMax;
```

```
22        boost::shared_ptr<Dataset> _ods;
23     };
24   }
25
26   #endif
```

Listing B.97: The source file of class DatasetNormalizer.

```
1    // cl/utilities/datasetnormalizer.cpp
2    #include<cl/utilities/datasetnormalizer.hpp>
3    #include<sstream>
4
5    namespace ClusLib {
6
7        DatasetNormalizer::DatasetNormalizer(const
8            boost::shared_ptr<Dataset> &ds): _ods(ds) {
9            ASSERT(_ods, "input_dataset_is_null");
10       }
11
12       void DatasetNormalizer::fill(boost::shared_ptr<Dataset> &ds) {
13           get_minmax();
14
15           boost::shared_ptr<Schema> schema(_ods->schema()->clone());
16           ds = boost::shared_ptr<Dataset>(new Dataset(schema));
17
18           for(Size i=0; i<_ods->size(); ++i) {
19               boost::shared_ptr<Record> rec(new Record(schema));
20               for(Size h=0; h<schema->size(); ++h) {
21                   if(! (*schema)[h]->can_cast_to_c()) {
22                       (*schema)[h]->set_d_val((*rec)[h],
23                           (*schema)[h]->get_d_val((*_ods)(i,h)));
24                   } else {
25                       Real dTemp = (
26                           (*schema)[h]->get_c_val((*_ods)(i,h)) -
27                           _dvMin[h]) / ( _dvMax[h] - _dvMin[h]);
28                       (*schema)[h]->set_c_val((*rec)[h], dTemp);
29                   }
30               }
31
32               if(schema->is_labelled()) {
33                   const boost::shared_ptr<DAttrInfo> &label =
34                       schema->labelInfo();
35                   label->set_d_val(rec->labelValue(),
36                       (*_ods)[i]->get_label());
37               }
38
39               const boost::shared_ptr<DAttrInfo> &id =
40                   schema->idInfo();
41               id->set_d_val(rec->idValue(), (*_ods)[i]->get_id());
42
43               ds->add(rec);
44           }
45       }
46
47       void DatasetNormalizer::get_minmax() {
48           boost::shared_ptr<Schema> schema = _ods->schema();
49           _dvMin.resize(schema->size());
50           _dvMax.resize(schema->size());
51
52           for(Size h=0; h<schema->size(); ++h) {
53               if(! (*schema)[h]->can_cast_to_c()) {
54                   continue;
55               }
56
57               Real dMin = MAX_REAL;
58               Real dMax = MIN_REAL;
```

```
                    Real dTemp;
                    for(Size i=0; i<_ods->size(); ++i) {
                        dTemp = (*schema)[h]->get_c_val((*_ods)(i,h));
                        if (dMin > dTemp ) {
                            dMin = dTemp;
                        }

                        if (dMax < dTemp ) {
                            dMax = dTemp;
                        }
                    }

                    if( dMax - dMin < EPSILON) {
                        dMax = dMin + 1.0;
                    }

                    _dvMin[h] = dMin;
                    _dvMax[h] = dMax;
                }
            }
        }
```

B.8.6 Class `DatasetReader`

Listing B.98: The header file of class `DatasetReader`.

```
// cl/utilities/datasetreader.hpp
#ifndef CLUSLIB_DATASETREADER_HPP
#define CLUSLIB_DATASETREADER_HPP

#include<cl/types.hpp>
#include<cl/datasets/dataset.hpp>
#include<cl/utilities/dataadapter.hpp>
#include<vector>
#include<string>
#include<boost/shared_ptr.hpp>

namespace ClusLib {

    class DatasetReader : public DataAdapter {
    public:
        DatasetReader(const std::string& fileName);
        void fill(boost::shared_ptr<Dataset>& ds);

    private:
        void createSchema();
        void fillData();
        boost::shared_ptr<Record> createRecord(
            const std::vector<std::string>& val);

        std::vector<std::string> split(const std::string&);

        std::string _fileName;
        Size _labelColumn;
        Size _idColumn;
        Size _numColumn;

        boost::shared_ptr<Schema> _schema;
        boost::shared_ptr<Dataset> _ds;
    };

}

#endif
```

Listing B.99: The source file of class `DatasetReader`.

```cpp
// cl/utilities/datasetreader.cpp
#include<cl/utilities/datasetreader.hpp>
#include<boost/tokenizer.hpp>
#include<boost/algorithm/string.hpp>
#include<boost/lexical_cast.hpp>
#include<fstream>
#include<iostream>

namespace ClusLib {

    DatasetReader::DatasetReader(const std::string& fileName)
        : _fileName(fileName), _labelColumn(Null<Size>()),
          _idColumn(Null<Size>()), _numColumn(0) {
    }

    void DatasetReader::fill(boost::shared_ptr<Dataset>& ds) {
        createSchema();
        fillData();
        if(_idColumn == Null<Size>()) {
            for(Size i=0; i<_ds->size(); ++i) {
                _schema->set_id((*_ds)[i],
                    boost::lexical_cast<std::string>(i));
            }
        }
        ds = _ds;
    }

    void DatasetReader::fillData() {
        std::ifstream file;
        std::string line;
        file.open(_fileName.c_str());
        ASSERT(file.good(), "can_not_open_file_" << _fileName);

        _ds = boost::shared_ptr<Dataset>(new Dataset(_schema));
        std::vector<std::string> temp;
        std::string ms, id;
        while(getline(file,line)){
            boost::trim(line);
            if(line.empty()){
                break;
            }
            temp = split(line);
            boost::shared_ptr<Record> pr = createRecord(temp);
            _ds->add(pr);
        }
    }

    boost::shared_ptr<Record> DatasetReader::createRecord(
        const std::vector<std::string>& val) {
        boost::shared_ptr<Record> rec =
            boost::shared_ptr<Record>(new Record(_schema));
        ASSERT(_numColumn == val.size(), "length_does_not_match");
        std::string label, id;
        Size j = 0;
        Size s;
        for(Size i=0;i<val.size();++i){
            if(i == _labelColumn) {
                label = val[i];
                continue;
            }
            if(i == _idColumn) {
                id = val[i];
                continue;
            }
            switch((*_schema)[j]->type()){
                case Continuous:
```

```cpp
            if(val[i]==""){
                (*_schema)[j]->set_c_val((*rec)[j], 0.0);
            }else{
                (*_schema)[j]->set_c_val((*rec)[j],
                    boost::lexical_cast<Real>(val[i]));
            }
            break;
        case Discrete:
            s = (*_schema)[j]->cast_to_d().add_value(
                val[i]);
            (*_schema)[j]->set_d_val((*rec)[j], s);
            break;
        }
        ++j;
    }

    if(_labelColumn != Null<Size>()) {
        _schema->set_label(rec, label);
    }

    if(_idColumn != Null<Size>()) {
        _schema->set_id(rec, id);
    }

    return rec;
}

void DatasetReader::createSchema() {
    size_t ind = _fileName.find_last_of('.');
    ASSERT(ind != std::string::npos,
        _fileName << " invalid file name");
    std::string schemaFile =
        _fileName.substr(0,ind+1) + "names";

    std::ifstream file;
    std::string line;
    file.open(schemaFile.c_str());
    ASSERT(file.good(), "can not open file " << schemaFile);

    bool bTag = false;
    while(getline(file,line)){
        ind = line.find("///:");
        if(ind!=std::string::npos){
            bTag = true;
            break;
        }
    }
    ASSERT(bTag,
        "Invalid names file (no ///:) " << schemaFile);
    std::vector<std::string> temp;
    _schema = boost::shared_ptr<Schema>(new Schema());
    _schema->idInfo() = boost::shared_ptr<DAttrInfo>(
        new DAttrInfo("Identifier"));
    Size nLine = 0;
    bool bClass = false;
    bool bId = false;
    while(getline(file,line)){
        boost::trim(line);
        if(line.empty()){
            break;
        }
        temp = split(line);
        ASSERT(temp.size()==2,"invalid schema line "<<line);
        if(temp[1]=="Class"){
            if (!bClass) {
                _schema->labelInfo() =
                    boost::shared_ptr<DAttrInfo>(
```

```
                        new DAttrInfo("Membership"));
                    bClass = true;
                    _labelColumn = nLine;
                } else {
                    FAIL("schema can not have two class columns");
                }
            } else if (temp[1] == "Continuous") {
                _schema->add(boost::shared_ptr<CAttrInfo>(
                    new CAttrInfo(temp[0])));
            } else if (temp[1] == "Discrete") {
                _schema->add(boost::shared_ptr<DAttrInfo>(
                    new DAttrInfo(temp[0])));
            } else if (temp[1] == "RecordID") {
                if (!bId) {
                    bId = true;
                    _idColumn = nLine;
                } else {
                    FAIL("schema can not have two id columns");
                }
            } else {
                FAIL("invalid type " << temp[1]
                    << " note that type name is case sensitive");
            }

            ++nLine;
        }

        _numColumn = nLine;
        file.close();
    }

    std::vector<std::string> DatasetReader::split(
        const std::string& str) {
        std::vector<std::string> ret;
        boost::char_separator<char> sep(",", "",
            boost::keep_empty_tokens);
        boost::tokenizer<boost::char_separator<char> >
            tokens(str, sep);
        for(boost::tokenizer<boost::char_separator<char>
            >::iterator it = tokens.begin();
            it != tokens.end(); ++it) {
            std::string temp = *it;
            boost::trim(temp);
            ret.push_back(temp);
        }

        return ret;
    }
}
```

B.8.7 Class Dendrogram

Listing B.100: The header file of class Dendrogram.

```
// cl/utilities/dendrogram.hpp
#ifndef CLUSLIB_DENDROGRAM_HPP
#define CLUSLIB_DENDROGRAM_HPP

#include<cl/types.hpp>
#include<sstream>
#include<string>

namespace ClusLib {
```

```
11      class Dendrogram {
12      public:
13          Dendrogram();
14          void setbox(Real x1, Real y1, Real x2, Real y2);
15          void drawDot(Real x, Real y);
16          void drawCircle(Real x, Real y, Real r);
17          void drawLine(Real x1, Real y1, Real x2, Real y2);
18          void drawText(Real x, Real y, const std::string &txt);
19          void save(const std::string &filename) const;
20
21      private:
22          std::stringstream _ss;
23          Real _x1;
24          Real _y1;
25          Real _x2;
26          Real _y2;
27      };
28
29  }
30
31  #endif
```

Listing B.101: The source file of class **Dendrogram**.

```
1   // cl/utilities/dendrogram.cpp
2   #include<cl/utilities/dendrogram.hpp>
3   #include<fstream>
4   #include<ctime>
5   #include<iomanip>
6
7   namespace ClusLib {
8
9       Dendrogram::Dendrogram()
10          : _x1(0), _y1(0), _x2(100), _y2(100) {
11      }
12
13      void Dendrogram::setbox(Real x1, Real y1, Real x2, Real y2) {
14          _x1 = x1;
15          _y1 = y1;
16          _x2 = x2;
17          _y2 = y2;
18      }
19
20      void Dendrogram::drawDot(Real x, Real y) {
21          _ss << "%_Dot\n";
22          _ss << " 3 slw ";
23          _ss << " 1 slc ";
24          _ss << " 0 slj ";
25          _ss << "n "
26              << x << " " << y << " "
27              << "m "
28              << x << " " << y << " "
29              << "1 0.0000 0.0000 0.0000 srgb stroke"
30              << std::endl;
31      }
32
33      void Dendrogram::drawCircle(Real x, Real y, Real r) {
34          _ss << "%_Ellipse\n";
35          _ss << " 0.5 slw ";
36          _ss << " 1 slc ";
37          _ss << " 0 slj " << std::endl;
38          _ss << "gs " << x << " " << y << " tr";
39          _ss << " n " << r << " 0 m 0 0 "
40              << r << " 0.0 360.0 arc ";
41          _ss << " 0.0000 0.0000 0.0000 srgb";
42          _ss << " stroke gr" << std::endl;
```

420   *Data Clustering in C++: An Object-Oriented Approach*

```cpp
43     }
44
45     void Dendrogram::drawLine(Real x1, Real y1,
46         Real x2, Real y2) {
47         _ss << "%_Line\n";
48         _ss << " _0.5_slw_";
49         _ss << " _1_slc_";
50         _ss << " _0_slj_";
51         _ss << "n_"
52             << x1 << "_" << y1 << "_"
53             << "m_"
54             << x2 << "_" << y2 << "_"
55             << "1_0.0000_0.0000_0.0000_srgb_stroke"
56             << std::endl;
57     }
58
59     void Dendrogram::drawText(Real x, Real y,
60         const std::string &txt) {
61         _ss << "%_Text\n";
62         _ss << "gs_/Times-Roman_ff_8_scf_sf";
63         _ss << "_" << x- 7 - txt.size()*3 << "_" << y-3 << "_m";
64         _ss << "_(" << txt << ")"
65             << "_0.0000_0.0000_0.0000_srgb"
66             << "_sh_gr" << std::endl;
67
68     }
69
70     void Dendrogram::save(const std::string &filename) const {
71         std::ofstream file( filename.c_str() );
72
73         file << "%!PS-Adobe-2.0_EPSF-2.0" << std::endl;
74         file << "%%Title:_" << filename << std::endl;
75         file << "%%Creator:_ClusLib_" << std::endl;
76         file << "%%CreationDate:_June_23,_2010_" <<std::endl;
77         file << "%%BoundingBox:_" << std::setprecision( 8 )
78             << _x1 << "_" << _y1 << "_"
79             << _x2 << "_" << _y2 << std::endl;
80         file << "%Magnification:_1.0000" << std::endl;
81         file << "%%EndComments" << std::endl;
82         file << std::endl;
83         file << "/cp_{closepath}_bind_def" << std::endl;
84         file << "/ef_{eofill}_bind_def" << std::endl;
85         file << "/gr_{grestore}_bind_def" << std::endl;
86         file << "/gs_{gsave}_bind_def" << std::endl;
87         file << "/sa_{save}_bind_def" << std::endl;
88         file << "/rs_{restore}_bind_def" << std::endl;
89         file << "/l_{lineto}_bind_def" << std::endl;
90         file << "/m_{moveto}_bind_def" << std::endl;
91         file << "/rm_{rmoveto}_bind_def" << std::endl;
92         file << "/n_{newpath}_bind_def" << std::endl;
93         file << "/s_{stroke}_bind_def" << std::endl;
94         file << "/sh_{show}_bind_def" << std::endl;
95         file << "/slc_{setlinecap}_bind_def" << std::endl;
96         file << "/slj_{setlinejoin}_bind_def" << std::endl;
97         file << "/slw_{setlinewidth}_bind_def" << std::endl;
98         file << "/srgb_{setrgbcolor}_bind_def" << std::endl;
99         file << "/rot_{rotate}_bind_def" << std::endl;
100        file << "/sc_{scale}_bind_def" << std::endl;
101        file << "/sd_{setdash}_bind_def" << std::endl;
102        file << "/ff_{findfont}_bind_def" << std::endl;
103        file << "/sf_{setfont}_bind_def" << std::endl;
104        file << "/scf_{scalefont}_bind_def" << std::endl;
105        file << "/sw_{stringwidth}_bind_def" << std::endl;
106        file << "/sd_{setdash}_bind_def" << std::endl;
107        file << "/tr_{translate}_bind_def" << std::endl;
108        file << "_0.5_setlinewidth" << std::endl;
109        file << _ss.str() <<std::endl;
```

```
              file << "showpage" << std::endl;
              file << "%%Trailer" << std::endl;
              file << "%%EOF" << std::endl;
              file.close();
           }
      }
```

B.8.8 Class nnMap

Listing B.102: The header file of class nnMap.

```cpp
// cl/utilities/nnmap.hpp
#ifndef CLUSLIB_NNMAP_HPP
#define CLUSLIB_NNMAP_HPP

#include<cl/types.hpp>
#include<cl/errors.hpp>
#include<map>
#include<algorithm>

namespace ClusLib {

    typedef std::pair<Size,Size> nnPair;

    class compare_a {
    public:
        bool operator()(const nnPair &a, const nnPair &b) const {
            Size amin = std::min(a.first, a.second);
            Size amax = std::max(a.first, a.second);
            Size bmin = std::min(b.first, b.second);
            Size bmax = std::max(b.first, b.second);

            if(amin < bmin) {
                return true;
            } else if (amin == bmin ){
                if (amax < bmax) {
                    return true;
                } else {
                    return false;
                }
            } else {
                return false;
            }
        }
    };

    class compare_b {
    public:
        bool operator()(const nnPair &a, const nnPair &b) const {
            if(a.first < b.first) {
                return true;
            } else if (a.first == b.first ){
                if (a.second < b.second) {
                    return true;
                } else {
                    return false;
                }
            } else {
                return false;
            }
        }
    };

    template<class T, class C>
```

```cpp
class nnMap {
public:
    typedef typename std::map<nnPair, T, C>::value_type
        value_type;
    typedef typename std::map<nnPair, T, C>::iterator
        iterator;
    typedef typename std::map<nnPair, T, C>::const_iterator
        const_iterator;

    std::pair<iterator, bool>
        add_item(Size i, Size j, T item);
    bool contain_key(Size i, Size j) const;
    T& operator()(Size i, Size j);
    const T& operator()(Size i, Size j) const;
    void clear();

    iterator begin();
    iterator end();
    const_iterator begin() const;
    const_iterator end() const;
private:
    std::map<nnPair, T, C> _map;
};

typedef nnMap<Real, compare_a> iirMapA;
typedef nnMap<Size, compare_b> iiiMapB;

class compare_iir {
public:
    bool operator()(const iirMapA::value_type& a,
            const iirMapA::value_type& b) {
        if(a.second < b.second) {
            return true;
        }
        return false;
    }
};

template<typename T, typename C>
inline std::pair<typename nnMap<T,C>::iterator, bool>
    nnMap<T,C>::add_item(Size i, Size j, T item) {
    std::pair<iterator, bool> ret =
        _map.insert(std::pair<nnPair, T>(nnPair(i,j), item));

    return ret;
}

template<typename T, typename C>
inline bool nnMap<T,C>::contain_key(Size i, Size j)
    const {
    const_iterator it;
    it = _map.find(nnPair(i,j));
    if(it != _map.end() ){
        return true;
    } else {
        return false;
    }
}

template<typename T, typename C>
inline T& nnMap<T,C>::operator()(Size i, Size j) {
    iterator it;
    it = _map.find(nnPair(i,j));
    if(it != _map.end() ){
        return it->second;
    } else {
        FAIL("Can_not_find_key_("<<i<<","<<j<<")_in_nnMap");
```

```
                }
        }

        template<typename T, typename C>
        inline const T& nnMap<T,C>::operator()(Size i, Size j)
                const {
            const_iterator it;
            it = _map.find(nnPair(i,j));
            if(it != _map.end() ){
                return it->second;
            } else {
                FAIL("Can not find key ("<<i<<","<<j<<") in nnMap");
            }
        }

        template<typename T, typename C>
        inline typename nnMap<T,C>::iterator nnMap<T,C>::begin() {
            return _map.begin();
        }

        template<typename T, typename C>
        inline typename nnMap<T,C>::iterator nnMap<T,C>::end() {
            return _map.end();
        }

        template<typename T, typename C>
        inline typename nnMap<T,C>::const_iterator nnMap<T,C>::begin()
                const {
            return _map.begin();
        }

        template<typename T, typename C>
        inline typename nnMap<T,C>::const_iterator nnMap<T,C>::end()
                const {
            return _map.end();
        }

        template<typename T, typename C>
        inline void nnMap<T,C>::clear() {
            _map.clear();
        }
}
#endif
```

B.8.9 Matrix Functions

Listing B.103: Declarations of matrix functions.

```
// cl/utilities/matrix.hpp
#ifndef CLUSLIB_CHOLESKY_HPP
#define CLUSLIB_CHOLESKY_HPP

#include<cl/types.hpp>
#include<boost/numeric/ublas/vector.hpp>
#include<boost/numeric/ublas/vector_proxy.hpp>
#include<boost/numeric/ublas/matrix.hpp>
#include<boost/numeric/ublas/matrix_proxy.hpp>
#include<boost/numeric/ublas/symmetric.hpp>
#include<boost/numeric/ublas/triangular.hpp>
#include<boost/numeric/ublas/io.hpp>
#include<boost/numeric/ublas/operation.hpp>

namespace ClusLib {
```

```
      namespace ublas = boost::numeric::ublas;

      Size chol(const ublas::symmetric_matrix<Real>& A,
                ublas::triangular_matrix<Real>& L);

      Size triangular_matrix_inverse(
          const ublas::triangular_matrix<Real>& L,
          ublas::triangular_matrix<Real>& iL);
}

#endif
```

Listing B.104: Implementation of matrix function.

```
// cl/utilities/matrix.cpp
#include<cl/utilities/matrix.hpp>
#include<cl/errors.hpp>
#include<cl/types.hpp>
#include<cmath>

namespace ClusLib {

    Size chol(const ublas::symmetric_matrix<Real>& A,
              ublas::triangular_matrix<Real>& L) {
        using namespace ublas;
        ASSERT(A.size1() == A.size2(), "matrix_A_is_not_square");
        ASSERT(L.size1() == L.size2(), "matrix_L_is_not_square");
        ASSERT(A.size1() == L.size1(),
            "matrix_A_and_matrix_L_have_different_dimensions");

        const Size n = A.size1();
        for (Size k=0; k < n; k++) {
            double qL_kk = A(k,k) - inner_prod(
                project( row(L, k), range(0, k) ),
                project( row(L, k), range(0, k) ) );

            if (qL_kk <= 0) {
                return 1 + k;
            }

            double L_kk = sqrt( qL_kk );
            L(k,k) = L_kk;

            matrix_column<triangular_matrix<Real> > clk(L, k);
            project( clk, range(k+1, n) ) = (
                project( column(A, k), range(k+1, n) ) - prod(
                    project(L, range(k+1, n), range(0, k)),
                    project(row(L, k), range(0, k) ) ) ) / L_kk;
        }

        return 0;
    }

    Size triangular_matrix_inverse(
            const ublas::triangular_matrix<Real>& L,
            ublas::triangular_matrix<Real>& iL) {
        using namespace ublas;
        ASSERT(L.size1() == L.size2(), "matrix_L_is_not_square");
        ASSERT(iL.size1() == iL.size2(),
            "matrix_iL_is_not_square");
        ASSERT(L.size1() == iL.size1(),
            "matrix_L_and_matrix_iL_have_different_dimensions");

        const Size n = L.size1();
        for (Size k=0; k < n; k++) {
            if (std::fabs(L(k,k)) < EPSILON) {
```

```
53                    return 1 + k;
54                }
55            }
56
57            for (Size k=0; k < n; k++) {
58                iL(k,k) = 1 / L(k,k);
59
60                for(Size j=k+1; j<n; ++j){
61                    iL(j, k) = - inner_prod(
62                        project(row(L, j), range(k, j)),
63                        project(column(iL, k), range(k, j)) )
64                        / L(j,j);
65                }
66            }
67
68            return 0;
69        }
70
71  }
```

B.8.10 Null Types

Listing B.105: The header file of null types.

```
1   // cl/utilities/null.hpp
2   #ifndef CLUSLIB_NULL_HPP
3   #define CLUSLIB_NULL_HPP
4
5   #include<cl/types.hpp>
6
7   namespace ClusLib {
8
9       template <class Type>
10      class Null;
11
12      template <>
13      class Null<Integer> {
14      public:
15          Null() {}
16          operator Integer() const {
17              return Integer(NULL_INTEGER);
18          }
19      };
20
21      template <>
22      class Null<Size> {
23      public:
24          Null() {}
25          operator Size() const {
26              return Size(NULL_SIZE);
27          }
28      };
29
30      template<>
31      class Null<Real> {
32      public:
33          Null() {}
34          operator Real() const {
35              return Real(NULL_REAL);
36          }
37      };
38  }
39
40  #endif
```

B.9 Files in Folder examples
B.9.1 Makefile

Listing B.106: The Makefile.am file in the directory examples.
```
1   AM_CPPFLAGS = -I${top_srcdir} -I${top_builddir}
2
3   SUBDIRS = \
4           agglomerative \
5           cmean \
6           diana \
7           fsc \
8           gkmode \
9           gmc \
10          kmean \
11          kprototype
12
13  EXTRA_DIST = \
14          container/container1.cpp \
15          container/container2.cpp \
16          dataset/dataset.cpp \
17          dataset/datasetout.txt \
18          datasetgenerator/datasetgenerator.cpp \
19          datasetgenerator/9points.txt \
20          datasetreader/datasetreader.cpp \
21          datasetreader/datasetreaderout.txt \
22          dummy/dummy.cpp \
23          dummy/out3.txt \
24          dummy/out5.txt \
25          mpikmean/mpikmean.hpp \
26          mpikmean/mpikmean.cpp \
27          mpikmean/mpimain.cpp \
28          nnmap/nnmap.cpp \
29          nnmap/nnmapout.txt
30
31  examples: $(SUBDIRS)
32
33  $(SUBDIRS):
34          $(MAKE) -C $@ examples
```

B.9.2 Agglomerative Hierarchical Algorithms

Listing B.107: The Makefile.am file in the directory agglomerative.
```
1   AM_CPPFLAGS = -I${top_srcdir} -I${top_builddir}
2
3   noinst_PROGRAMS = agglomerative
4
5   agglomerative_SOURCES = agglomerative.cpp
6   agglomerative_LDADD = ../../cl/libClusLib.la
7   agglomerative_LDFLAGS = -l${BOOST_PROGRAM_OPTIONS_LIB}
```

Listing B.108: Program to illustrate agglomerative hierarchical algorithms.
```
1   // examples/agglomerative/agglomerative.cpp
2   #include<cl/cluslib.hpp>
3
```

```cpp
#include<boost/timer.hpp>
#include<iostream>
#include<fstream>
#include<iomanip>
#include<boost/program_options.hpp>

using namespace ClusLib;
using namespace boost::program_options;
using namespace std;

int main(int ac, char* av[]){
    try{
        options_description desc("Allowed options");
        desc.add_options()
            ("help", "produce help message")
            ("method", value<string>()->default_value("single"),
                "method (single, complete, gaverage, wgaverage, \
                 centroid, median, ward)")
            ("datafile", value<string>(), "the data file")
            ("p", value<Size>()->default_value(50),
                "maximum number of nodes to show in dendrogram")
            ("maxclust", value<Size>()->default_value(3),
                "maximum number of clusters");

        variables_map vm;
        store(parse_command_line(ac, av, desc), vm);
        notify(vm);

        if (vm.count("help") || ac==1) {
            cout << desc << "\n";
            return 1;
        }

        string datafile;
        if (vm.count("datafile")) {
            datafile = vm["datafile"].as<string>();
        } else {
            cout << "Please provide a data file\n";
            return 1;
        }

        string method;
        if (vm.count("method") ){
            method = vm["method"].as<string>();
        }

        Size maxclust;
        if (vm.count("maxclust")) {
            maxclust = vm["maxclust"].as<Size>();
        }
        Size p;
        if (vm.count("p")) {
            p = vm["p"].as<Size>();
        }

        DatasetReader reader(datafile);
        boost::shared_ptr<Dataset> ds;
        reader.fill(ds);

        std::cout<<*ds<<std::endl;

        boost::shared_ptr<Algorithm> ca;
        boost::shared_ptr<Distance> dist(new EuclideanDistance());
        if (method == "single" ) {
            ca = boost::shared_ptr<Algorithm>(new Single());
        } else if (method == "complete") {
            ca = boost::shared_ptr<Algorithm>(new Complete());
```

```cpp
        } else if (method == "gaverage") {
            ca = boost::shared_ptr<Algorithm>(new Average());
        } else if (method == "wgaverage") {
            ca = boost::shared_ptr<Algorithm>(new Weighted());
        } else if (method == "centroid") {
            ca = boost::shared_ptr<Algorithm>(new Centroid());
        } else if (method == "median") {
            ca = boost::shared_ptr<Algorithm>(new Median());
        } else if (method == "ward") {
            ca = boost::shared_ptr<Algorithm>(new Ward());
        } else {
            FAIL("method " << method << " is not available ");
        }

        Arguments &Arg = ca->getArguments();
        Arg.ds = ds;
        Arg.distance = dist;

        boost::timer t;
        t.restart();
        ca->clusterize();
        double seconds = t.elapsed();
        std::cout<<"completed in "<<seconds<<" seconds"
            <<std::endl;

        std::string prefix;
        size_t ind = datafile.find_last_of('.');
        if(ind != std::string::npos) {
            prefix = datafile.substr(0,ind);
        } else {
            prefix = datafile;
        }
        prefix += "-" + method + "-";

        const Results& Res = ca->getResults();

        HClustering hc =
            boost::any_cast<HClustering>(Res.get("hc"));
        hc.save(prefix + "dendrogram.eps",p);
        JoinValueVisitor jv;
        hc.root()->accept(jv);

        std::string jvfile = prefix + "joinValues.csv";
        std::ofstream of;
        of.open(jvfile.c_str());
        of<<jv;
        of.close();

        PClustering pc = hc.get_pc(maxclust);
        std::cout<<pc<<std::endl;
        pc.save(prefix + "pcsummary.txt");

        return 0;
    } catch (std::exception& e) {
        std::cout<<e.what()<<std::endl;
        return 1;
    } catch (...) {
        std::cout<<"unknown error"<<std::endl;
        return 2;
    }
}
```

B.9.3 A Divisive Hierarchical Algorithm

Listing B.109: The Makefile.am file in the directory diana.

```
1   AM_CPPFLAGS = -I${top_srcdir} -I${top_builddir}
2
3   noinst_PROGRAMS = diana
4
5   diana_SOURCES = diana.cpp
6   diana_LDADD = ../../cl/libClusLib.la
7   diana_LDFLAGS = -l${BOOST_PROGRAM_OPTIONS_LIB}
```

Listing B.110: Program to illustrate the DIANA algorithm.

```cpp
// examples/diana/diana.cpp
#include<cl/cluslib.hpp>

#include<boost/timer.hpp>
#include<iostream>
#include<fstream>
#include<iomanip>
#include<boost/program_options.hpp>

using namespace ClusLib;
using namespace boost::program_options;
using namespace std;

int main(int ac, char* av[]){
    try{
        options_description desc("Allowed options");
        desc.add_options()
            ("help", "produce help message")
            ("datafile", value<string>(), "the data file")
            ("p", value<Size>()->default_value(50),
             "maximum number of nodes to show in dendrogram")
            ("maxclust", value<Size>()->default_value(3),
             "maximum number of clusters");

        variables_map vm;
        store(parse_command_line(ac, av, desc), vm);
        notify(vm);

        if (vm.count("help") || ac==1) {
            cout << desc << "\n";
            return 1;
        }

        string datafile;
        if (vm.count("datafile")) {
            datafile = vm["datafile"].as<string>();
        } else {
            cout << "Please provide a data file\n";
            return 1;
        }

        Size maxclust;
        if (vm.count("maxclust")) {
            maxclust = vm["maxclust"].as<Size>();
        }
        Size p;
        if (vm.count("p")) {
            p = vm["p"].as<Size>();
        }

        DatasetReader reader(datafile);
```

```cpp
52      boost::shared_ptr<Dataset> ds;
53      reader.fill(ds);
54
55      std::cout<<*ds<<std::endl;
56
57      boost::shared_ptr<Algorithm> ca(new Diana());
58      boost::shared_ptr<Distance> dist(new EuclideanDistance());
59
60      Arguments &Arg = ca->getArguments();
61      Arg.ds = ds;
62      Arg.distance = dist;
63
64      boost::timer t;
65      t.restart();
66      ca->clusterize();
67      double seconds = t.elapsed();
68      std::cout<<"completed in "<<seconds<<" seconds"
69          <<std::endl;
70
71      std::string prefix;
72      size_t ind = datafile.find_last_of('.');
73      if(ind != std::string::npos ) {
74          prefix = datafile.substr(0,ind);
75      } else {
76          prefix = datafile;
77      }
78      prefix += "-diana-";
79
80      const Results& Res = ca->getResults();
81
82      HClustering hc =
83          boost::any_cast<HClustering>(Res.get("hc"));
84      hc.save(prefix + "dendrogram.eps",p);
85      JoinValueVisitor jv;
86      hc.root()->accept(jv);
87
88      std::string jvfile = prefix + "joinValues.csv";
89      std::ofstream of;
90      of.open(jvfile.c_str());
91      of<<jv;
92      of.close();
93
94      PClustering pc = hc.get_pc(maxclust);
95      std::cout<<pc<<std::endl;
96      pc.save(prefix + "pcsummary.txt");
97
98      return 0;
99  } catch (std::exception& e) {
100     std::cout<<e.what()<<std::endl;
101     return 1;
102 } catch (...) {
103     std::cout<<"unknown error"<<std::endl;
104     return 2;
105 }
106 }
```

B.9.4 The k-means Algorithm

Listing B.111: The Makefile.am file in the directory kmean.

```
1  AM_CPPFLAGS = -I${top_srcdir} -I${top_builddir}
2
3  noinst_PROGRAMS = kmean
4
```

```
5   kmean_SOURCES = kmean.cpp
6   kmean_LDADD  = ../../cl/libClusLib.la
7   kmean_LDFLAGS = -l${BOOST_PROGRAM_OPTIONS_LIB}
```

Listing B.112: Program to illustrate the k-means algorithm.

```cpp
1   // examples/kmean/kmean.cpp
2   #include<cl/cluslib.hpp>
3
4   #include<boost/timer.hpp>
5   #include<boost/program_options.hpp>
6   #include<iostream>
7   #include<sstream>
8   #include<iomanip>
9
10  using namespace ClusLib;
11  using namespace std;
12  using namespace boost::program_options;
13
14  int main(int ac, char* av[]){
15      try{
16          options_description desc("Allowed options");
17          desc.add_options()
18              ("help", "produce help message")
19              ("datafile", value<string>(), "the data file")
20              ("k", value<Size>()->default_value(3),
21                  "number of clusters")
22              ("seed", value<Size>()->default_value(1),
23                  "seed used to choose random initial centers")
24              ("maxiter", value<Size>()->default_value(100),
25                  "maximum number of iterations")
26              ("numrun", value<Size>()->default_value(1),
27                  "number of runs");
28
29          variables_map vm;
30          store(parse_command_line(ac, av, desc), vm);
31          notify(vm);
32
33          if (vm.count("help") || ac==1) {
34              cout << desc << "\n";
35              return 1;
36          }
37
38          string datafile;
39          if (vm.count("datafile")) {
40              datafile = vm["datafile"].as<string>();
41          } else {
42              cout << "Please provide a data file\n";
43              return 1;
44          }
45
46          Size numclust = vm["k"].as<Size>();
47          Size maxiter = vm["maxiter"].as<Size>();
48          Size numrun = vm["numrun"].as<Size>();
49          Size seed = vm["seed"].as<Size>();
50
51          DatasetReader reader(datafile);
52          boost::shared_ptr<Dataset> ds;
53          reader.fill(ds);
54          std::cout<<*ds<<std::endl;
55
56          boost::shared_ptr<EuclideanDistance>
57              ed(new EuclideanDistance());
58
59
60          boost::timer t;
```

```cpp
61          t.restart();
62
63          Results Res;
64          Real avgiter = 0.0;
65          Real avgerror = 0.0;
66          Real dMin = MAX_REAL;
67          Real error;
68          for(Size i=1; i<=numrun; ++i) {
69              Kmean ca;
70              Arguments &Arg = ca.getArguments();
71              Arg.ds = ds;
72              Arg.distance = ed;
73              Arg.insert("numclust", numclust);
74              Arg.insert("maxiter", maxiter);
75              Arg.insert("seed", seed);
76              if (numrun == 1) {
77                  Arg.additional["seed"] = seed;
78              } else {
79                  Arg.additional["seed"] = i;
80              }
81
82              ca.clusterize();
83
84              const Results &tmp = ca.getResults();
85              avgiter += boost::any_cast<Size>(tmp.get("numiter"));
86              error = boost::any_cast<Real>(tmp.get("error"));
87              avgerror += error;
88              if (error < dMin) {
89                  dMin = error;
90                  Res = tmp;
91              }
92          }
93          avgiter /= numrun;
94          avgerror /= numrun;
95
96          double seconds = t.elapsed();
97          std::cout<<"completed in "<<seconds<<" seconds"
98              <<std::endl;
99
100         PClustering pc =
101             boost::any_cast<PClustering>(Res.get("pc"));
102
103         std::cout<<pc<<std::endl;
104         std::cout<<"Number of runs: "<<numrun<<std::endl;
105         std::cout<<"Average number of iterations: "
106             <<avgiter<<std::endl;
107         std::cout<<"Average error: "<<avgerror<<std::endl;
108         std::cout<<"Best error: "<<dMin<<std::endl;
109
110         std::string prefix;
111         size_t ind = datafile.find_last_of('.');
112         if(ind != std::string::npos ) {
113             prefix = datafile.substr(0,ind);
114         } else {
115             prefix = datafile;
116         }
117         std::stringstream ss;
118         ss<<prefix<<"-kmean-k"<<numclust<<"-s"<<seed<<".txt";
119         pc.save(ss.str());
120
121         return 0;
122     } catch (std::exception& e) {
123         std::cout<<e.what()<<std::endl;
124         return 1;
125     } catch (...){
126         std::cout<<"unknown error"<<std::endl;
127         return 2;
```

```
128      }
129  }
```

B.9.5 The c-means Algorithm

Listing B.113: The Makefile.am file in the directory cmean.

```
1  AM_CPPFLAGS = -I${top_srcdir} -I${top_builddir}
2
3  noinst_PROGRAMS = cmean
4
5  cmean_SOURCES = cmean.cpp
6  cmean_LDADD = ../../cl/libClusLib.la
7  cmean_LDFLAGS = -l${BOOST_PROGRAM_OPTIONS_LIB}
```

Listing B.114: Program to illustrate the c-means algorithm.

```cpp
1   // examples/cmean/cmean.cpp
2   #include<cl/cluslib.hpp>
3
4   #include<boost/timer.hpp>
5   #include<boost/program_options.hpp>
6   #include<iostream>
7   #include<fstream>
8   #include<iomanip>
9
10  using namespace ClusLib;
11  using namespace boost::program_options;
12
13  int main(int ac, char* av[]){
14      try{
15          options_description desc("Allowed options");
16          desc.add_options()
17              ("help", "produce help message")
18              ("datafile", value<std::string>(), "the data file")
19              ("k", value<Size>()->default_value(3),
20               "number of clusters")
21              ("seed", value<Size>()->default_value(1),
22               "seed used to choose random initial centers")
23              ("maxiter", value<Size>()->default_value(100),
24               "maximum number of iterations")
25              ("numrun", value<Size>()->default_value(1),
26               "number of runs")
27              ("epsilon",value<Real>()->default_value(1e-6, "1e-6"),
28               "epsilon")
29              ("alpha",value<Real>()->default_value(2.1,"2.1"),
30               "alpha")
31              ("threshold",
32               value<Real>()->default_value(1e-12,"1e-12"),
33               "Objective function tolerance");
34
35          variables_map vm;
36          store(parse_command_line(ac, av, desc), vm);
37          notify(vm);
38
39          if (vm.count("help") || ac==1) {
40              std::cout << desc << "\n";
41              return 1;
42          }
43
44          std::string datafile;
45          if (vm.count("datafile")) {
```

```cpp
             datafile = vm["datafile"].as<std::string>();
         } else {
             std::cout << "Please provide a data file \n";
             return 1;
         }

         Size numclust = vm["k"].as<Size>();
         Size maxiter = vm["maxiter"].as<Size>();
         Size numrun = vm["numrun"].as<Size>();
         Size seed = vm["seed"].as<Size>();
         Real alpha = vm["alpha"].as<Real>();
         Real epsilon = vm["epsilon"].as<Real>();
         Real threshold = vm["threshold"].as<Real>();

         DatasetReader reader(datafile);
         boost::shared_ptr<Dataset> ds;
         reader.fill(ds);

         std::cout<<*ds<<std::endl;

         boost::timer t;
         t.restart();

         Results Res;
         Real avgiter = 0.0;
         Real avgerror = 0.0;
         Real dMin = MAX_REAL;
         Real error;
         for(Size i=1; i<=numrun; ++i) {
             Cmean ca;
             Arguments &Arg = ca.getArguments();
             Arg.ds = ds;
             Arg.insert("alpha", alpha);
             Arg.insert("epsilon", epsilon);
             Arg.insert("threshold", threshold);
             Arg.insert("numclust", numclust);
             Arg.insert("maxiter", maxiter);
             Arg.insert("seed", seed);
             if (numrun == 1) {
                 Arg.additional["seed"] = seed;
             } else {
                 Arg.additional["seed"] = i;
             }

             ca.clusterize();

             const Results &tmp = ca.getResults();
             avgiter += boost::any_cast<Size>(tmp.get("numiter"));
             error = boost::any_cast<Real>(tmp.get("dObj"));
             avgerror += error;
             if (error < dMin) {
                 dMin = error;
                 Res = tmp;
             }
         }
         avgiter /= numrun;
         avgerror /= numrun;

         double seconds = t.elapsed();
         std::cout<<"completed in "<<seconds<<" seconds"
             <<std::endl;

         PClustering pc =
             boost::any_cast<PClustering>(Res.get("pc"));
         std::cout<<pc<<std::endl;
         Size numiter = boost::any_cast<Size>(Res.get("numiter"));
         error = boost::any_cast<Real>(Res.get("dObj"));
```

```
114            std::cout<<"\nNumber_of_run:_"<<numrun<<std::endl;
115            std::cout<<"Average_number_of_iterations:_"
116                <<avgiter<<std::endl;
117            std::cout<<"Average_error:_"<<avgerror<<std::endl;
118            std::cout<<"Number_of_iterations_for_the_best_case:_"
119                <<numiter<<std::endl;
120            std::cout<<"Best_error:_"<<error<<std::endl;
121
122            boost::numeric::ublas::matrix<Real> fcm =
123                boost::any_cast<boost::numeric::ublas::matrix<Real> >(
124                Res.get("fcm"));
125            std::cout<<"\nFuzzy_cluster_memberships_of_"
126                <<"the_first_5_records:"<<std::endl;
127            for(Size i=0;i<fcm.size1();++i){
128                std::cout<<"Record_"<<i;
129                for(Size j=0;j<fcm.size2();++j){
130                    std::cout<<",_"<<fcm(i,j);
131                }
132                std::cout<<std::endl;
133                if (i==4 && fcm.size1()>4) {
134                    std::cout<<"..."<<std::endl;
135                    break;
136                }
137            }
138
139            return 0;
140        } catch (std::exception& e) {
141            std::cout<<e.what()<<std::endl;
142            return 1;
143        } catch (...){
144            std::cout<<"unknown_error"<<std::endl;
145            return 2;
146        }
147    }
```

B.9.6 The k-prototypes Algorithm

Listing B.115: The Makefile.am file in the directory kprototype.

```
1  AM_CPPFLAGS = -I${top_srcdir} -I${top_builddir}
2
3  noinst_PROGRAMS = kprototype
4
5  kprototype_SOURCES = kprototype.cpp
6  kprototype_LDADD = ../../cl/libClusLib.la
7  kprototype_LDFLAGS = -l${BOOST_PROGRAM_OPTIONS_LIB}
```

Listing B.116: Program to illustrate the k-prototype algorithm.

```
1  // examples/kprototype/kprototype.cpp
2  #include<cl/cluslib.hpp>
3
4  #include<boost/timer.hpp>
5  #include<boost/program_options.hpp>
6  #include<iostream>
7  #include<fstream>
8  #include<iomanip>
9
10 using namespace ClusLib;
11 using namespace std;
12 using namespace boost::program_options;
```

```cpp
int main(int ac, char* av[]){
    try{
        options_description desc("Allowed options");
        desc.add_options()
            ("help", "produce help message")
            ("datafile", value<string>(), "the data file")
            ("normalize", value<string>()->default_value("no"),
             "normalize the data or not")
            ("k", value<Size>()->default_value(3),
             "number of clusters")
            ("beta", value<Real>()->default_value(1),
             "balance weight for distance")
            ("seed", value<Size>()->default_value(1),
             "seed used to choose random initial centers")
            ("maxiter", value<Size>()->default_value(100),
             "maximum number of iterations")
            ("numrun", value<Size>()->default_value(1),
             "number of runs");

        variables_map vm;
        store(parse_command_line(ac, av, desc), vm);
        notify(vm);

        if (vm.count("help") || ac==1) {
            cout << desc << "\n";
            return 1;
        }

        string datafile;
        if (vm.count("datafile")) {
            datafile = vm["datafile"].as<string>();
        } else {
            cout << "Please provide a data file\n";
            return 1;
        }

        Real beta = vm["beta"].as<Real>();
        Size numclust = vm["k"].as<Size>();
        Size maxiter = vm["maxiter"].as<Size>();
        Size numrun = vm["numrun"].as<Size>();
        Size seed = vm["seed"].as<Size>();
        string normalize = vm["normalize"].as<string>();

        DatasetReader reader(datafile);
        boost::shared_ptr<Dataset> ds;
        reader.fill(ds);

        if( normalize != "no" ) {
            boost::shared_ptr<Dataset> ods = ds;
            DatasetNormalizer dn(ods);
            dn.fill(ds);
        }

        std::cout<<*ds<<std::endl;

        boost::shared_ptr<Distance> dist(new MixedDistance(beta));

        boost::timer t;
        t.restart();

        Results Res;
        Real avgiter = 0.0;
        Real avgerror = 0.0;
        Real dMin = MAX_REAL;
        Real error;
        for(Size i=1; i<=numrun; ++i) {
```

```cpp
80              Kprototype ca;
81              Arguments &Arg = ca.getArguments();
82              Arg.ds = ds;
83              Arg.distance = dist;
84              Arg.insert("numclust", numclust);
85              Arg.insert("maxiter", maxiter);
86              Arg.insert("seed", seed);
87              if (numrun == 1) {
88                  Arg.additional["seed"] = seed;
89              } else {
90                  Arg.additional["seed"] = i;
91              }
92
93              ca.clusterize();
94
95              const Results &tmp = ca.getResults();
96              avgiter += boost::any_cast<Size>(tmp.get("numiter"));
97              error = boost::any_cast<Real>(tmp.get("error"));
98              avgerror += error;
99              if (error < dMin) {
100                 dMin = error;
101                 Res = tmp;
102             }
103         }
104         avgiter /= numrun;
105         avgerror /= numrun;
106
107         double seconds = t.elapsed();
108         std::cout<<"completed in "<<seconds<<" seconds"
109             <<std::endl;
110
111         PClustering pc =
112             boost::any_cast<PClustering>(Res.get("pc"));
113
114         std::cout<<pc<<std::endl;
115         std::cout<<"Number of runs: "<<numrun<<std::endl;
116         std::cout<<"Average number of iterations: "
117             <<avgiter<<std::endl;
118         std::cout<<"Average error: "<<avgerror<<std::endl;
119         std::cout<<"Best error: "<<dMin<<std::endl;
120
121         return 0;
122     } catch (std::exception& e) {
123         std::cout<<e.what()<<std::endl;
124         return 1;
125     } catch (...) {
126         std::cout<<"unknown error"<<std::endl;
127         return 2;
128     }
129 }
```

B.9.7 The Genetic k-modes Algorithm

Listing B.117: The Makefile.am file in the directory gkmode.

```
1  AM_CPPFLAGS = -I${top_srcdir} -I${top_builddir}
2
3  noinst_PROGRAMS = gkmode
4
5  gkmode_SOURCES = gkmode.cpp
6  gkmode_LDADD = ../../cl/libClusLib.la
7  gkmode_LDFLAGS = -l${BOOST_PROGRAM_OPTIONS_LIB}
```

Listing B.118: Program to illustrate the genetic k-modes algorithm.

```cpp
// examples/gkmode/gkmode.cpp
#include<cl/cluslib.hpp>

#include<boost/timer.hpp>
#include<boost/program_options.hpp>
#include<iostream>
#include<fstream>
#include<iomanip>

using namespace ClusLib;
using namespace std;
using namespace boost::program_options;

int main(int ac, char* av[]){
    try{
        options_description desc("Allowed options");
        desc.add_options()
            ("help", "produce_help_message")
            ("datafile", value<string>(), "the_data_file")
            ("k", value<Size>()->default_value(3),
             "number_of_clusters")
            ("numpop", value<Size>()->default_value(50),
             "number_of_chromosomes_in_the_population")
            ("maxgen", value<Size>()->default_value(100),
             "maximum_number_of_generations")
            ("c", value<Real>()->default_value(1.5),
             "parameter_c")
            ("cm", value<Real>()->default_value(1.5),
             "parameter_c_m")
            ("pm", value<Real>()->default_value(0.2,"0.2"),
             "mutation_probability");

        variables_map vm;
        store(parse_command_line(ac, av, desc), vm);
        notify(vm);

        if (vm.count("help") || ac==1) {
            cout << desc << "\n";
            return 1;
        }

        string datafile;
        if (vm.count("datafile")) {
            datafile = vm["datafile"].as<string>();
        } else {
            cout << "Please_provide_a_data_file\n";
            return 1;
        }

        Size numclust = vm["k"].as<Size>();
        Size numpop = vm["numpop"].as<Size>();
        Size maxgen = vm["maxgen"].as<Size>();
        Real c = vm["c"].as<Real>();
        Real cm = vm["cm"].as<Real>();
        Real pm = vm["pm"].as<Real>();

        DatasetReader reader(datafile);
        boost::shared_ptr<Dataset> ds;
        reader.fill(ds);
        std::cout<<*ds<<std::endl;

        GKmode ca;
        Arguments &Arg = ca.getArguments();
        Arg.ds = ds;
        Arg.insert("numclust", numclust);
        Arg.insert("numpop", numpop);
```

```
67              Arg.insert("maxgen", maxgen);
68              Arg.insert("c", c);
69              Arg.insert("cm", cm);
70              Arg.insert("pm", pm);
71
72              boost::timer t;
73              t.restart();
74
75              ca.clusterize();
76
77              double seconds = t.elapsed();
78              std::cout<<"completed in "<<seconds<<" seconds"
79                  <<std::endl;
80
81              const Results& Res = ca.getResults();
82
83              PClustering pc =
84                  boost::any_cast<PClustering>(Res.get("pc"));
85              std::cout<<pc<<std::endl;
86
87              return 0;
88          } catch (std::exception& e) {
89              std::cout<<e.what()<<std::endl;
90              return 1;
91          } catch (...) {
92              std::cout<<"unknown error"<<std::endl;
93              return 2;
94          }
95      }
```

B.9.8 The FSC Algorithm

Listing B.119: The Makefile.am file in the directory fsc.

```
1  AM_CPPFLAGS = -I${top_srcdir} -I${top_builddir}
2
3  noinst_PROGRAMS = fsc
4
5  fsc_SOURCES = fsc.cpp
6  fsc_LDADD = ../../cl/libClusLib.la
7  fsc_LDFLAGS = -l${BOOST_PROGRAM_OPTIONS_LIB}
```

Listing B.120: Program to illustrate the FSC algorithm.

```
1   // examples/fsc/fsc.cpp
2   #include<cl/cluslib.hpp>
3
4   #include<boost/timer.hpp>
5   #include<boost/program_options.hpp>
6   #include<iostream>
7   #include<fstream>
8   #include<iomanip>
9
10  using namespace ClusLib;
11  using namespace boost::program_options;
12
13  int main(int ac, char* av[]){
14      try{
15          options_description desc("Allowed options");
16          desc.add_options()
17              ("help", "produce help message")
18              ("datafile", value<std::string>(), "the data file")
```

```
            ("k" , value<Size>()->default_value(3) ,
            "number_of_clusters")
            ("seed" , value<Size>()->default_value(1) ,
            "seed_used_to_choose_random_initial_centers")
            ("maxiter" , value<Size>()->default_value(100) ,
            "maximum_number_of_iterations")
            ("numrun" , value<Size>()->default_value(1) ,
            "number_of_runs")
            ("epsilon",value<Real>()->default_value(0) ,
            "epsilon")
            ("alpha",value<Real>()->default_value(2.1,"2.1") ,
            "alpha")
            ("threshold",
            value<Real>()->default_value(1e-12,"1e-12") ,
            "Objective_function_tolerance");

        variables_map vm;
        store(parse_command_line(ac, av, desc), vm);
        notify(vm);

        if (vm.count("help") || ac==1) {
            std::cout << desc << "\n";
            return 1;
        }

        std::string datafile;
        if (vm.count("datafile")) {
            datafile = vm["datafile"].as<std::string>();
        } else {
            std::cout << "Please_provide_a_data_file\n";
            return 1;
        }

        Size numclust = vm["k"].as<Size>();
        Size maxiter = vm["maxiter"].as<Size>();
        Size numrun = vm["numrun"].as<Size>();
        Size seed = vm["seed"].as<Size>();
        Real alpha = vm["alpha"].as<Real>();
        Real epsilon = vm["epsilon"].as<Real>();
        Real threshold = vm["threshold"].as<Real>();

        DatasetReader reader(datafile);
        boost::shared_ptr<Dataset> ds;
        reader.fill(ds);

        std::cout<<*ds<<std::endl;

        boost::timer t;
        t.restart();

        Results Res;
        Real avgiter = 0.0;
        Real avgerror = 0.0;
        Real dMin = MAX_REAL;
        Real error;
        for(Size i=1; i<=numrun; ++i) {
            FSC ca;
            Arguments &Arg = ca.getArguments();
            Arg.ds = ds;
            Arg.insert("alpha", alpha);
            Arg.insert("epsilon", epsilon);
            Arg.insert("threshold", threshold);
            Arg.insert("numclust", numclust);
            Arg.insert("maxiter", maxiter);
            Arg.insert("seed", seed);
            if (numrun == 1) {
                Arg.additional["seed"] = seed;
```

```
 86             } else {
 87                 Arg.additional["seed"] = i;
 88             }
 89
 90             ca.clusterize();
 91
 92             const Results &tmp = ca.getResults();
 93             avgiter += boost::any_cast<Size>(tmp.get("numiter"));
 94             error = boost::any_cast<Real>(tmp.get("dObj"));
 95             avgerror += error;
 96             if (error < dMin) {
 97                 dMin = error;
 98                 Res = tmp;
 99             }
100         }
101         avgiter /= numrun;
102         avgerror /= numrun;
103
104         double seconds = t.elapsed();
105         std::cout<<"completed in "<<seconds<<" seconds"
106             <<std::endl;
107
108         PClustering pc =
109             boost::any_cast<PClustering>(Res.get("pc"));
110         std::cout<<pc<<std::endl;
111         Size numiter =
112             boost::any_cast<Size>(Res.get("numiter"));
113         error = boost::any_cast<Real>(Res.get("dObj"));
114         const SubspaceCluster *p;
115         std::cout<<"Attribute Weights: "<<std::endl;
116         for(Size k=0;k<pc.size();++k) {
117             p = dynamic_cast<const SubspaceCluster*>(pc[k].get());
118             std::cout<<"Cluster "<<k;
119             for(Size j=0;j<p->w().size();++j) {
120                 std::cout<<", "<<p->w(j);
121             }
122             std::cout<<std::endl;
123         }
124         std::cout<<"\nNumber of run: "<<numrun<<std::endl;
125         std::cout<<"Average number of iterations: "
126             <<avgiter<<std::endl;
127         std::cout<<"Average error: "<<avgerror<<std::endl;
128         std::cout<<"Number of iterations for the best case: "
129             <<numiter<<std::endl;
130         std::cout<<"Best error: "<<error<<std::endl;
131
132         return 0;
133     } catch (std::exception& e) {
134         std::cout<<e.what()<<std::endl;
135         return 1;
136     } catch (...) {
137         std::cout<<"unknown error"<<std::endl;
138         return 2;
139     }
140 }
```

B.9.9 The Gaussian Mixture Clustering Algorithm

Listing B.121: The Makefile.am file in the directory gmc.

```
1  AM_CPPFLAGS = -I${top_srcdir} -I${top_builddir}
2
3  noinst_PROGRAMS = gmc
4
```

```
5   gmc_SOURCES = gmc.cpp
6   gmc_LDADD = ../../cl/libClusLib.la
7   gmc_LDFLAGS = -l${BOOST_PROGRAM_OPTIONS_LIB}
```

Listing B.122: Program to illustrate the Gaussian mixture clustering algorithm.

```
1   // examples/gmc/gmc.cpp
2   #include<cl/cluslib.hpp>
3
4   #include<boost/timer.hpp>
5   #include<boost/program_options.hpp>
6   #include<iostream>
7   #include<fstream>
8   #include<iomanip>
9
10  using namespace ClusLib;
11  using namespace boost::program_options;
12
13  int main(int ac, char* av[]){
14      try{
15          options_description desc("Allowed options");
16          desc.add_options()
17              ("help", "produce help message")
18              ("datafile", value<std::string>(), "the data file")
19              ("k", value<Size>()->default_value(3),
20               "number of clusters")
21              ("seed", value<Size>()->default_value(1),
22               "seed used to choose random initial centers")
23              ("maxiter", value<Size>()->default_value(100),
24               "maximum number of iterations")
25              ("numrun", value<Size>()->default_value(1),
26               "number of runs")
27              ("threshold", value<Real>()->default_value(1e-10),
28               "Likelihood tolerance")
29              ("epsilon", value<Real>()->default_value(0.0),
30               "Regularization parameter");
31
32          variables_map vm;
33          store(parse_command_line(ac, av, desc), vm);
34          notify(vm);
35
36          if (vm.count("help") || ac==1) {
37              std::cout << desc << "\n";
38              return 1;
39          }
40
41          std::string datafile;
42          if (vm.count("datafile")) {
43              datafile = vm["datafile"].as<std::string>();
44          } else {
45              std::cout << "Please provide a data file\n";
46              return 1;
47          }
48
49          Size numclust = vm["k"].as<Size>();
50          Size maxiter = vm["maxiter"].as<Size>();
51          Size numrun = vm["numrun"].as<Size>();
52          Size seed = vm["seed"].as<Size>();
53          Real threshold = vm["threshold"].as<Real>();
54          Real epsilon = vm["epsilon"].as<Real>();
55
56          if(numrun ==0 ){
57              return 1;
58          }
59
```

```
     DatasetReader reader(datafile);
     boost::shared_ptr<Dataset> ds;
     reader.fill(ds);

     std::cout<<*ds<<std::endl;

     boost::timer t;
     t.restart();

     Results Res;
     Real avgiter = 0.0;
     Real avgll = 0.0;
     Real maxll = MIN_REAL;
     for(Size i=1; i<=numrun; ++i) {
         GMC ca;
         Arguments &Arg = ca.getArguments();
         Arg.ds = ds;
         Arg.insert("numclust", numclust);
         Arg.insert("maxiter", maxiter);
         if(numrun == 1) {
             Arg.insert("seed", seed);
         } else {
             Arg.insert("seed", i);
         }
         Arg.insert("epsilon", epsilon);
         Arg.insert("threshold", threshold);

         ca.clusterize();
         const Results &tmp = ca.getResults();
         Real ll = boost::any_cast<Real>(tmp.get("ll"));
         avgll += ll;
         if (ll > maxll) {
             maxll = ll;
             Res = tmp;
         }
         avgiter += boost::any_cast<Size>(tmp.get("numiter"));
     }
     avgiter /= numrun;
     avgll /= numrun;

     double seconds = t.elapsed();
     std::cout<<"completed_in_"<<seconds<<"_seconds"
         <<std::endl;

     PClustering pc =
         boost::any_cast<PClustering>(Res.get("pc"));
     std::cout<<pc<<std::endl;

     pc.save("iris.txt");

     std::vector<Real> p = boost::any_cast<std::vector<Real> >(
         Res.get("p"));
     ublas::matrix<Real> mu =
         boost::any_cast<ublas::matrix<Real> >(Res.get("mu"));
     std::cout<<"Component_size:_"<<std::endl;
     for(Size i=0; i<p.size(); ++i){
         std::cout<<"Cluster_"<<i<<":_"<<p[i]<<std::endl;
     }
     std::cout<<"\nCluster_Center:_"<<std::endl;
     for(Size i=0; i<p.size(); ++i){
         std::cout<<"Center_"<<i<<"_"
             <<ublas::row(mu,i)<<std::endl;
     }
     std::cout<<"\nNumber_of_runs:_"<<numrun<<std::endl;
     std::cout<<"Average_number_of_iterations:_"
         <<avgiter<<std::endl;
     std::cout<<"Average_likelihood:_"
```

```
                <<avgll<<std::endl;
    std::cout<<"Best_likelihood:_"<<maxll<<std::endl;
    std::cout<<"Number_of_iterations_for_the_best_case:_"
        <<boost::any_cast<Size>(Res.get("numiter"))
        <<std::endl;

    return 0;
} catch (std::exception& e) {
    std::cout<<e.what()<<std::endl;
    return 1;
} catch (...){
    std::cout<<"unknown_error"<<std::endl;
    return 2;
}
}
```

B.9.10 A Parallel k-means Algorithm

Listing B.123: The header file of the parallel k-means algorithm.

```
// examples/mpikmean/mpikmean.hpp
#ifndef CLUSLIB_MPIKMEAN_HPP
#define CLUSLIB_MPIKMEAN_HPP

#include<cl/cluslib.hpp>
#include<boost/mpi.hpp>
#include<boost/serialization/vector.hpp>
#include<boost/timer.hpp>

namespace ClusLib {

    template<typename T>
    struct vplus {
        std::vector<T> operator()(const std::vector<T> &x,
            const std::vector<T> &y) {
            std::vector<T> result = x;
            for(size_t i=0; i<x.size(); ++i) {
                result[i] += y[i];
            }
            return result;
        }
    };

    class MPIKmean: public Algorithm {
    protected:
        void setupArguments();
        void performClustering() const;
        void fetchResults() const;
        virtual void initialization() const;
        virtual void iteration() const;
        virtual Real dist(Size i, Size j) const;

        mutable std::vector<Real> _centers;
        mutable std::vector<Real> _data;
        mutable Size _numObj;
        mutable Size _numAttr;
        mutable std::vector<Size> _CM;

        mutable std::vector<boost::shared_ptr<CenterCluster> >
            _clusters;
        mutable Real _error;
        mutable Size _numiter;

        Size _numclust;
```

```
45          Size _maxiter;
46          Size _seed;
47          boost::mpi::communicator _world;
48      };
49
50  }
51
52
53  #endif
```

Listing B.124: The source file of the parallel k-means algorithm.

```
1   // examples/mpikmean/mpikmean.cpp
2   #include<iostream>
3   #include<boost/random.hpp>
4   #include<cmath>
5   #include<mpikmean.hpp>
6   #include<boost/serialization/vector.hpp>
7
8   namespace ClusLib {
9
10      void MPIKmean::performClustering() const {
11          initialization();
12          iteration();
13      }
14
15      void MPIKmean::setupArguments() {
16          _numclust = boost::any_cast<Size>(
17              _arguments.get("numclust"));
18
19          _maxiter = boost::any_cast<Size>(
20              _arguments.get("maxiter"));
21          ASSERT(_maxiter>0, "invalid _maxiter");
22
23          _seed = boost::any_cast<Size>(
24              _arguments.get("seed"));
25          ASSERT(_seed>0, "invalid _seed");
26
27          if(_world.rank()==0) {
28              Algorithm::setupArguments();
29              ASSERT(_ds->is_numeric(), "dataset _is _not _numeric");
30
31              ASSERT(_numclust>=2 && _numclust<=_ds->size(),
32                  "invalid _numclust");
33          }
34
35      }
36
37      void MPIKmean::fetchResults() const {
38          std::vector<Real> error(1, 0.0), totalerror(1);
39          for(Size i=0;i<_numObj;++i) {
40              error[0] += dist(i,_CM[i]);
41          }
42
43          reduce(_world, error, totalerror, vplus<Real>(), 0);
44
45          if(_world.rank() == 0) {
46              boost::shared_ptr<Schema> schema = _ds->schema();
47              PClustering pc;
48              for(Size i=0;i<_numclust;++i){
49                  for(Size j=0; j<_numAttr; ++j) {
50                      (*schema)[j]->set_c_val(
51                          (*_clusters[i]->center())[j],
52                          _centers[i*_numAttr+j]);
53                  }
54                  pc.add(_clusters[i]);
```

```cpp
55              }
56
57              for(Size i=0; i<_CM.size(); ++i) {
58                  _clusters[_CM[i]]->add((*_ds)[i]);
59              }
60
61              _results.CM = _CM;
62              _results.insert("pc", boost::any(pc));
63
64              _error = totalerror[0];
65              _results.insert("error", boost::any(_error));
66              _results.insert("numiter", boost::any(_numiter));
67          }
68      }
69
70      void MPIKmean::iteration() const {
71          std::vector<Size> nChanged(1,1);
72
73          _numiter = 1;
74          while(nChanged[0] > 0) {
75              nChanged[0] = 0;
76              Size s;
77              Real dMin,dDist;
78              std::vector<Size> nChangedLocal(1,0);
79              std::vector<Real> newCenters(_numclust*_numAttr,0.0);
80              std::vector<Size> newSize(_numclust,0);
81
82              for(Size i=0;i<_numObj;++i) {
83                  dMin = MAX_REAL;
84                  for(Size k=0;k<_numclust;++k) {
85                      dDist = dist(i, k);
86                      if (dMin > dDist) {
87                          dMin = dDist;
88                          s = k;
89                      }
90                  }
91
92                  for(Size j=0; j<_numAttr; ++j) {
93                      newCenters[s*_numAttr+j] +=
94                              _data[i*_numAttr+j];
95                  }
96                  newSize[s] +=1;
97
98                  if (_CM[i] != s){
99                      _CM[i] = s;
100                     nChangedLocal[0]++;
101                 }
102             }
103
104             all_reduce(_world, nChangedLocal, nChanged,
105                         vplus<Size>());
106             all_reduce(_world, newCenters, _centers,
107                         vplus<Real>());
108             std::vector<Size> totalSize(_numclust,0);
109             all_reduce(_world, newSize, totalSize, vplus<Size>());
110
111             for(Size k=0; k<_numclust; ++k) {
112                 for(Size j=0; j<_numAttr; ++j) {
113                     _centers[k*_numAttr+j] /= totalSize[k];
114                 }
115             }
116
117             ++_numiter;
118             if (_numiter > _maxiter){
119                 break;
120             }
121         }
```

```cpp
            if(_world.rank() > 0) {
                _world.send(0,0,_CM);
            } else {
                for(Size p=1; p<_world.size(); ++p) {
                    std::vector<Size> msg;
                    _world.recv(p,0,msg);
                    for(Size j=0; j<msg.size(); ++j) {
                        _CM.push_back(msg[j]);
                    }
                }
            }
        }

        void MPIKmean::initialization() const {
            Size numRecords;
            Size rank = _world.rank();

            if (rank == 0) {
                numRecords = _ds->size();
                _numAttr = _ds->num_attr();
                _centers.resize(_numclust * _numAttr);

                std::vector<Integer> index(numRecords,0);
                for(Size i=0;i<index.size();++i){
                    index[i] = i;
                }

                boost::shared_ptr<Schema> schema = _ds->schema();
                boost::minstd_rand generator(_seed);
                for(Size i=0;i<_numclust;++i){
                    boost::uniform_int<> uni_dist(0,numRecords-i-1);
                    boost::variate_generator<boost::minstd_rand&,
                        boost::uniform_int<> >
                        uni(generator,uni_dist);
                    Integer r = uni();
                    boost::shared_ptr<Record> cr = boost::shared_ptr
                        <Record>(new Record(*(*_ds)[r]));
                    boost::shared_ptr<CenterCluster> c =
                        boost::shared_ptr<CenterCluster>(
                            new CenterCluster(cr));
                    c->set_id(i);
                    _clusters.push_back(c);
                    for(Size j=0; j<_numAttr; ++j) {
                        _centers[i*_numAttr + j] =
                            (*schema)[j]->get_c_val((*_ds)(r,j));
                    }
                    index.erase(index.begin()+r);
                }

            }

            boost::mpi::broadcast(_world, _centers, 0);
            boost::mpi::broadcast(_world, numRecords, 0);
            boost::mpi::broadcast(_world, _numAttr, 0);

            Size nDiv = numRecords / _world.size();
            Size nRem = numRecords % _world.size();

            if(rank == 0) {
                boost::shared_ptr<Schema> schema = _ds->schema();
                _numObj = (nRem >0) ? nDiv+1: nDiv;
                _data.resize(_numObj * _numAttr);
                _CM.resize(_numObj);
                for(Size i=0; i<_numObj; ++i) {
                    for(Size j=0; j<_numAttr; ++j) {
                        _data[i*_numAttr +j] =
```

```
                            (*schema)[j]->get_c_val((*_ds)(i, j));
                }
            }

            Size nCount = _numObj;
            for(Size p=1; p<_world.size(); ++p) {
                Size s = (p< nRem) ? nDiv +1 : nDiv;
                std::vector<Real> dv(s*_numAttr);
                for(Size i=0; i<s; ++i) {
                    for(Size j=0; j<_numAttr; ++j) {
                        dv[i*_numAttr+j] =
                            (*schema)[j]->get_c_val(
                                (*_ds)(i+nCount,j));
                    }
                }
                nCount += s;
                _world.send(p, 0, dv);
            }
        } else {
            _numObj = (rank < nRem) ? nDiv+1: nDiv;
            _CM.resize(_numObj);
            _world.recv(0,0,_data);
        }
    }

    Real MPIKmean::dist(Size i, Size j) const {
        Real dDist = 0.0;
        for(Size h=0; h<_numAttr; ++h) {
            dDist += std::pow(_data[i*_numAttr + h]
                      - _centers[j*_numAttr + h], 2.0);
        }
        return std::pow(dDist, 0.5);
    }
}
```

Listing B.125: Program to illustrate the parallel k-means algorithm.

```
// examples/mpikmean/mpimain.cpp
#include<cl/cluslib.hpp>
#include<mpikmean.hpp>

#include<boost/timer.hpp>
#include<boost/mpi.hpp>
#include<boost/program_options.hpp>
#include<iostream>
#include<sstream>
#include<iomanip>
#include<functional>

using namespace ClusLib;
using namespace std;
using namespace boost::program_options;
namespace mpi=boost::mpi;

int main(int ac, char* av[]){
    try{
        mpi::environment env(ac, av);
        mpi::communicator world;

        options_description desc("Allowed options");
        desc.add_options()
            ("help", "produce_help_message")
            ("datafile", value<string>(), "the_data_file")
            ("k", value<Size>()->default_value(3),
             "number_of_clusters")
            ("seed", value<Size>()->default_value(1),
```

```
                    "seed_used_to_choose_random_initial_centers")
                    ("maxiter", value<Size>()->default_value(100),
                    "maximum_number_of_iterations")
                    ("numrun", value<Size>()->default_value(1),
                    "number_of_runs");

        variables_map vm;
        store(parse_command_line(ac, av, desc), vm);
        notify(vm);

        if (vm.count("help") || ac==1) {
            cout << desc << "\n";
            return 1;
        }

        Size numclust = vm["k"].as<Size>();
        Size maxiter = vm["maxiter"].as<Size>();
        Size numrun = vm["numrun"].as<Size>();
        Size seed = vm["seed"].as<Size>();

        string datafile;
        if (vm.count("datafile")) {
            datafile = vm["datafile"].as<string>();
        } else {
            cout << "Please_provide_a_data_file\n";
            return 1;
        }

        boost::shared_ptr<Dataset> ds;

        if (world.rank() ==0) {
            DatasetReader reader(datafile);
            reader.fill(ds);
            std::cout<<*ds<<std::endl;
        }

        boost::timer t;
        t.restart();

        Results Res;
        Real avgiter = 0.0;
        Real avgerror = 0.0;
        Real dMin = MAX_REAL;
        Real error;

        for(Size i=1; i<=numrun; ++i) {
            MPIKmean ca;
            Arguments &Arg = ca.getArguments();
            Arg.ds = ds;
            Arg.insert("numclust", numclust);
            Arg.insert("maxiter", maxiter);
            Arg.insert("seed", seed);
            if (numrun == 1) {
                Arg.additional["seed"] = seed;
            } else {
                Arg.additional["seed"] = i;
            }

            ca.clusterize();

            if(world.rank() == 0) {
                const Results &tmp = ca.getResults();
                avgiter +=
                    boost::any_cast<Size>(tmp.get("numiter"));
                error = boost::any_cast<Real>(tmp.get("error"));
                avgerror += error;
                if (error < dMin) {
```

```
                        dMin = error;
                        Res = tmp;
                    }
                }
            }

            double seconds = t.elapsed();
            if(world.rank() == 0) {
                avgiter /= numrun;
                avgerror /= numrun;

                std::cout<<"completed in "<<seconds
                    <<" seconds"<<std::endl;
                std::cout<<"number of processes: "
                    <<world.size()<<std::endl;

                PClustering pc =
                    boost::any_cast<PClustering>(Res.get("pc"));

                std::cout<<pc<<std::endl;
                std::cout<<"Number of runs: "<<numrun<<std::endl;
                std::cout<<"Average number of iterations: "
                    <<avgiter<<std::endl;
                std::cout<<"Average error: "<<avgerror<<std::endl;
                std::cout<<"Best error: "<<dMin<<std::endl;

                std::string prefix;
                size_t ind = datafile.find_last_of('.');
                if(ind != std::string::npos ) {
                    prefix = datafile.substr(0,ind);
                } else {
                    prefix = datafile;
                }
                std::stringstream ss;
                ss<<prefix<<"-kmean-k"<<numclust<<"-s"<<seed<<".txt";
                pc.save(ss.str());
            }

            return 0;
        } catch (std::exception& e) {
            std::cout<<e.what()<<std::endl;
            return 1;
        } catch (...) {
            std::cout<<"unknown error"<<std::endl;
            return 2;
        }
    }
```

B.10 Files in Folder test-suite

B.10.1 Makefile

Listing B.126: The Makefile.am file in the directory test-suite.

```
CL_TESTS = \
    cluslibtestsuite.cpp \
    attrinfo.hpp attrinfo.cpp \
    dataset.hpp dataset.cpp \
```

```
5        distance.hpp distance.cpp \
6        nnmap.hpp nnmap.cpp \
7          matrix.hpp matrix.cpp \
8        schema.hpp schema.cpp
9
10   AM_CPPFLAGS = -I${top_srcdir} -I${top_builddir}
11
12   bin_PROGRAMS = cluslib-test-suite
13   cluslib_test_suite_SOURCES = ${CL_TESTS}
14   cluslib_test_suite_LDADD = ${top_builddir}/cl/libClusLib.la
```

B.10.2 The Master Test Suite

Listing B.127: The source file of the master test suite.

```cpp
1  // test-suite/cluslibtestsuite.cpp
2  #include<boost/test/included/unit_test_framework.hpp>
3
4  #include<iostream>
5  #include"attrinfo.hpp"
6  #include"matrix.hpp"
7  #include"dataset.hpp"
8  #include"nnmap.hpp"
9  #include"distance.hpp"
10 #include"schema.hpp"
11
12 using namespace boost::unit_test_framework;
13
14 test_suite* init_unit_test_suite(int, char* []) {
15     std::string header = "Testing ClusLib";
16     std::string rule = std::string(header.length(),'=');
17
18     BOOST_MESSAGE(rule);
19     BOOST_MESSAGE(header);
20     BOOST_MESSAGE(rule);
21
22     test_suite* test = BOOST_TEST_SUITE("ClusLib test suite");
23
24     test->add(AttrInfoTest::suite());
25     test->add(DatasetTest::suite());
26     test->add(nnMapTest::suite());
27     test->add(DistanceTest::suite());
28     test->add(MatrixTest::suite());
29     test->add(SchemaTest::suite());
30
31     return test;
32 }
```

B.10.3 Test of `AttrInfo`

Listing B.128: The header file of class `AttrInfoTest`.

```cpp
1  // test-suite/attrinfo.hpp
2  #ifndef TEST_ATTRINFO_HPP
3  #define TEST_ATTRINFO_HPP
4
5  #include<boost/test/unit_test.hpp>
6
7  class AttrInfoTest {
8  public:
```

```
 9        static void testDAttrInfo();
10        static void testCAttrInfo();
11        static boost::unit_test_framework::test_suite* suite();
12   };
13
14   #endif
```

Listing B.129: The source file of class `AttrInfoTest`.

```
 1   // test-suite/attrinfo.cpp
 2   #include"attrinfo.hpp"
 3
 4   #include<cl/datasets/cattrinfo.hpp>
 5   #include<cl/datasets/dattrinfo.hpp>
 6   #include<sstream>
 7
 8   using namespace ClusLib;
 9   using namespace boost::unit_test_framework;
10
11   void AttrInfoTest::testDAttrInfo() {
12       std::stringstream ss;
13
14       DAttrInfo nai("Nominal_A");
15       nai.add_value("A");
16       nai.add_value("B");
17       nai.add_value("A");
18       ss<<nai.name()<<" has "<<nai.num_values()
19          <<" values"<<std::endl;
20
21       ss<<" B: "<<nai.str_to_int("B")<<std::endl;
22       ss<<" A: "<<nai.str_to_int("A")<<std::endl;
23       ss<<" 0: "<<nai.int_to_str(0)<<std::endl;
24       ss<<" 1: "<<nai.int_to_str(1)<<std::endl;
25
26       DAttrInfo naib("Nominal_B");
27       if(nai==naib)
28           ss<<nai.name()<<"="<<naib.name()<<std::endl;
29       if(nai!=naib)
30           ss<<nai.name()<<"!="<<naib.name()<<std::endl;
31       if(naib.can_cast_to_d())
32           ss<<naib.name()<<" can_cast_to_discrete"<<std::endl;
33       else
34           ss<<naib.name()<<" can_not_cast_to_discrete"<<std::endl;
35
36       AttrValue nava, navb;
37       nai.set_unknown(nava);
38       BOOST_CHECK(nai.is_unknown(nava));
39
40       nai.set_d_val(nava,0);
41       nai.set_d_val(navb,1);
42       ss<<"nava navb distance: "<<nai.distance(nava,navb)
43          <<std::endl;
44       ss<<"nava has value: "<<nai.get_d_val(nava)<<std::endl;
45       ss<<"navb has value: "<<nai.get_d_val(navb)<<std::endl;
46
47       BOOST_MESSAGE(ss.str());
48   }
49
50   void AttrInfoTest::testCAttrInfo() {
51       std::stringstream ss;
52
53       CAttrInfo rai("Real_A");
54       CAttrInfo raib("Real_B");
55       CAttrInfo raic("Real_B");
56
57       if(raic==raib)
```

```
58        ss<<raic.name()<<"="<<raib.name()<<std::endl;
59     if(rai!=raib)
60        ss<<rai.name()<<"!="<<raib.name()<<std::endl;
61
62     if(rai.can_cast_to_c())
63        ss<<rai.name()<<" can_cast_to_continuous"<<std::endl;
64     else
65        ss<<rai.name()<<" can_not_cast_to_continuous"<<std::endl;
66
67     AttrValue rava, ravb;
68     rai.set_unknown(rava);
69     BOOST_CHECK(rai.is_unknown(rava));
70
71     rai.set_c_val(rava,2.0);
72     rai.set_c_val(ravb,4.0);
73
74     ss<<"rava_ravb_distance: "<<rai.distance(rava,ravb)
75        <<std::endl;
76     ss<<"rava_has_value: "<<rai.get_c_val(rava)<<std::endl;
77     ss<<"ravb_has_value: "<<rai.get_c_val(ravb)<<std::endl;
78
79     BOOST_MESSAGE(ss.str());
80  }
81
82  test_suite* AttrInfoTest::suite() {
83     test_suite* suite = BOOST_TEST_SUITE(
84        "Testing Attribute Infomation");
85
86     suite->add(BOOST_TEST_CASE(&AttrInfoTest::testDAttrInfo));
87     suite->add(BOOST_TEST_CASE(&AttrInfoTest::testCAttrInfo));
88
89     return suite;
90  }
```

B.10.4 Test of `Dataset`

Listing B.130: The header file of class `DatasetTest`.

```
1  // test-suite/dataset.hpp
2  #ifndef TEST_DATASET_HPP
3  #define TEST_DATASET_HPP
4
5  #include<boost/test/unit_test.hpp>
6
7  class DatasetTest {
8  public:
9      static void testDataset();
10     static boost::unit_test_framework::test_suite* suite();
11 };
12
13 #endif
```

Listing B.131: The source file of class `DatasetTest`.

```
1  // test-suite/dataset.cpp
2  #include"dataset.hpp"
3  #include<cl/datasets/dataset.hpp>
4  #include<cl/utilities/datasetreader.hpp>
5  #include<sstream>
6  #include<fstream>
7  #include<iostream>
8
```

```
9    using namespace ClusLib;
10   using namespace boost::unit_test_framework;
11
12   void DatasetTest::testDataset() {
13       BOOST_MESSAGE("Testing_Dataset");
14
15       std::string fileName("../../Data/iris.data");
16       DatasetReader reader(fileName);
17       boost::shared_ptr<Dataset> ds;
18       reader.fill(ds);
19
20       std::cout<<"Num_records:_"<<ds->size()<<std::endl
21           <<"Num_attributes:_"<<ds->size()<<std::endl
22           <<"Num_categories:_"
23           <<ds->schema()->labelInfo()->num_values()<<std::endl;
24
25
26       Dataset ds2 = *ds;
27       std::cout<<"Num_records:_"<<ds2.size()<<std::endl
28           <<"Num_attributes:_"<<ds2.size()<<std::endl
29           <<"Num_categories:_"
30           <<ds2.schema()->labelInfo()->num_values()<<std::endl;
31
32   }
33
34   test_suite* DatasetTest::suite() {
35       test_suite* suite = BOOST_TEST_SUITE("Testing_Dataset");
36
37       suite->add(BOOST_TEST_CASE(&DatasetTest::testDataset));
38
39       return suite;
40   }
```

B.10.5 Test of Distance

Listing B.132: The header file of class `DistanceTest`.

```
1    // test-suite/distance.hpp
2    #ifndef TEST_DISTANCE_HPP
3    #define TEST_DISTANCE_HPP
4
5    #include<boost/test/unit_test.hpp>
6
7    class DistanceTest {
8    public:
9        static void testEuclidean();
10       static void testMahalanobis();
11       static boost::unit_test_framework::test_suite* suite();
12   };
13
14   #endif
```

Listing B.133: The source file of class `DistanceTest`.

```
1    // test-suite/distance.cpp
2    #include"distance.hpp"
3
4    #include<cl/datasets/dataset.hpp>
5    #include<cl/distances/euclideandistance.hpp>
6    #include<cl/distances/mahalanobisdistance.hpp>
7    #include<cl/distances/minkowskidistance.hpp>
8    #include<cl/utilities/datasetreader.hpp>
```

```cpp
#include<sstream>
#include<fstream>

using namespace ClusLib;
using namespace boost::unit_test_framework;

void DistanceTest::testEuclidean() {
    BOOST_MESSAGE("Testing Euclidean");

    std::string fileName("../../Data/iris.data");
    DatasetReader reader(fileName);
    boost::shared_ptr<Dataset> ds;
    reader.fill(ds);

    EuclideanDistance ed;
    std::vector<Real> dist(ds->size());
    for(Size i=0;i<ds->size();++i){
        dist[i] = ed((*ds)[0],(*ds)[i]);
    }

    std::stringstream ss;
    for(Size i=0;i<dist.size();++i){
        ss<<"distance between records 0 and "<<(*ds)[i]->get_id()
            <<": "<<dist[i]<<'\n';
    }


    std::ofstream out;
    out.open("irisdist.txt");
    out<<ss.str();
    out.close();
}

void DistanceTest::testMahalanobis() {
    BOOST_MESSAGE("Testing Mahalanobis");

    std::string fileName("../../Data/bezdekIris.data");
    DatasetReader reader(fileName);
    boost::shared_ptr<Dataset> ds;
    reader.fill(ds);

    boost::shared_ptr<Schema> schema = ds->schema();
    ublas::matrix<Real> data;
    ublas::vector<Real> mu;

    data.resize(ds->size(), ds->num_attr());
    mu.resize(ds->num_attr());
    for(Size j=0; j<mu.size(); ++j) {
        mu(j) = 0.0;
    }
    for(Size i=0; i<ds->size(); ++i) {
        for(Size j=0; j<ds->num_attr(); ++j) {
            data(i,j) = (*schema)[j]->get_c_val((*ds)(i,j));
            mu(j) += data(i,j);
        }
    }

    ublas::symmetric_matrix<Real> cov;
    cov.resize(ds->num_attr(), ds->num_attr());
    for(Size i=0; i<ds->num_attr(); ++i) {
        for(Size j=0; j<=i; ++j) {
            cov(i, j) = ( inner_prod( column(data, i),
                column(data, j)) - mu(i)*mu(j) / ds->size())
                / (ds->size() - 1.0);
        }
    }
```

```
76        std::cout<<"cov:_"<<std::endl;
77        std::cout<<cov<<std::endl;
78
79        boost::shared_ptr<Record> r(new Record(ds->schema()));
80        for(Size i=0; i<r->size(); ++i) {
81            (*schema)[i]->set_c_val((*r)[i], mu(i)/ds->size());
82        }
83
84        MahalanobisDistance md(cov);
85        std::vector<Real> dist(ds->size());
86        for(Size i=0;i<ds->size();++i){
87            dist[i] = md(r,(*ds)[i]);
88        }
89
90        std::stringstream ss;
91        for(Size i=0;i<dist.size();++i){
92            ss<<"distance_between_records_0_and_"<<(*ds)[i]->get_id()
93                <<":_"<<dist[i]<<'\n';
94        }
95
96
97        std::ofstream out;
98        out.open("iris_mahal.txt");
99        out<<ss.str();
100       out.close();
101   }
102
103   test_suite* DistanceTest::suite() {
104       test_suite* suite = BOOST_TEST_SUITE("Testing_Distances");
105
106       suite->add(BOOST_TEST_CASE(&DistanceTest::testEuclidean));
107       suite->add(BOOST_TEST_CASE(&DistanceTest::testMahalanobis));
108
109       return suite;
110   }
```

B.10.6 Test of nnMap

Listing B.134: The header file of class **nnMapTest**.

```
1   // test-suite/nnmap.hpp
2   #ifndef TEST_NNMAP_HPP
3   #define TEST_NNMAP_HPP
4
5   #include<boost/test/unit_test.hpp>
6
7   class nnMapTest {
8   public:
9       static void testiirMapA();
10      static void testiiiMapB();
11      static boost::unit_test_framework::test_suite* suite();
12  };
13
14  #endif
```

Listing B.135: The source file of class **nnMapTest**.

```
1   // test-suite/nnmap.cpp
2   #include"nnmap.hpp"
3   #include<cl/datasets/dataset.hpp>
4   #include<cl/utilities/datasetreader.hpp>
5   #include<cl/utilities/nnmap.hpp>
```

```cpp
#include<cl/distances/euclideandistance.hpp>
#include<sstream>
#include<fstream>
#include<iostream>

using namespace ClusLib;
using namespace boost::unit_test_framework;

void nnMapTest::testiirMapA() {
    BOOST_MESSAGE("Testing_iirMapA");

    std::string fileName("../../Data/bezdekIris.data");
    DatasetReader reader(fileName);
    boost::shared_ptr<Dataset> ds;
    reader.fill(ds);

    EuclideanDistance ed;
    iirMapA dm;
    Size n = ds->size();
    for(Size i=0;i<n;++i){
        for(Size j=i+1;j<n;++j){
            dm.add_item(i,j,ed((*ds)[i],(*ds)[j]));
        }
    }

    std::stringstream ss;
    for(Size i=0;i<n;++i){
        for(Size j=i+1;j<n;++j){
            ss<<i+1<<","<<j+1<<","<<dm(j,i)<<std::endl;
        }
    }

    std::ofstream out("irisdm.csv");
    out<<ss.str();
    out.close();
}

void nnMapTest::testiiiMapB() {
    BOOST_MESSAGE("Testing_iiiMapB");

    iiiMapB dm;
    Size n = 10;
    for(Size i=0;i<n;++i){
        for(Size j=0;j<n;++j){
            dm.add_item(i,j,i+j);
        }
    }

    std::stringstream ss;
    for(Size i=0;i<n;++i){
        for(Size j=0;j<n;++j){
            ss<<i+1<<","<<j+1<<","<<dm(j,i)<<std::endl;
        }
    }

    std::ofstream out("iirmapb.csv");
    out<<ss.str();
    out.close();
}

test_suite* nnMapTest::suite() {
    test_suite* suite = BOOST_TEST_SUITE("Testing_nnMap");

    suite->add(BOOST_TEST_CASE(&nnMapTest::testiirMapA));
    suite->add(BOOST_TEST_CASE(&nnMapTest::testiiiMapB));

    return suite;
```

```
73  }
```

B.10.7 Test of Matrices

Listing B.136: The header file of class `MatrixTest`.

```
 1  // test-suite/matrix.hpp
 2  #ifndef TEST_MATRIX_HPP
 3  #define TEST_MATRIX_HPP
 4
 5  #include<boost/test/unit_test.hpp>
 6
 7  class MatrixTest {
 8  public:
 9      static void testCholesky();
10      static void testTriangularInverse();
11      static boost::unit_test_framework::test_suite* suite();
12  };
13
14  #endif
```

Listing B.137: The source file of class `MatrixTest`.

```
 1  // test-suite/matrix.cpp
 2  #include"matrix.hpp"
 3
 4  #include<cl/utilities/matrix.hpp>
 5  #include<boost/numeric/ublas/io.hpp>
 6  #include<sstream>
 7  #include<fstream>
 8
 9  using namespace ClusLib;
10  using namespace boost::unit_test_framework;
11  using namespace boost::numeric::ublas;
12
13  void MatrixTest::testCholesky() {
14      BOOST_MESSAGE("Testing_Cholesky");
15
16      symmetric_matrix<Real> A(3,3);
17      A(0,0) = 3.1325;
18      A(1,0) = 0.9748; A(1,1) = 1.4862;
19      A(2,0) = -0.7613; A(2,1) = -0.8402; A(2,2) = 0.7390;
20
21      triangular_matrix<Real> L(3,3);
22      Size k = chol(A,L);
23
24      BOOST_CHECK(k==0);
25
26      std::cout<<"A:_\n";
27      for(Size i=0; i<A.size1(); ++i) {
28          std::cout<<row(A, i)<<std::endl;
29      }
30      std::cout<<"L:_\n";
31      for(Size i=0; i<L.size1(); ++i) {
32          std::cout<<row(L, i)<<std::endl;
33      }
34  }
35
36  void MatrixTest::testTriangularInverse() {
37      BOOST_MESSAGE("Testing_Triangular_Inverse");
38
39      triangular_matrix<Real> L(4,4);
```

```
          L(0,0) = 0.8281;
          L(1,0) =-0.0512; L(1,1) = 0.4328;
          L(2,0) = 1.5389; L(2,1) = -0.5794; L(2,2) = 0.6421;
          L(3,0) = 0.6235; L(3,1) = -0.2072; L(3,2) = 0.3365;
          L(3,3)=0.1900;

          triangular_matrix<Real> iL(4,4);
          Size k = triangular_matrix_inverse(L,iL);

          BOOST_CHECK(k==0);

          std::cout<<"L:_\n";
          for(Size i=0; i<L.size1(); ++i) {
              std::cout<<row(L, i)<<std::endl;
          }
          std::cout<<"iL:_\n";
          for(Size i=0; i<iL.size1(); ++i) {
              std::cout<<row(iL, i)<<std::endl;
          }
      }

      test_suite* MatrixTest::suite() {
          test_suite* suite = BOOST_TEST_SUITE("Testing_Matrix");

          suite->add(BOOST_TEST_CASE(&MatrixTest::testCholesky));
          suite->add(BOOST_TEST_CASE(
              &MatrixTest::testTriangularInverse));

          return suite;
      }
```

B.10.8 Test of Schema

Listing B.138: The header file of class SchemaTest.

```
// test-suite/nnmap.hpp
#ifndef TEST_NNMAP_HPP
#define TEST_NNMAP_HPP

#include<boost/test/unit_test.hpp>

class nnMapTest {
public:
    static void testiirMapA();
    static void testiiiMapB();
    static boost::unit_test_framework::test_suite* suite();
};

#endif
```

Listing B.139: The source file of class SchemaTest.

```
// test-suite/nnmap.cpp
#include"nnmap.hpp"
#include<cl/datasets/dataset.hpp>
#include<cl/utilities/datasetreader.hpp>
#include<cl/utilities/nnmap.hpp>
#include<cl/distances/euclideandistance.hpp>
#include<sstream>
#include<fstream>
#include<iostream>
```

```cpp
using namespace ClusLib;
using namespace boost::unit_test_framework;

void nnMapTest::testiirMapA() {
    BOOST_MESSAGE("Testing iirMapA");

    std::string fileName("../../Data/bezdekIris.data");
    DatasetReader reader(fileName);
    boost::shared_ptr<Dataset> ds;
    reader.fill(ds);

    EuclideanDistance ed;
    iirMapA dm;
    Size n = ds->size();
    for(Size i=0;i<n;++i){
        for(Size j=i+1;j<n;++j){
            dm.add_item(i,j,ed((*ds)[i],(*ds)[j]));
        }
    }

    std::stringstream ss;
    for(Size i=0;i<n;++i){
        for(Size j=i+1;j<n;++j){
            ss<<i+1<<","<<j+1<<","<<dm(j,i)<<std::endl;
        }
    }

    std::ofstream out("irisdm.csv");
    out<<ss.str();
    out.close();
}

void nnMapTest::testiiiMapB() {
    BOOST_MESSAGE("Testing iiiMapB");

    iiiMapB dm;
    Size n = 10;
    for(Size i=0;i<n;++i){
        for(Size j=0;j<n;++j){
            dm.add_item(i,j,i+j);
        }
    }

    std::stringstream ss;
    for(Size i=0;i<n;++i){
        for(Size j=0;j<n;++j){
            ss<<i+1<<","<<j+1<<","<<dm(j,i)<<std::endl;
        }
    }

    std::ofstream out("iirmapb.csv");
    out<<ss.str();
    out.close();
}

test_suite* nnMapTest::suite() {
    test_suite* suite = BOOST_TEST_SUITE("Testing nnMap");

    suite->add(BOOST_TEST_CASE(&nnMapTest::testiirMapA));
    suite->add(BOOST_TEST_CASE(&nnMapTest::testiiiMapB));

    return suite;
}
```

Appendix C

Software

C.1 An Introduction to Makefiles

In this appendix, we give a basic introduction to makefiles, which tell the program **make** what to do. For a detailed introduction to makefiles, readers are referred to Vaughan et al. (2010), Mecklenburg (2004), and Calcote (2010).

A *makefile* is a file that contains a set of rules used by **make** to build an application. The first rule in the file is the default rule. Usually, a rule consists of three parts: the target, its prerequisites, and one or more commands to execute. The format of a rule is:

```
target1 target2: prerequisite1 prerequisite2
    command1
    command2
```

The target is the file that we want to create or a label representing a command script. The prerequisites are files on which the target depends. The commands are shell commands used to create the target from those prerequisites. If the prerequisites do not exist, then **make** will refuse to work. Also note that a command must start with a tab character. The pound sign # is the comment character for **make**.

C.1.1 Rules

There are several kinds of rules in **make**: explicit rules, pattern rules, and suffix rules. Pattern rules and suffix rules are also called implicit rules. Explicit rules specify particular files as targets and prerequisites. Unlike explicit rules, pattern rules use wildcards rather than explicit filenames to specify targets and prerequisites. Suffix rules are built-in rules of **make**, which enable **make** to handle common tasks by knowing the file types.

When a target is a label rather than a file, we call it a phony target. For example, `clean` is a standard phony target:

```
.PHONY: clean
clean:
    rm -f *.o *~
```

461

The target .PHONY is a special target which tells make that a target is not a real file. Other examples of standard phony targets include all and install.

C.1.2 Variables

A makefile will define some variables. In general, a variable has the syntax:

$(variable-name)

For example, the command to compile a C++ program is contained in the variable COMPILE.cc.

The program make also defines a set of automatic variables which are set by make after a rule is matched. Table C.1 shows several very useful automatic variables. Each of these automatic variables has two variants by appending a "D" to the symbol or appending an "F" to the symbol. For example, $(@D) returns only the directory portion of the value; While $(@F) returns only the file portion of the value.

Variable	Description
$@	The filename of the target.
$%	The filename element of an archive member specification.
$<	The filename of the first prerequisite.
$?	The filenames of all prerequisites that are newer than the target, separated by black spaces.
$^	The filenames of all prerequisites with duplicate filenames removed.
$+	The filenames of all prerequisites with duplicates.

TABLE C.1: Some automatic variables in make.

Usually, the sources of an application are in a number of directories. For example, header files are in a directory named "include" and implementation files are in a directory named "src". Unless we direct it, make will not find these files. In this case, we can use VPATH and vpath to tell make to search for files in different directories. For example, to tell make to look in the directory "src", we can add a VPATH assignment to the makefile:

VPATH = src

The vpath feature is more sophisticated than the VPATH feature. The syntax of vpath is:

vpath pattern directory-list

For example, the following rules

vpath %.h include
vpath %.cpp src

tell **make** to search header files in the directory "include" and source files in "src".

We can create makefiles by hand. However, makefiles are usually created by GNU Autotools as makefiles created manually are error prone. In Section 5.3 we introduce how to use GNU Autotools to create makefiles automatically.

C.2 Installing Boost

Most of the Boost libraries are header-only libraries. Installing these header-only libraries is very simple. One can just download the package of the Boost libraries from Boost's website and uncompress the package to disk. To use these header-only libraries, one lets the program know where the Boost headers are.

Several Boost libraries are automatic linking libraries. For example, the program options library is an automatic linking library. These libraries require separate compilation. After the package of Boost libraries is downloaded and uncompressed, one has to compile these automatic linking libraries into binaries before using them.

C.2.1 Boost for Windows

To install Boost on Windows systems, we first download the Boost package for windows. The release version of Boost as of June 30, 2010 is 1.43.0. We download the Boost package `boost_1_43_0.zip` from `http://www.boost.org` and extract the zip file[1] to a local folder, for example, `C:\`. Then we see all Boost files in the folder `C:\boost_1_43_0\`.

Use the following steps to build the Boost program options library using Visual C++ and install it.

(a) Open a "Command Prompt" window and execute the command

   `cd C:\boost_1_43_0`

(b) Execute the command

   `bootstrap`

   to build `bjam`;

---

[1] You might need to unblock the zip file by right clicking the zip file, clicking "Properties" in the pop-up window, clicking the "Unblock" button in the "General" tab, and clicking the "Apply" button. Unblocking the zip file will speed the extracting process. The above process is for Windows XP. For other Windows systems, the steps might be different or not necessary.

(c) Execute the command

.\bjam --with-program_options install

to build and install the program options library.

The third step takes a while to complete. Once all the steps are finished, all the header-only libraries are also installed. These libraries are installed to the default locations, which is `C:\Boost\include` and `C:\Boost\lib`.

C.2.2 Boost for Cygwin or Linux

In Cygwin, one has the option to install Boost libraries and binaries using the Cygwin intaller. However, the Boost libraries and binaries included in Cygwin are usually older than the latest Boost release. In this section, we describe how to install the latest Boost libraries for Cygwin.

To install Boost libraries in Cygwin, download the Boost package for Unix. That is, download the file `boost_1_43_0.tar.gz`. Save the file to `C:\cygwin\usr\tmp` and follow the following steps to build and install the Boost program options library.

(a) Open a "Cygwin Bash Shell" window and execute the command

cd /usr/tmp;

(b) Execute the command

tar xvfz boost_1_43_0.tar.gz

to extract the files;

(c) Execute the command

cd boost_1_43_0

and

sh bootstrap.sh

to build `bjam`;

(d) Execute the command

./bjam --with-program_options install

to build and install the Boost program options library.

Once the above steps are completed, the header files will be installed to `/usr/local/include` and the libraries will be installed to `/usr/local/lib`. Installing Boost libraries in Linux such as Ubuntu is very similar to that in Cygwin. Note that you need to use `sudo` to install the Boost libraries in Ubuntu.

C.3 Installing Cygwin

Installing Cygwin on Windows systems is very easy. One can follow the following steps to install Cygwin.

(a) Download the setup.exe file from http://www.cygwin.com;

(b) Double click the setup program and click the "Next" button;

(c) Check "Install from Internet" and click "Next";

(d) Click "Next" in the "Choose Installation Directory" window;

(e) Click "Next" in the "Select Local Package Directory" window;

(f) Click "Next" in the "Select Connection Type" window;

(g) Choose a download site in the "Choose Download Site(s)" window and click "Next";

(h) Expand the "Devel" section and choose "autoconf", "automake", "gcc-g++", "gdb", "libboost", "libtool", and "make";

(i) Expand the "Editors" section and choose "vim" (or other editors you like);

(j) Click "Next";

(k) Click "Finish."

Note that some download sites have the latest Boost binaries and some download sites may have old Boost binaries.

C.4 Installing GMP

GMP is a free library for arbitrary precision arithmetic, operating on signed integers, rational numbers, and floating point numbers. In this appendix, we describe how to install GMP in Cygwin. One can use the same installation instruction to install GMP in UNIX-like platforms such as Ubuntu.

To install GMP, we first download the GMP library from the website http://gmplib.org/[2]. As of this writing, the latest version of GMP was gmp-5.0.1.tar.bz2. Save this file to a local folder such as C:\ and install the library as follows.

---

[2] If the link is out of date, one can find the link by searching "GNU Multiple Precision Arithmetic Library" in a search engine.

(a) Run "Cygwin Bash" and uncompress the file using the following command:

```
cd C:
tar xvfj gmp-5.0.1.tar.bz2
```

(b) Configure GMP for compilation:

```
cd gmp-5.0.1
./configure
```

(c) Build GMP:

```
make
```

(d) Install GMP:

```
make install
```

This will install the GMP library to the default locations in Cygwin. After installation, you can find the GMP header `gmp.h` at `/usr/local/include` and the libraries at `/usr/local/lib`.

Note that uncompressing the library in Cygwin bash is necessary. Uncompressing the file using some Windows program such as WinZip will corrupt the configuration files such as `configure` and `config.sub`.

C.5 Installing MPICH2 and Boost MPI

MPICH2 is an implementation of the messaging passing interface (MPI) standard. The source code can be downloaded from the website http://www.mcs.anl.gov/research/projects/mpich2/[3]. As of August 2010, the latest version of the package was `mpich2-1.2.1p1.tar.gz`. Once you download the package to local disk, you can use the following steps to install MPICH2 in Cygwin or Linux systems.

(a) Uncompress the file using the following command:

```
tar xvfz mpich2-1.2.1p1.tar.gz
```

(b) Configure MPICH2 for compilation:

---

[3] If the link is no longer available, one can search for MPICH using a search engine to find the latest link.

```
cd mpich2-1.2.1p1
./configure
```

(c) Build MPICH2:

```
make
```

(d) Install MPICH2:

```
make install
```

It takes a while to compile MPICH2. Once the above steps are completed, the header files will be installed to `/usr/local/include` and the libraries will be installed to `/usr/local/lib`.

Boost MPI library requires separate compilation. However, installing Boost MPI is a little bit different from installing Boost program options shown in Section C.2. To install Boost MPI, proceed as follows.

(a) Execute the command

```
tar xvfz boost_1_43_0.tar.gz
```

to extract the files;

(b) Execute the command

```
cd boost_1_43_0
```

and

```
sh bootstrap.sh
```

to build `bjam`;

(c) Add "`using mpi ;`" to the file `project-config.jam` (line 19) as shown in Listing C.1. The file is located in the root folder of the Boost package (i.e., `boost_1_43_0`).

(d) Execute the command

```
./bjam --with-program_options --with-mpi install
```

to build and install the Boost program options library and the Boost MPI library.

Listing C.1: The `project-config.jam` file.

```
# Boost.Build Configuration
# Automatically generated by bootstrap.sh

import option ;
import feature ;

# Compiler configuration. This definition will be used unless
# you already have defined some toolsets in your user-config.jam
# file.
if ! gcc in [ feature.values <toolset> ]
{
    using gcc ;
}

project : default-build <toolset>gcc ;

path-constant ICU_PATH : /usr ;

using mpi ;

# List of --with-<library> and --without-<library>
# options. If left empty, all libraries will be built.
# Options specified on the command line completely
# override this variable.
libraries = --with-program_options --with-mpi ;

# These settings are equivivalent to corresponding command-line
# options.
option.set prefix : /usr/local ;
option.set exec-prefix : /usr/local ;
option.set libdir : /usr/local/lib ;
option.set includedir : /usr/local/include ;
```

Once the above steps are completed, the header files will be installed to /usr/local/include and the libraries will be installed to /usr/local/lib. Note that the Boost serialization library is also installed.

Bibliography

Aggarwal, C., Wolf, J., Yu, P., Procopiuc, C., and Park, J. (1999). Fast algorithms for projected clustering. In *Proceedings of the 1999 ACM SIGMOD International Conference on Management of Data*, pages 61–72. ACM Press.

Aggarwal, C. and Yu, P. (2000). Finding generalized projected clusters in high dimensional spaces. In Chen, W., Naughton, J. F., and Bernstein, P. A., editors, *Proceedings of the 2000 ACM SIGMOD International Conference on Management of Data, May 16-18, 2000, Dallas, TX*, volume 29, pages 70–81. ACM.

Agrawal, R., Gehrke, J., Gunopulos, D., and Raghavan, P. (1998). Automatic subspace clustering of high dimensional data for data mining applications. In *SIGMOD Record ACM Special Interest Group on Management of Data*, pages 94–105, New York, NY, ACM Press.

Al-Sultan, K. and Fedjki, C. (1997). A tabu search-based algorithm for the fuzzy clustering problem. *Pattern Recognition*, 30(12):2023–2030.

Alsabti, K., Ranka, S., and Singh, V. (1998). An efficient k-means clustering algorithm. In *Proceedings of IPPS/SPDP Workshop on High Performance Data Mining*, Orlando, FL.

Anderberg, M. (1973). *Cluster Analysis for Applications*. Academic Press, New York.

Banfield, C. (1976). Statistical algorithms: Algorithm AS 102: Ultrametric distances for a single linkage dendrogram. *Applied Statistics*, 25(3):313–315.

Banfield, J. and Raftery, A. (1993). Model-based gaussian and non-gaussian clustering. *Biometrics*, 49(3):803–821.

Baraldi, A. and Blonda, P. (1999a). A survey of fuzzy clustering algorithms for pattern recognition. i. *IEEE Transactions on Systems, Man and Cybernetics, Part B*, 29(6):778–785.

Baraldi, A. and Blonda, P. (1999b). A survey of fuzzy clustering algorithms for pattern recognition. ii. *IEEE Transactions on Systems, Man and Cybernetics, Part B*, 29(6):786–801.

Basak, J. and Krishnapuram, R. (2005). Interpretable hierarchical clustering by constructing an unsupervised decision tree. *IEEE Transactions on Knowledge and Data Engineering*, 17(1):121–132.

Bellman, R., Kalaba, R., and Zadeh, L. (1966). Abstraction and pattern classification. *Journal of Mathematical Analysis and Applications*, 2:581–586.

Berkhin, P. (2002). Survey of clustering data mining techniques. Technical report, Accrue Software, San Jose, CA.

Berry, M. and Linoff, G. (2000). *Mastering Data Mining*. John Wiley & Sons, New York.

Beyer, K., Goldstein, J., Ramakrishnan, R., and Shaft, U. (1999). When is "nearest neighbor" meaningful? In Beeri, C. and Buneman, P., editors, *Proceedings of 7th International Conference on Database Theory*, volume 1540 of *Lecture Notes in Computer Science*, pages 217–235. Springer, New York.

Bezdek, J. (1974). *Fuzzy mathematics in pattern classification*. PhD thesis, Cornell University, Ithaca, NY.

Bezdek, J. (1981a). *Pattern Recognition with Fuzzy Objective Function Algorithms*. Plenum, New York.

Bezdek, J. (1981b). *Pattern Recognition with Fuzzy Objective Function Algorithms*. Kluwer Academic Publishers, Norwell, MA.

Bezdek, J., Hathaway, R., Sabin, M., and Tucker, W. (1992). Convergence theory for fuzzy c-means: Counterexamples and repairs. In Bezdek, J. and Pal, S., editors, *Fuzzy Models for Pattern Recognition: Methods that Search for Approximate Structures in Data*, pages 138–142. IEEE Press.

Bijnen, E. (1973). *Cluster Analysis: Survey and evaluation of techniques*. Tilburg University Press, The Netherlands.

Binder, D. (1978). Bayesian cluster analysis. *Biometrika*, 65(1):31–38.

Bobrowski, L. and Bezdek, J. (1991). c-means clustering with the l_1 and l_∞ norms. *IEEE Transactions on Systems, Man and Cybernetics*, 21(3):545–554.

Bock, H. (1989). Probabilistic aspects in cluster analysis. In Opitz, O., editor, *Conceptual and Numerical Analysis of Data*, pages 12–44, Augsburg, FRG. Springer-Verlag.

Bock, H. (1996). Probabilistic models in cluster analysis. *Computational Statistics and Data Analysis*, 23(1):5–28.

Booch, G., Maksimchuk, R., Engel, M., Young, B., Conallen, J., and Houston, K. (2007). *Object-Oriented Analysis and Design with Applications*. Addison-Wesley Professional, Upper Saddle River, NJ, 3rd edition.

Calcote, J. (2010). *Autotools: A Practical Guide To GNU Autoconf, Automake, and Libtool*. No Starch Press, San Francisco, CA.

Cameron, D., Rosenblatt, B., and Raymond, E. (1996). *Learning GNU Emacs*. O'Reilly & Associates, Inc., Sebastopol, CA, 2nd edition.

Campello, R., Hruschka, E., and Alves, V. (2009). On the efficiency of evolutionary fuzzy clustering. *Journal of Heuristics*, 15(1):43–75.

Cao, Y. and Wu, J. (2002). Projective ART for clustering data sets in high dimensional spaces. *Neural Networks*, 15(1):105–120.

Cao, Y. and Wu, J. (2004). Dynamics of projective adaptive resonance theory model: The foundation of PART algorithm. *IEEE Transactions on Neural Networks*, 15(2):245–260.

Carmichael, J., George, J., and Julius, R. (1968). Finding natural clusters. *Systematic Zoology*, 17(2):144–150.

Casselman, B. (2004). *Mathematical Illustrations: A Manual of Geometry and PostScript*. Cambridge University Press, Cambridge.

Celeux, G. and Govaert, G. (1995). Gaussian parsimonious clustering models. *Pattern Recognition*, 28(5):781–793.

Chang, J. and Jin, D. (2002). A new cell-based clustering method for large, high-dimensional data in data mining applications. In *SAC '02: Proceedings of the 2002 ACM symposium on Applied computing*, pages 503–507. ACM Press, New York.

Chaturvedi, A., Green, P., and Carroll, J. (2001). k-modes clustering. *Journal of Classification*, 18(1):35–55.

Cheng, C., Fu, A., and Zhang, Y. (1999). Entropy-based subspace clustering for mining numerical data. In *Proceedings of the Fifth ACM SIGKDD International Conference on Knowledge Discovery and Data Mining*, pages 84–93. ACM Press, New York.

Cheng, Y. (1995). Mean shift, mode seeking, and clustering. *IEEE Transactions on Pattern Analysis and Machine Intelligence*, 17(8):790–799.

Cherkassky, V. and Mulier, F. (1998). *Learning from Data: Concepts, Theory, and Methods*. John Wiley & Sons, New York, NY.

Clatworthy, J., Buick, D., Hankins, M., Weinman, J., and Horne, R. (2005). The use and reporting of cluster analysis in health psychology: A review. *British Journal of Health Psychology*, 10(3):329–358.

Comaniciu, D. and Meer, P. (1999). Mean shift analysis and applications. In *The Proceedings of the Seventh IEEE International Conference on ComputerVision*, volume 2, pages 1197–1203. IEEE.

Comaniciu, D. and Meer, P. (2002). Mean shift: A robust approach toward feature space analysis. *IEEE Transactions on Pattern Analysis and Machine Intelligence*, 24(5):603–619.

Cormack, R. (1971). A review of classification. *Journal of the Royal Statistical Society. Series A (General)*, 134(3):321–367.

Cuesta-Albertos, J., Gordaliza, A., and Matrán, C. (1997). Trimmed k-means: An attempt to robustify quantizers. *The Annals of Statistics*, 25(2):553–576.

Dash, M., Liu, H., and Xu, X. (2001). '1+1>2': Merging distance and density based clustering. In *Proceedings of the Seventh International Conference on Database Systems for Advanced Applications, 2001*, pages 32 –39, Hong Kong, China.

Day, N. (1969). Estimating the components of a mixture of normal distributions. *Biometrika*, 56(3):463–474.

Deitel, P. and Deitel, H. (2006). *C++ how to program*. Prentice Hall Press, Upper Saddle River, NJ, 5th edition.

Deitel, P. and Deitel, H. (2009). *C++ for Programmers*. Prentice Hall Press, Upper Saddle River, NJ, 1st edition.

Dempster, A., Laird, N., and Rubin, D. (1977). Maximum likelihood from incomplete data via the EM algorithm. *Journal of the Royal Statistical Society. Series B (Methodological)*, 39(1):1–38.

Deng, Z., Choi, K.-S., Chung, F.-L., and Wang, S. (2010). Enhanced soft subspace clustering integrating within-cluster and between-cluster information. *Pattern Recognition*, 43(3):767–781.

Devroye, L. (1986). *Non-Uniform Random Variate Generation*. Springer-Verlag, New York.

Ding, C. and Zha, H. (2010). *Spectral Clustering, Ordering and Ranking*. Springer-Verlag, New York.

Döring, C., Lesot, M.-J., and Kruse, R. (2006). Data analysis with fuzzy clustering methods. *Computational Statistics and Data Analysis*, 51(1):192–214.

DuBien, J. and Warde, W. (1979). A mathematical comparison of the members of an infinite family of agglomerative clustering algorithms. *The Canadian Journal of Statistics*, 7:29–38.

Duda, R. and Hart, P. (1973). *Pattern Classification and Scene Analysis.* John Willey & Sons, New Yotk.

Duda, R., Hart, P., and Stork, D. (2001). *Pattern Classification.* John Wiley & Sons, New York, NY, 2nd edition.

Dunn, J. (1974a). A fuzzy relative of the ISODATA process and its use in detecting compact well-separated clusters. *Journal of Cybernetics,* 3(3):32–57.

Dunn, J. (1974b). Well separated clusters and optimal fuzzy partitions. *Journal of Cybernetics,* 4:95–104.

Duran, B. and Odell, P. (1974). *Cluster Analysis — A Survey,* volume 100 of *Lecture Notes in Economics and Mathematical Systems.* Springer-Verlag, Berlin, Heidelberg, New York.

Edwards, A. and Cavalli-Sforza, L. (1965). A method for cluster analysis. *Biometrics,* 21(2):362–375.

Eisen, M., Spellman, P., Brown, P., and Botstein, D. (1998). Cluster analysis and display of genome-wide expression patterns. *Proceedings of the National Academy of Sciences of the United States of America,* 95(25):14863–14868.

El-Sonbaty, Y., Ismail, M., and Farouk, M. (2004). An efficient density based clustering algorithm for large databases. In *16th IEEE International Conference on Tools with Artificial Intelligence, 2004. ICTAI 2004,* pages 673–677.

Ester, M., Kriegel, H., Sander, J., and Xu, X. (1996). A density-based algorithm for discovering clusters in large spatial databases with noise. In Simoudis, E., Han, J., and Fayyad, U., editors, *Second International Conference on Knowledge Discovery and Data Mining,* pages 226–231, Portland, Oregon. AAAI Press.

Everitt, B. (1993). *Cluster Analysis.* Halsted Press, New York, Toronto, 3rd edition.

Everitt, B., Landau, S., and Leese, M. (2001). *Cluster Analysis.* Oxford University Press, New York, 4th edition.

Faber, V. (1994). Clustering and the continuous k-means algorithm. *Los Alamos Science,* 22:138–144.

Filippone, M., Camastra, F., Masulli, F., and Rovetta, S. (2008). A survey of kernel and spectral methods for clustering. *Pattern Recognition,* 41(1):176–190.

Fisher, R. (1936). The use of multiple measurements in taxonomic problems. *Annual Eugenics,* 7 (Part II):179–188.

Foggia, P., Percannella, G., Sansone, C., and Vento, M. (2007). A graph-based clustering method and its applications. In *Proceedings of the 2nd International Conference on Advances in Brain, Vision and Artificial Intelligence*, pages 277–287, Berlin, Heidelberg. Springer-Verlag.

Foggia, P., Percannella, G., Sansone, C., and Vento, M. (2009). Benchmarking graph-based clustering algorithms. *Image and Vision Computing*, 27(7):979–988.

Fraley, C. and Raftery, A. (1998). How many clusters? Which clustering method? Answers via model-based cluster analysis. *The Computer Journal*, 41(8):578–588.

Frank, A. and Asuncion, A. (2010). UCI machine learning repository.

Fukunaga, K. (1990). *Introduction to Statistical Pattern Recognition*. Computer Science and Scientific Computing. Academic Press, Inc., San Diego, CA, 2nd edition.

Fukunaga, K. and Hostetler, L. (1975). The estimation of the gradient of a density function, with applications in pattern recognition. *IEEE Transactions on Information Theory*, 21(1):32–40.

Fukushima, K. (1975). Cognitron: A self-organizing multi-layered neural network. *Biological Cybernetics*, 20:121–136.

Fukuyama, Y. and Sugeno, M. (1989). A new method of choosing the number of clusters for the fuzzy c-means method. In *Proceedings of 5th Fuzzy Syst. Symp.*, pages 247–250.

Gaber, M., Zaslavsky, A., and Krishnaswamy, S. (2005). Mining data streams: A review. *ACM SIGMOD Record*, 34(2):18–26.

Gamma, E., Helm, R., Johnson, R., and Vlissides, J. (1994). *Design Patterns: Elements of Reusable Object-Oriented Software*. Addison-Wesley Professional, Upper Saddle River, NJ.

Gan, G. (2007). Subspace Clustering Based on Fuzzy Models and Mean Shifts. PhD thesis, Department of Mathematics and Statistics, York University, Toronto, ON.

Gan, G., Ma, C., and Wu, J. (2007). *Data Clustering: Theory, Algorithms, and Applications*, volume 20 of *ASA-SIAM Series on Statistics and Applied Probability*. SIAM Press, SIAM, Philadelphia, ASA, Alexandria, VA.

Gan, G. and Wu, J. (2004). Subspace clustering for high dimensional categorical data. *ACM SIGKDD Explorations Newsletter*, 6(2):87–94.

Gan, G. and Wu, J. (2008). A convergence theorem for the fuzzy subspace clustering (fsc) algorithm. *Pattern Recognition*, 41(6):1939–1947.

Gan, G., Wu, J., and Yang, Z. (2006a). A fuzzy subspace algorithm for clustering high dimensional data. In Li, X., Zaiane, O., and Li, Z., editors, *Lecture Notes in Artificial Intelligence*, volume 4093, pages 271–278. Springer.

Gan, G., Wu, J., and Yang, Z. (2006b). PARTCAT: A subspace clustering algorithm for high dimensional categorical data. In *IJCNN '06. International Joint Conference on Neural Networks*, pages 4406–4412. IEEE.

Gan, G., Yang, Z., and Wu, J. (2005). A genetic k-modes algorithm for clustering categorical data. In Li, X., Wang, S., and Dong, Z., editors, *Proceedings on Advanced Data Mining and Applications: First International Conference, ADMA 2005, Wuhan, China*, volume 3584 of *Lecture Notes in Artificial Intelligence*, pages 195–202. Springer-Verlag GmbH.

Ganti, V., Gehrke, J., and Ramakrishnan, R. (1999). CACTUS: Clustering categorical data using summaries. In Chaudhuri, S. and Madigan, D., editors, *Proceedings of the Fifth ACM SIGKDD International Conference on Knowledge Discovery and Data Mining*, pages 73–83, N.Y. ACM Press.

Gath, I. and Geva, A. (1989). Unsupervised optimal fuzzy clustering. *IEEE Transactions on Pattern Analysis and Machine Intelligence*, 11(7):773–780.

Gibson, D., Kleinberg, J., and Raghavan, P. (2000). Clustering categorical data: An approach based on dynamical systems. *The VLDB Journal*, 8(3-4):222–236.

Glover, F., Taillard, E., and de Werra, D. (1993). A user's guide to tabu search. *Annals of Operations Research*, 41:3–28.

Goil, S., Nagesh, H., and Choudhary, A. (1999). MAFIA: Efficient and scalable subspace clustering for very large datasets. Technical Report CPDC-TR-9906-010, Center for Parallel and Distributed Computing, Department of Electrical & Computer Engineering, Northwestern University.

Gordon, A. (1987). A review of hierarchical classification. *Journal of the Royal Statistical Society. Series A (General)*, 150(2):119–137.

Gordon, A. (1996). Hierarchical classification. In Arabie, P., Hubert, L., and Soete, G., editors, *Clustering and Classification*, pages 65–121, River Edge, NJ, USA. World Scientific.

Gordon, A. (1999). *Classification*. Chapman & Hall/CRC, 2nd edition.

Gower, J. (1971). A general coefficient of similarity and some of its properties. *Biometrics*, 27(4):857–874.

Gower, J. and Ross, G. (1969). Minimum spanning trees and single linkage cluster analysis. *Applied Statistics*, 18(1):54–64.

Grabusts, P. and Borisov, A. (2002). Using grid-clustering methods in data classification. In *2002. PARELEC '02. Proceedings. International Conference on Parallel Computing in Electrical Engineering*, pages 425–426, Latvia. IEEE.

Gropp, W., Lusk, E., and Skjellum, A. (1999). *Using MPI: Portable parallel programming with the message-passing interface*. MIT Press, Cambridge, MA, 2nd edition.

Grossberg, S. (1976a). Adaptive pattern classification and universal recoding: I. Parallel development and coding of neural feature detectors. *Biological Cybernetics*, 23:121–134.

Grossberg, S. (1976b). Adaptive pattern recognition and universal encoding ii: Feedback, expectation, olfaction, and illusions. *Biological Cybernetics*, 23:187–202.

Guha, S., Rastogi, R., and Shim, K. (1998). CURE: An efficient clustering algorithm for large databases. In *Proceedings of the 1998 ACM SIGMOD International Conference on Management of Data*, pages 73–84. ACM Press.

Guha, S., Rastogi, R., and Shim, K. (2000). ROCK: A robust clustering algorithm for categorical attributes. *Information Systems*, 25(5):345–366.

Gürsoy, A. (2004). Data decomposition for parallel k-means clustering. In *Parallel Processing and Applied Mathematics*, volume 3019 of *Lecture Notes in Computer Science*, pages 241–248. Springer Berlin/Heidelberg.

Gürsoy, A. and Cengiz, I. (2001). Parallel pruning for k-means clustering on shared memory architectures. In *Euro-Par 2001 Parallel Processing*, volume 2150 of *Lecture Notes in Computer Science*, pages 321–325. Springer Berlin/Heidelberg.

Hai, N. and Susumu, H. (2005). Performances of parallel clustering algorithm for categorical and mixed data. In *Parallel and Distributed Computing: Applications and Technologies*, volume 3320 of *Lecture Notes in Computer Science*, pages 53–89. Springer Berlin/Heidelberg.

Halkidi, M., Batistakis, Y., and Vazirgiannis, M. (2002a). Cluster validity methods: part I. *ACM SIGMOD Record*, 31(2).

Halkidi, M., Batistakis, Y., and Vazirgiannis, M. (2002b). Clustering validity checking methods: part II. *ACM SIGMOD Record*, 31(3).

Hartigan, J. (1975). *Clustering Algorithms*. John Wiley & Sons, Toronto.

Hartuv, E. and Shamir, R. (2000). A clustering algorithm based on graph connectivity. *Information Processing Letters*, 76(4–6):175–181.

Hathaway, R. and Bezdek, J. (1984). Local convergence of the fuzzy c-means algorithms. *Pattern Recogniztion*, 19(6):477–480.

Hathaway, R., Bezdek, J., and Tucker, W. (1987). An improved covergence theorem for the fuzzy c-means clustering algorithms. In Bezdek, J., editor, *Analysis of Fuzzy Information*, volume III, pages 123–131. CRC Press, Inc.

Holland, J. (1975). *Adaptation in Natural and Artificial Systems*. University of Michigan Press, Ann Arbor, MI.

Holman, E. (1992). Statistical properties of large published classifications. *Journal of Classification*, 9(2):187–210.

Holzner, S. (2006). *Design Patterns For Dummies*. John Wiley & Sons, Inc., New York, NY.

Höppner, F., Klawonn, F., Kruse, R., and Runkler, T. (1999). *Fuzzy Cluster Analysis: Methods for Classification, Data Analysis and Image Recognition*. Wiley, New York.

Hua, K., Lang, S., and Lee, W. (1994). A decomposition-based simulated annealing technique for data clustering. In ACM, editor, *Proceedings of the Thirteenth ACM SIGACT-SIGMOD-SIGART Symposium on Principles of Database Systems, May 24–26, 1994, Minneapolis, MN*, volume 13, pages 117–128, New York, NY. ACM Press.

Huang, Z. (1997a). Clustering large data sets with mixed numeric and categorical values. In *Knowledge Discovery and Data Mining: Techniques and Applications*. World Scientific, Singapore.

Huang, Z. (1997b). A fast clustering algorithm to cluster very large categorical data sets in data mining. In *SIGMOD Workshop on Research Issues on Data Mining and Knowledge Discovery*, Tucson, AZ.

Huang, Z. (1998). Extensions to the k-means algorithm for clustering large data sets with categorical values. *Data Mining and Knowledge Discovery*, 2(3):283–304.

Huang, Z. and Ng, M. (1999). A fuzzy k-modes algorithm for clustering categorical data. *IEEE Transactions on Fuzzy Systems*, 7(4):446–452.

Ichino, M. (1988). General metrics for mixed features—the Cartesian space theory for pattern recognition. In *Proceedings of the 1988 IEEE International Conference on Systems, Man, and Cybernetics*, volume 1, pages 494–497.

Ichino, M. and Yaguchi, H. (1994). Generalized Minkowski metrics for mixed feature-type data analysis. *IEEE Transactions on Systems, Man and Cybernetics*, 24(4):698–708.

Jain, A. (2010). Data clustering: 50 years beyond k-means. *Pattern Recognition Letters*, 31(8):651–666.

Jain, A. and Dubes, R. (1988). *Algorithms for Clustering Data*. Prentice Hall, Englewood Cliffs, New Jersey.

Jain, A., Duin, R., and Mao, J. (2000). Statistical pattern recognition: A review. *IEEE Transactions on Pattern Analysis and Machine Intelligence*, 22(1):4–37.

Jain, A., Murty, M., and Flynn, P. (1999). Data clustering: A review. *ACM Computing Surveys*, 31(3):264–323.

Jambu, M. (1978). *Classification automatique pour l'analyse de données*. Dunod, Paris.

Jardine, C., Jardine, N., and Sibson, R. (1967). The structure and construction of taxonomic hierarchies. *Mathematical Biosciences*, 1(2):173–179.

Jiang, D., Tang, C., and Zhang, A. (2004). Cluster analysis for gene expression data: A survey. *IEEE Transactions on Knowledge and Data Engineering*, 16(11):1370–1386.

Jiménez, D. and Vidal, V. (2005). Parallel implementation of information retrieval clustering models. In *High Performance Computing for Computational Science — VECPAR 2004*, volume 3402 of *Lecture Notes in Computer Science*, pages 129–141. Springer Berlin/Heidelberg.

Jing, L., Ng, M., and Huang, J. (2007). An entropy weighting k-means algorithm for subspace clustering of high-dimensional sparse data. *IEEE Transactions on Knowledge and Data Engineering*, 19(8):1026–1041.

Johnson, S. (1967). Hierarchical clustering schemes. *Psychometrika*, 32(3):241–254.

Judd, D., McKinley, P., and Jain, A. (1998). Large-scale parallel data clustering. *IEEE Transactions on Pattern Analysis and Machine Intelligence*, 20(8):871–876.

Karlsson, B. (2005). *Beyond the C++ Standard Library: An Introduction to Boost*. Addison-Wesley Professional, Upper Saddle River, NJ.

Karypis, G., Han, E., and Kumar, V. (1999). Chameleon: Hierarchical clustering using dynamic modeling. *Computer*, 32(8):68–75.

Kaufman, L. and Rousseeuw, P. (1990). *Finding Groups in Data—An Introduction to Cluster Analysis*. Wiley series in probability and mathematical statistics. John Wiley & Sons, Inc., New York.

Keim, D. and Hinneburg, A. (1999). Optimal grid-clustering: Towards breaking the curse of dimensionality in high-dimensional clustering. In *Proceedings of the 25th International Conference on Very Large Data Bases (VLDB '99)*, pages 506–517, San Francisco. Morgan Kaufmann.

Kohavi, R. (1995). A study of cross-validation and bootstrap for accuracy estimation and model selection. In *Proceedings of the 14th International Joint Conference on Artificial Intelligence*, pages 1137–1143, San Francisco, CA. Morgan Kaufmann Publishers Inc.

Kohavi, R., Sommerfield, D., and Dougherty, J. (1998). Data mining using MLC++: A machine learning library in C++. *International Journal of Artificial Intelligence Tools*, 6(4):537–566.

Kriegel, H.-P., Kröger, P., and Zimek, A. (2009). Clustering high-dimensional data: A survey on subspace clustering, pattern-based clustering, and correlation clustering. *ACM Transactions on Knowledge Discovery from Data (TKDD)*, 3(1):1–58.

Krishna, K. and Narasimha, M. (1999). Genetic k-means algorithm. *IEEE Transactions on Systems, Man and Cybernetics, Part B*, 29(3):433–439.

Kwok, T., Smith, K., Lozano, S., and Taniar, D. (2002). Parallel fuzzy c-means clustering for large data sets. In *Euro-Par 2002 Parallel Processing*, volume 2400 of *Lecture Notes in Computer Science*, pages 27–58. Springer, Berlin/Heidelberg.

Lamb, L. and Robbins, A. (1998). *Learning the vi Editor*. O'Reilly & Associates, Inc., Sebastopol, CA, 6th edition.

Lance, G. and Williams, W. (1967a). A general theory of classificatory sorting strategies I. Hierarchical systems. *The Computer Journal*, 9(4):373–380.

Lance, G. and Williams, W. (1967b). A general theory of classificatory sorting strategies II. Clustering systems. *The Computer Journal*, 10(3):271–277.

Lasater, C. (2007). *Design Patterns*. Wordware Publishing, Inc., Plano, TX.

Legendre, L. and Legendre, P. (1983). *Numerical Ecology*. Elsevier Scientific, New York.

Leung, Y., Zhang, J.-S., and Xu, Z.-B. (2000). Clustering by scale-space filtering. *IEEE Transactions on Pattern Analysis and Machine Intelligence*, 22:1396–1410.

Li, C. and Biswas, G. (2002). Unsupervised learning with mixed numeric and nominal data. *IEEE Transactions on Knowledge and Data Engineering*, 14:673–690.

Likas, A. and Verbeek, N. V. J. (2003). The global k-means clustering algorithm. *Pattern Recognition*, 36(2):45–461.

Lin, N., Chang, C., Chueh, H.-E., Chen, H.-J., and Hao, W.-H. (2008). A deflected grid-based algorithm for clustering analysis. *WSEAS Transactions on Computers*, 7(4):125–132.

Lippman, S., Lajoie, J., and Moo, B. (2005). *C++ Primer*. Addison-Wesley Professional, Upper Saddle River, NJ, 4th edition.

Liu, B., Xia, Y., and Yu, P. (2000). Clustering through decision tree construction. In *Proceedings of the Ninth International Conference on Information and Knowledge Management*, pages 20–29, McLean, VA. ACM Press.

Lorr, M. (1983). *Cluster Analysis for Social Scientists*. The Jossey-Bass Social and Behavioral Science Series. Jossey-Bass, San Francisco, Washington, London.

Luxburg, U. (2007). A tutorial on spectral clustering. *Statistics and Computing*, 17(4):395–416.

Macnaughton-Smith, P., Williams, W., Dale, M., and Mockett, L. (1964). Dissimilarity analysis: A new technique of hierarchical sub-division. *Nature*, 202:1034–1035.

Macqueen, J. (1967). Some methods for classification and analysis of multivariate observations. In Cam, L. and Neyman, J., editors, *Proceedings of the 5th Berkeley Symposium on Mathematical Statistics and Probability*, volume 1, pages 281–297, Berkely, CA. University of California Press.

Maksimchuk, R. and Naiburg, E. (2005). *UML for Mere Mortals*. Addison-Wesley Professional, Boston, MA.

Mao, J. and Jain, A. (1996). A self-organizing network for hyperellipsoidal clustering (hec). *IEEE Transactions on Neural Networks*, 7(1):16–29.

Martinez, W. and Martinez, A. (2005). *Exploratory Data Analysis with MATLAB*. Computer Science and Data Analysis. Chapman & Hall/CRC, Boca Raton, FL.

McLachlan, G. and Krishnan, T. (1997). *The EM Algorithm and Extensions*. Wiley, NY.

Mecklenburg, R. (2004). *Managing Projects with GNU Make*. O'Reilly Media, Inc., 3rd edition.

Meng, X. and van Dyk, D. (1997). The EM algorithm—An old folk-song sung to a fast new tune. *Journal of the Royal Statistical Society. Series B (Methodological)*, 59(3):511–567.

Meyers, S. (1997). *Effective C++: 50 Specific Ways to Improve Your Programs and Design*. Addison-Wesley Professional, Upper Saddle River, NJ, 2nd edition.

Michaud, P. (1997). Clustering techniques. *Future Generation Computer Systems*, 13(2-3):135–147.

Milligan, G. (1979). Ultrametric hierarchical clustering algorithms. *Psychometrika*, 44:343–346.

Mirkin, B. (2005). *Clustering for Data Mining: A Data Recovery Approach*. Computer Science and Data Analysis Series. Chapman & Hall/CRC, Boca Raton, FL.

Miyamoto, S., Ichihashi, H., and Honda, K. (2008). *Algorithms for Fuzzy Clustering: Methods in c-Means Clustering with Applications*. Springer-Verlag, Berlin, Heidelberg.

Mladenović, N. and Hansen, P. (1997). Variable neighborhood search. *Computers and Operations Research*, 24(11):1097–1100.

Morrison, D. (1967). Measurement problems in cluster analysis. *Management Science (Series B, Managerial)*, 13(12):B775–B780.

Müller, E., Günnemann, S., Assent, I., and Seidl, T. (2009). Evaluating clustering in subspace projections of high dimensional data. *Proceedings of the VLDB Endowment*, 2(1):1270–1281.

Murtagh, F. (1983). A survey of recent advances in hierarchical clustering algorithms. *The Computer Journal*, 26(4):354–359.

Murtagh, F. (1984). Counting dendrograms: A survey. *Discrete Applied Mathematics*, 7(2):191–199.

Nagesh, H., Goil, S., and Choudhary, A. (2001). Adaptive grids for clustering massive data sets. In *First SIAM International Conference on Data Mining*, Chicago, IL.

Ng, M. and Wong, J. (2002). Clustering categorical data sets using tabu search techniques. *Pattern Recognition*, 35(12):2783–2790.

Orlóci, L. (1967). An agglomerative method for classification of plant communities. *Journal of Ecology*, 55:193–205.

Osherove, R. (2009). *The Art of Unit Testing: With Examples in .Net*. Manning Publications, Greenwich, CT.

Othman, F., Abdullah, R., Rashid, N., and Salam, R. (2005). Parallel k-means clustering algorithm on DNA dataset. In *Parallel and Distributed Computing: Applications and Technologies*, volume 3320 of *Lecture Notes in Computer Science*, pages 1–34. Springer Berlin/Heidelberg.

Park, N. and Lee, W. (2007). Grid-based subspace clustering over data streams. In *Proceedings of the Sixteenth ACM Conference on Information and Knowledge Management*, pages 801–810, New York, NY, USA. ACM.

Parsons, L., Haque, E., and Liu, H. (2004). Subspace clustering for high dimensional data: A review. *SIGKDD, Newsletter of the ACM Special Interest Group on Knowledge Discovery and Data Mining*, 6(1):90–105.

Patrikainen, A. and Meila, M. (2006). Comparing subspace clusterings. *IEEE Transactions on Knowledge and Data Engineering*, 18(7):902–916.

Pelleg, D. and Moore, A. (1999). Accelerating exact k-means algorithms with geometric reasoning. In *Proceedings of the Fifth ACM SIGKDD International Conference on Knowledge Discovery and Data Mining*, pages 277–281, San Diego, California, United States. ACM Press.

Phillips, S. (2002). Acceleration of k-means and related clustering algorithms. In Mount, D. and Stein, C., editors, *ALENEX: International Workshop on Algorithm Engineering and Experimentation, LNCS*, volume 2409, pages 166–177, San Francicsco, CA. Springer-Verlag Heidelberg.

Press, W., Flannery, B., Teukolsky, S., and Vetterling, W. (1992). *Numerical Recipes in C: The Art of Scientific Computing*. Cambridge University Press, Cambridge, 2nd edition.

Procopiuc, C., Jones, M., Agarwal, P., and Murali, T. (2002). A Monte Carlo algorithm for fast projective clustering. In *Proceedings of the 2002 ACM SIGMOD International Conference on Management of Data*, pages 418–427. ACM Press.

Qiu, B.-Z., Li, X.-L., and Shen, J.-Y. (2007). Grid-based clustering algorithm based on intersecting partition and density estimation. In *Proceedings of the 2007 International Conference on Emerging Technologies in Knowledge Discovery and Data Mining*, pages 368–377, Berlin, Heidelberg. Springer-Verlag.

Rohlf, F. (1974). Algorithm 81: Dendrogram plot. *The Computer Journal*, 17(1):89–91.

Rose, K., Gurewitz, E., and Fox, G. (1990). Statistical mechanics and phase transitions in clustering. *Physical Review Letters*, 65(8):945–948.

Rumelhart, D. and Zipser, D. (1986). Feature discovery by competitive learning. In *Parallel Distributed Processing: Explorations in the Microstructure of Cognition, Vol. 1: Foundations*, pages 151–193, Cambridge, MA. MIT Press.

Rummel, R. (1970). *Applied Factor Analysis*. Northwestern University Press, Evanston, IL.

Ruspini, E. (1969). A new approach to clustering. *Information and Control*, 15:22–32.

Salton, G. and McGill, M. (1983). *Introduction to Modern Information Retrieval*. McGraw-Hill, New York, Tokyo.

Sander, J., Ester, M., Kriegel, H., and Xu, X. (1998). Density-based clustering in spatial databases: The algorithm gdbscan and its applications. *Data Mining and Knowledge Discovery*, 2(2):169–194.

Schikuta, E. (1996). Grid-clustering: A efficient hierarchical clustering method for very large data sets. In *1996, Proceedings of the 13th International Conference on Pattern Recognition*, volume 2, pages 101–105, Vienna, Austria. IEEE.

Schikuta, E. and Erhart, M. (1997). The BANG-clustering system: Grid-based data analysis. In Liu, X., Cohen, P., and Berthold, M., editors, *Lecture Notes in Computer Science*, volume 1280, pages 513–524, Berlin/Heidelberg. Springer-Verlag.

Scoltock, J. (1982). A survey of the literature of cluster analysis. *The Computer Journal*, 25(1):130–134.

Scott, A. and Symons, M. (1971). Clustering methods based on likelihood ratio criteria. *Biometrics*, 27(2):387–397.

Selim, S. and Ismail, M. (1984). k-means-type algorithms: A generalized convergence theorem and characterization of local optimality. *IEEE Transactions on Pattern Analysis and Machine Intelligence*, 6(1):81–87.

Selim, S. and Ismail, M. (1986). Fuzzy c-means: Optimality of solutions and effective termination of the algorithm. *Pattern Recognition*, 19(6):651–663.

Shalloway, A. and Trott, J. (2001). *Design Patterns Explained: A New Perspective on Object-Oriented Design*. Addison-Wesley Professional, Upper Saddle River, NJ.

Sheikholeslami, G., Chatterjee, S., and Zhang, A. (2000). WaveCluster: a wavelet-based clustering approach for spatial data in very large databases. *The VLDB Journal*, 8(3–4):289–304.

Sibson, R. (1973). SLINK: An optimally efficient algorithm for the single link cluster method. *The Computer Journal*, 16(1):30–34.

Sintes, T. (2001). *Sams Teach Yourself Object Oriented Programming in 21 Days*. Sams, Indianapolis, Indiana, 2nd edition.

Sokal, R. and Sneath, P. (1963). *Principles of Numerical Taxonomy*. W.H. Freeman, San Francisco, CA.

Sokal, R. and Sneath, P. (1973). *Numerical Taxonomy: The Principles and Practice of Numerical Classification*. W.H. Freeman, San Francisco, CA.

Song, Y., Chen, W., Bai, H., Lin, C., and Chang, E. (2008). Parallel spectral clustering. In *Machine Learning and Knowledge Discovery in Databases*, volume 5212 of *Lecture Notes in Computer Science*, pages 374–389. Springer Berlin/Heidelberg.

Stroustrup, B. (1994). *The Design and Evolution of C++*. ACM Press/Addison-Wesley Publishing Co., New York, NY.

Teboulle, M. (2007). A unified continuous optimization framework for center-based clustering methods. *The Journal of Machine Learning Research*, 8:65–102.

Theodoridis, S. and Koutroubas, K. (1999). *Pattern Recognition*. Academic Press, London.

Valente de Oliveira, J. and Pedrycz, W. (2007). *Advances in Fuzzy Clustering and its Applications*. John Wiley & Sons, Inc., New York, NY.

van Rijsbergen, C. (1970). Algorithm 47: A clustering algorithm. *The Computer Journal*, 13(1):113–115.

Vanjak, Z. and Mornar, V. (2001). General object-oriented framework for iterative optimization algorithms. *Journal of Computing and Information Technology*, 9(3).

Vaughan, G., Elliston, B., Tromey, T., and Taylor, I. (2010). *GNU Autoconf, Automake, and Libtool*. Pearson Education, Upper Saddle River, NJ.

Wang, L. and Wang, Z. (2003). CUBN: A clustering algorithm based on density and distance. In *2003 International Conference on Machine Learning and Cybernetics*, pages 108–112.

Wang, W., Yang, J., and Muntz, R. (1997). STING: A statistical information grid approach to spatial data mining. In Jarke, M., Carey, M., Dittrich, K., Lochovsky, F., and Jeusfeld, P. L. M., editors, *Twenty-Third International Conference on Very Large Data Bases*, pages 186–195, Athens, Greece. Morgan Kaufmann.

Ward, Jr., J. (1963). Hierarchical grouping to optimize an objective function. *Journal of the American Statistical Association*, 58(301):236–244.

Ward, Jr., J. and Hook, M. (1963). Application of an hierarchical grouping procedure to a problem of grouping profiles. *Educational and Psychological Measurement*, 23(1):69–81.

Willett, P. (1988). Recent trends in hierarchical document clustering: A criticial review. *Information Processing and Management*, 24(5):577–597.

Wishart, D. (2002). k-means clustering with outlier detection, mixed variables and missing values. In Schwaiger, M. and Opitz, O., editors, *Exploratory Data Analysis in Empirical Research*, pages 216–226. Springer, New York.

Wolfe, J. (1970). Pattern clustering by multivariate mixture analysis. *Multivariate Behavioral Research*, 5:329–350.

Woo, K. and Lee, J. (2002). *FINDIT: A fast and intelligent subspace clustering* algorithm using dimension voting. PhD thesis, Korea Advanced Institute of Science and Technology, Department of Electrical Engineering and Computer Science.

Xu, R. and Wunsch II, D. (2005). Survey of clustering algorithms. *IEEE Transactions on Neural Networks*, 16(3):645–678.

Xu, R. and Wunsch, II, D. (2009). *Clustering*. Wiley-IEEE Press, Hoboken, New Jersey.

Xu, X., Jäger, J., and Kriegel, H. (1999). A fast parallel clustering algorithm for large spatial databases. *Data Mining and Knowledge Discovery*, 3(3):263–290.

Yang, J., Wang, W., Wang, H., and Yu, P. (2002). δ-clusters: Capturing subspace correlation in a large data set. *Proceedings. 18th International Conference on Data Engineering*, pages 517–528.

Yang, M. (1993). A survey of fuzzy clustering. *Mathematical and Computer Modelling*, 18(11):1–16.

Yeung, K., Medvedovic, M., and Bumgarner, R. (2003). Clustering gene-expression data with repeated measurements. *Genome Biology*, 4(5):G34.1–17.

Zadeh, L. (1965). Fuzzy sets. *Information and Control*, 8:338–353.

Zaiane, O. and Lee, C. (2002). Clustering spatial data in the presence of obstacles: A density-based approach. In *Proceedings. International Database Engineering and Applications Symposium, 2002*, pages 214–223.

Zhang, B. (2003). Comparison of the performance of center-based clustering algorithms. In *Proceedings of the 7th Pacific-Asia conference on Advances in Knowledge Discovery and Data Mining*, pages 63–74, Berlin, Heidelberg. Springer-Verlag.

Zhang, B., Hsu, M., and Dayal, U. (2001). k-harmonic means — A spatial clustering algorithm with boosting. In *Proceedings of the First International Workshop on Temporal, Spatial, and Spatio-Temporal Data Mining—Revised Papers*, volume 2007 of *Lecture Notes in Computer Science*, pages 31–45, London, UK. Springer-Verlag.

Zhang, B. and Srihari, S. (2003). Properties of binary vector dissimilarity measures. Technical report, CEDAR, Department of Computer Science & Engineering, University at Buffalo, State University of New York. http://www.cedar.buffalo.edu/papers/publications.html.

Zhang, T., Ramakrishnan, R., and Livny, M. (1996). BIRCH: An efficient data clustering method for very large databases. In *Proceedings of the 1996 ACM SIGMOD International Conference on Management of Data*, pages 103–114. ACM Press.

Zhang, Y., Wang, W., Zhang, X., and Li, Y. (2008). A cluster validity index for fuzzy clustering. *Information Sciences*, 178(4):1205–1218.

Zhao, W., Ma, H., and He, Q. (2009). Parallel k-means clustering based on MapReduce. In *Proceedings of the 1st International Conference on Cloud Computing*, pages 674–679, Berlin, Heidelberg. Springer-Verlag.

Zhao, Y. and Song, J. (2001). GDILC: A grid-based density-isoline clustering algorithm. In *International Conferences on Info-tech and Info-net, 2001. Proceedings. ICII 2001*, volume 3, pages 140–145, Beijing, China. IEEE.

Zhong, S. and Ghosh, J. (2003). A unified framework for model-based clustering. *The Journal of Machine Learning Research*, 4:1001–1037.

Author Index

Abdullah, R., 310
Agarwal, P.K., 22
Aggarwal, C.C., 22, 147
Agrawal, R., 22
Alsabti, K., 320
Al-Sultan, K.S., 19
Alves, V.S., 23
Anderberg, M.R., 8, 139
Assent, I., 22
Asuncion, A., 258

Bai, H., 321
Banfield, C.F., 15
Banfield, J.D., 21
Baraldi, A., 22
Basak, J., 15
Batistakis, Y., 23
Bellman, R., 23
Berry, M.J.A., 4
Beyer, K.S., 22
Bezdek, J.C., 18, 23, 149, 241, 253
Binder, D.A., 21
Biswas, G., 15
Blonda, P., 22
Bobrowski, L., 18
Booch, G. 29
Bock, H.H., 5, 21, 131, 306
Borisov, A., 20
Botstein, D., 10
Brown, P.O., 10
Buick, D., 149
Bumgarner, R.E., 149

Calcote, J., 96, 105, 461
Camastra, F., 19
Cameron, D., 43

Campello, R.J., 23
Cao, Y., 22, 23
Carmichael, J.W., 5, 131
Carroll, J.D., 18, 265
Casselman, B., 177
Cavalli-Sforza, L.L., 14, 21
Celeux, G., 21, 291
Cengiz, I., 320
Chang, C., 20
Chang, E., 321
Chang, J., 22
Chatterjee, S., 20
Chaturvedi, A., 18, 265
Chen, H.-J., 20
Chen, W.Y., 321
Cheng, C., 22
Cheng, Y., 18
Cherkassky, V., 5
Choi, K.-S., 22
Choudhary, A., 20
Chueh, H.-E., 20
Chung, F.-L., 22
Clatworthy, J., 149
Comaniciu, D., 18
Conallen, J., 29
Cuesta-Albertos, J.A., 18

Dale, M.B., 217
Dash, M., 21
Day, N.E., 21
Dayal, U., 18
de Werra, D., 19
Deitel, H., 42, 56
Deitel, P., 42, 56
Dempster, A.P., 21
Deng, Z., 22

Devroye, L., 271
Ding, C., 19
Döring, C., 23
Dougherty, J., 103, 115
Dubes, R.C., 4, 139
DuBien, J.L., 186
Duda, R., 9, 167
Dunn, J.C., 18, 241

Edwards, A.W.F., 14, 21
Eisen, M.B., 10
Elliston, B., 96, 105, 461
El-Sonbaty, Y., 20
Engel, M.W., 29
Erhart, M., 20
Ester, M., 20
Everitt, B., 5, 6, 131, 139

Faber, V., 18
Farouk, M., 20
Fedjki, C.A., 19
Filippone, M., 19
Fisher, R.A., 15, 167
Flannery, B.P., 146
Flynn, P.J., 3
Foggia, P., 19
Fox, G.C., 18
Fraley, C., 21
Frank, A., 258
Fu, A.W., 22
Fukunaga, K., 18
Fukushima, K., 22
Fukuyama, Y., 23

Gamma, E., 57, 66, 69, 75
Gan, G., 3, 14, 279
Ganti, V., 19
Gath, I., 253
Gehrke, J., 19, 22
George, J.A., 5, 131
Geva, A.B., 253
Ghosh, J., 306
Gibson, D., 19
Glover, F., 19
Goil, S., 20
Goldstein, J., 22

Gordaliza, A., 18
Gordon, A.D., 14, 139, 187
Govaert, G., 21, 291
Gower, J.C., 10, 15, 143
Grabusts, P., 20
Green, P.E., 18, 265
Gropp, W., 307
Grossberg, S., 22
Guha, S., 15, 19, 147
Günnemann, S., 22
Gunopulos, D., 22
Gurewitz, E., 18
Gürsoy, A., 320

Hai, N., 320
Halkidi, M., 23
Han, E., 15, 19
Hankins, M., 149
Hansen, P., 19
Hao, W.-H., 20
Haque, E., 22
Hart, P., 9, 167
Hartuv, E., 19
Hathaway, R.J., 149
He, Q., 321
Helm, R., 57, 66, 69, 75
Hinneburg, A., 20, 21
Holland, J.H., 19, 265
Holman, E.W., 14
Holzner, S., 75
Honda, K., 253
Hook, M.E., 196
Höppner, 23
Horne, R., 149
Hostetler, L., 18
Houston, K.A., 29
Hruschka, E.R., 23
Hsu, M., 18
Hua, K.A., 19
Huang, J.Z., 22
Huang, Z., 10, 18, 142, 265

Ichihashi, H., 253
Ichino, M., 11, 143, 263
Ismail, M.A., 18, 20

Jäger, J., 20
Jain, A.K., 3, 5, 10, 18, 139, 144, 229, 310
Jambu, M., 14, 187
Jardine, C.J., 15
Jardine, N., 15
Jiménez, D., 320
Jin, D., 22
Jing, L., 22
Johnson, R., 57, 66, 69, 75
Johnson, S.C., 8, 14
Jones, M., 22
Judd, D., 310
Julius, R.S., 5, 131

Kalaba, R., 23
Karlsson, B., 99
Karypis, G., 15, 19
Kaufman, L., 10, 142, 152, 217
Keim, D., 20, 21
Klawonn, F., 23
Kleinberg, J., 19
Kohavi, R., 5, 103, 115
Koutroubas, K., 23
Kriegel, H.-P., 22, 27
Krishna, K., 19, 265
Krishnan, T., 292
Krishnapuram, R., 15
Kröger, P., 22
Kruse, R., 23
Kumar, V., 15, 19
Kwok, T., 320

Laird, N.M., 21
Lajoie, J., 44, 56
Lamb, L., 43
Lance, G.N., 13, 185
Landau, S., 5, 139
Lang, S.D., 19
Lasater, C.G., 66, 69, 75
Lee, C., 20
Lee, J., 22
Lee, W.K., 19
Lee, W.S., 20, 22
Leese, M., 5, 139

Legendre, L., 10, 139, 147
Legendre, P., 10, 139, 147
Lesot, M.-J., 23
Leung, Y., 15
Li, C., 15
Li, X.-L., 20
Li, Y., 23
Likas, A., 19
Lin, C.J., 321
Lin, N.P., 20
Linoff, G.S., 4
Lippman, S.B., 44, 56
Liu, B., 22
Liu, H., 21, 22
Livny, M., 15
Lorr, M., 5
Lozano, S., 320
Lusk, E., 307
Luxburg, U., 19

Ma, C., 3
Ma, H., 321
Macnaughton-Smith, P., 217
Macqueen, J.B., 149
Maksimchuk, R.A., 29, 38, 40
Mao, J., 10, 140, 144
Martinez, A.R., 21
Martinez, W.L., 21
Masulli, F., 19
Matrán, C., 18
McGill, M.J., 147
McKinley, P.K., 310
McLachlan, G.J., 292
Mecklenburg, R., 461
Medvedovic, M., 149
Meer, P., 18
Meila, M., 22
Meng, X., 21
Meyers, S., 56, 119
Michaud, P., 6
Milligan, G., 14, 187
Mirkin, B., 26
Miyamoto, S., 253
Mladenović, N., 19
Mockett, L.G., 217

Moo, B.E., 44, 56
Moore, A., 18
Mornar, V., 324
Morrison, D.G., 10, 147
Mulier, F., 5
Müller, E., 22
Muntz, R.R., 20
Murali, T.M., 22
Murtagh, F., 12, 187
Murty, M.N., 3

Nagesh, H., 20
Naiburg, E.J., 38, 40
Narasimha, M.M., 19, 265
Ng, M.K., 18, 22

Orlóci, L., 10, 147
Osherove, R., 112
Othman, F., 310

Park, J.S., 22, 147
Park, N.H., 20, 22
Parsons, L., 22
Patrikainen, A., 22
Pedrycz, W., 253
Pelleg, D., 18
Percannella, G., 19
Phillips, S.J., 18
Press, W.H., 146
Procopiuc, C., 22, 147

Qiu, B.-Z., 20

Raftery, A.E., 21
Raghavan, P., 19, 22
Ramakrishnan, R., 15, 19, 22
Ranka, S., 320
Rashid, N., 310
Rastogi, R., 15, 19, 147
Raymond, E., 43
Robbins, A., 43
Rohlf, F.J., 15
Rose, K., 18
Rosenblatt, B., 43
Ross, G.J.S., 15
Rousseeuw, P.J., 10, 142, 152, 217

Rovetta, S., 19
Rubin, D.B., 21
Rumelhart, D.E., 22
Rummel, R.J., 145
Runkler, T., 23
Ruspini, E.H., 23

Sabin, M.J., 149
Salam, R., 310
Salton, G., 147
Sander, J., 20
Sansone, C., 19
Schikuta, E., 20
Scoltock, J., 24
Scott, A.J., 21
Seidl, T., 22
Selim, S.Z., 18
Shaft, U., 22
Shalloway, A., 75
Shamir, R., 19
Sheikholeslami, G., 20
Shen, J.-Y., 20
Shim, K., 15, 19, 147
Sibson, R., 15
Singh, V., 320
Sintes, T., 41
Skjellum, A., 307
Smith, K., 320
Sneath, P.H.A., 139
Sokal, R.R., 139
Sommerfield, D., 103, 115
Song, J., 20
Song, Y., 321
Spellman, P.T., 10
Srihari, S.N., 8
Stork, D., 9
Stroustrup, B., 42
Sugeno, M., 23
Susumu, H., 320
Symons, M.J., 21

Taillard, E., 19
Taniar, D., 320
Taylor, I.L., 96, 105, 461
Teboulle, M., 17

Teukolsky, S.A., 146
Theodoridis, S., 23
Tromey, T., 96, 105, 461
Trott, J., 75
Tucker, W., 149

Valente de Oliveira, J., 253
van Rijsbergen, C.J., 15
van Dyk, D., 21
Vanjak, Z., 324
Vaughan, G.V., 96, 105, 461
Vazirgiannis, M., 23
Vento, M., 19
Verbeek, J.J., 19
Vetterling, W.T., 146
Vidal, V., 320
Vlassis, N., 19
Vlissides, J.M., 57, 66, 69, 75

Wang, H., 22
Wang, L., 21
Wang, S., 22
Wang, W., 20, 22, 23
Wang, Z., 21
Warde, W.D., 186
Ward, Jr., J.H., 196
Weinman, J., 149
Willett, P., 14
Williams, W.T., 13, 185, 217
Wishart, D., 10, 18, 143
Wolf, J.L., 22, 147
Wolfe, J.H., 21
Wong, J.C., 18
Woo, K., 22
Wu, J., 3, 14, 22, 23
Wunsch II, D., 4, 27

Xia, Y., 22
Xu, R., 4, 27
Xu, X., 20
Xu, Z.-B., 15

Young, B.J., 29
Yaguchi, H., 11, 143, 263
Yang, J., 20, 22
Yeung, K.Y., 149

Yu, P., 22

Zadeh, L.A., 23
Zaiane, O.R., 20
Zha, H., 19
Zhang, A., 20
Zhang, B., 8, 17, 18
Zhang, J.-S., 15
Zhang, T., 15
Zhang, X., 23
Zhang, Y., 22, 23
Zhao, W., 321
Zhao, Y., 20
Zhong, S., 306
Zimek, A., 22
Zipser, D., 22

Subject Index

Abstract class, *see* Class
Abstraction, 46
Accessor, 48
AC_CONFIG_AUX_DIR, 105
AC_CONFIG_HEADERS, 105
AC_CONFIG_SRCDIR, 105
AC_INIT, 105
AC_OUTPUT, 105
Activity diagram, 29, 38
Actor, 36
Agglomerative hierarchical clustering, 185
    centroid, 194
    complete linkage, 192
    geometric, 187
    Graph, 187
    group average, 193
    median, 195
    single linkage, 192
    Ward's, 196
    weighted group average, 194
Aggregation, 33
Algorithm, 151
all_reduce, 309
AM_INIT_AUTOMAKE, 105
Arguments, 149
Association, 33
    bidirectional, 35
    unidirectional, 35
Associative container, 77
Attribute, 3
AttrValue, 117
Autoconf, 105
Autoconf, 96
autoconf, 96

autoheader, 96
autom4te, 96
Automake, 105
autoreconf, 96
autoscan, 96
autoscan, 105
autoupdate, 96
Average, 193

B, 42
BCPL, 42
Behavior diagram, 29
Bidirectional association, *see* Association
binary_function, 139
BIRCH, 15
Boost, 86
Boost MPI, *see* MPI
Boost serialization, 308
broadcast, 309

CenterCluster, 132
Centroid, 12, 194
Chained cluster, *see* Cluster
Chameleon, 15
Cholesky decomposition, 146, 297
Chromosome, 265
Class
    abstract, 51
Class diagram, 29
Classification, 4
Classification approach, 292
Cluster, 3, 5, 131
    chained, 6
    compact, 5
Cluster, 131

Cluster analysis, 3
Cluster validity, 23
Clustering membership, 151
c-means, 241
Collective operation, 309
Communication diagram, 29
Compact cluster, *see* Cluster
Competitive learning, 22
`Complete`, 192
Complete linkage, 12, 192
Component diagram, 29
Composite pattern, *see* Design pattern
Composite structure diagram, 29
Composition, 33
`configure.ac`, 105
`configure.scan`, 105
Constructor, 47
Container, 77
Container adapter, 77
Continuous, 7
Covariant return type, 64
Crossover, 265
CURE, 15

Data clustering, 3, 5
    parallel, 310
Data mining, 4
    direct, 4
    indirect, 4
Data point, 3
`DatasetGenerator`, 170
`DatasetReader`, 167
DBSCAN, 20
Dendrogram, 15
Denpendency, 33
`Dendrogram`, 177
Dependency, 32
deployment diagram, 29
Design pattern, 57
    behavioral, 57
    composite, 61
    creational, 57
    prototype, 64
    singleton, 58
    strategy, 67
    structural, 57
    template method, 69
    visitor, 72
Diameter, 218
DIANA, 14, 217
`Diana`, 218
Discrete, 7
Dissimilarity, 8
Distance
    chord, 10
    city block, 9
    Euclidean, 4, 9, 141
    Mahalanobis, 9, 144
    Manhattan, 9
    maximum, 9
    Minkowski, 9, 140
    mixed, 143
    simple matching, 10, 142
    squared Euclidean, 9
`Distance`, 139
Distance function, 8
Divisive hierarchical clustering, 185

EM algorithm, 292
Encapsulated PostScript, *see* EPS
Encapsulation, 45
EPS, 177
`Error`, 111
E-step, 292
Euclidean distance, *see* Distance
Exception, 54
Exception handling, 54

FCM, 241
Feature, 3
Feature extraction, 3
Feature selection, 3
`__FILE__`, 112
Fitness value, 266
Forward declaration, 119
FSC, 279
`FSC`, 281
Fuzzy clustering, 4, 15
Fuzzy membership, 279

Subject Index

Fuzzy partition matrix, 241
Fuzzy subspace clustering, 279
fuzzy c-means, 241
fuzzy k-means, 253

Gaussian mixture model, 291
GDBSCAN, 20
GDILC, 20
Generalization, 33
Genetic algorithm, 265
Genetic operator, 265
Genetic k-modes, 265
Geometric hierarchical, 187
GKmode, 267
GMC, 293
Graph, 19
Graph hierarchical, 187
GRIDCLUS, 20
Group average, 12, 193

Hard clustering, 4, 15
Hierarchical, 4
Hierarchical clustering, 11, 135, 185
    Monotonic, 187

ifnames, 96
Inducer, 5
Inductive learning algorithm, 5
inline, 117
Interaction diagram, 29
Interaction overview diagram, 29
InternalNode, 135
Interval scale, 7
Inversion method, 271
Iterator, 77, 82
    input, 83
    output, 83

k-means, 18, 229
    parallel, 307
k-modes operator, 265
k-prototypes, 255

LAM/MPI, 307
Lance-Williams formula, 185, 186

LeafNode, 135
Libtool, 105
__LINE__, 112
Log-likelihood, 292
Loss function, 266
LW, 187

Mahalanobis distance, *see* Distance
Makefile, 461
Median, 12, 195
Message passing interface, 307
Metric, 8
Minkowski distance, *see* Distance
Mixed distance, *see* Distance
Mixture approach, 292
Mode, 265
Monothetic, 14
Monotonic hierarchical, 187
Monotonic inequality, 187
MPD, 309
mpd, 309
MPI, 307
    Boost, 307
MPI communicator, 308
MPI environment, 308
mpic++, 309
MPICH, 307
MPICH2, 466
MPIKmean, 311
mpirun, 309
M-step, 292
Multiplicity, 35
mutable, 152
Mutation, 265
Mutation probability, 266
Mutator, 48

Natural selection, 265
Nest link, 31
Node, 135
Nominal, 7

Object, 3, 41
Object diagram, 29
Object-oriented programming, 41

Observation, 3
OOP, 41
OpenMPI, 307
OptiGrid, 20
Optimistic flow, 36
Ordinal scale, 7

Package diagram, 29
Parallel data clustering, *see* Data clustering
Parallel *k*-means, *see* *k*-means
Partitional, 4
Partitional clustering, 15, 133
Pattern, 3
`PClustering`, 133
PDBSCAN, 20
Polymorphism, 50
    dynamic, 51
    inclusion, 50
    overloading, 50
    overriding, 50
    parametric, 50
    static, 51
Polythetic, 14
Population, 265
Positive definite, 295
Posterior probability, 294
Pragmatic flow, 36
Prototype pattern, *see* Design pattern

Ratio scale, 7
Realization, 33
Record, 3
`recv`, 309
`reduce`, 309
Regularization parameter, 295
`Results`, 151
ROCK, 15

`Schema`, 124
Segmentation analysis, 3
Semimetric, 8
`send`, 309
Sequence container, 77
Sequence diagram, 29

Serialization, 308
`shared_ptr`, 88
Simple matching distance, *see* Distance
`SimpleMatchingDistance`, 142
`Single`, 192
Single linkage, 12, 192
Singleton pattern, *see* Design pattern
Smart pointer, 87
Spectral clustering, 19
Standard Template Library, 48, 77
State machine diagram, 29
STING, 20
STL, 48
Strategy pattern, *see* Design pattern
`struct`, 314
Structure diagram, 29
`SubspaceCluster`, 132

Taxonomy analysis, 3
Template, 33
Template method, *see* Design pattern
`tokenizer`, 92
`typedef`, 110

Ultrametric, 8
UML, 29
UML diagram, 29
Unidirectional association, *see* Association
Unified Modeling Language, 29
Unsupervised classification, 3
Use case diagram, 29, 36

Variable, 3
`variant`, 89
Visibility, 31
Visitor pattern, *see* Design pattern

Ward, 12, 196
WaveCluster, 20
`Weighted`, 194
Weighted group average, 12, 194